BRITISH MEDICAL ASSOCIATION
1000662

Diagnostic Lymph Node Pathology

THIRD EDITION

Diagnostic Lymph Node Pathology

THIRD EDITION

Margaret Ashton-Key
Consultant Histopathologist
University Hospital Southampton NHS Foundation Trust
Honorary Senior Lecturer, University of Southampton, UK

Penny Wright
Consultant Histopathologist
Cambridge University Hospitals NHS Trust
Associate Lecturer, Cambridge University, UK

Dennis Wright
Emeritus Professor of Pathology
University of Southampton, UK

CRC Press
Taylor & Francis Group
Boca Raton London New York

CRC Press is an imprint of the
Taylor & Francis Group, an **informa** business

CRC Press
Taylor & Francis Group
6000 Broken Sound Parkway NW, Suite 300
Boca Raton, FL 33487-2742

Printed and bound in India by Replika Press Pvt. Ltd.

Printed on acid-free paper
Version Date: 20160317

International Standard Book Number-13: 978-1-4987-3269-7 (Pack - Book and Ebook)

Visit the Taylor & Francis Web site at
http://www.taylorandfrancis.com

and the CRC Press Web site at
http://www.crcpress.com

Contents

Preface to the third edition

When our new publishers invited us to update *Diagnostic Lymph Node Pathology* with a third edition we were happy to comply. This has enabled us to include a few new items, to improve some of the illustrations and to introduce new ones. The text and references have been updated. We have retained the chapter on the handling of lymph node biopsies and revised the chapter on needle biopsies. The format of the book remains as before with the use of high-quality illustrations and boxes to succinctly present the clinical, morphological, immunohistochemical and genetic features of each entity. The text is concise and is accompanied by relevant and up-to-date references. These features make the book more than an atlas. Our aim has been to create a readily available guide to the correct diagnosis and categorisation of lymph node pathology.

It is the custom in many administrations to refer biopsies diagnosed as malignant lymphoma to a panel of expert haematopathologists for confirmation of the diagnosis. Our hope is that this book will help general pathologists to identify cases in need of referral and to provide a readily available source of information as a bench book for trainees, for general pathologists and also for haematopathologists.

This book was originally conceived by Professor Anthony Leong. Sadly, Professor Leong died at the time the second edition of the book was published.

Preface to the second edition

Although it is only four years since this book was first published, our understanding of the biology and pathology of malignant lymphomas, and consequently the way we interpret lymph node biopsies, has changed to the point where a new edition has become necessary. An additional stimulus has been the publication of a new edition of the *WHO Classification of Tumours of Haematopoietic and Lymphoid Tissues*, in which a number of new entities are recognized, and some regrouping has taken place.

Our main aims remain the same: to provide general histopathologists and trainees with a systematic and logical way of approaching lymph node biopsies; to help them use diagnostic techniques wisely, in particular the bewildering variety of antibodies now available for immunohistochemistry; and to guide them through the maze of differential diagnoses, enabling them to reach an accurate diagnosis. If specialized haematopathologists are tempted to keep a copy hidden in a desk drawer for quick reference we would be more than gratified. We recognize the important role molecular techniques now play in providing supporting evidence, but, as before, the emphasis throughout is on morphology, supplemented by immunohistochemistry.

It is easy to become excited by new entities and concepts. While including these, we hope we have managed to preserve the balance of the text, without giving undue prominence to rarities. At the same time, we have attempted to provide a reasonably complete guide to lymph node pathology, so that the answer to most diagnostic problems will be found in these pages. We should emphasize that this is not a comprehensive guide to malignant haematopathology: only those conditions that involve lymph nodes, either primarily or secondarily, are discussed. This inevitably means that some important conditions, for example some forms of extranodal T-cell lymphoma, are either not considered or receive only brief mention.

The general format of the book has been retained and the layout has, we think, been improved. For quick reference, the boxes containing concise clinical, morphological and immunohistochemical features of each entity have been preserved and updated. A number of illustrations have been replaced and additional ones included. The layout of the illustrations has been changed to avoid boxes that overflow onto two pages, a major criticism of the previous edition. The references given are by no means comprehensive. We have included those that we regard as being seminal, useful reviews or evidence of recent significant advances, and we hope the interested reader will find these a useful starting point.

Since the previous edition the use of needle core biopsies has increased substantially. In expert hands, they provide a rapid and increasingly reliable means of diagnosis, with minimal inconvenience and discomfort for the patient. Chapter 11 has been expanded to reflect this and to provide a guide to their interpretation and limitations.

We hope our efforts have resulted in a useful and practical guide to a subject that remains endlessly challenging and fascinating.

Preface to the first edition

Haematopathology has become the subject of specialist reporting in many countries in the developed world. This is seen as a necessary evolution and a consequence of the increasing complexity of the subject, the need for sophisticated ancillary investigations in some cases and the fundamental need for an accurate diagnosis on which to base further patient management. Nevertheless, most lymph node biopsies will land on the desks of general pathologists who will need to make the judgement as to whether the pathology is that of a reactive or neoplastic process and whether referral is necessary. We have aimed this book at general pathologists and trainees, although we hope that dedicated haematopathologists may find some gems between its covers. In light of our target readership we have placed our main emphasis on morphology rather than molecular techniques.

A number of authors have tried to base lymph node diagnosis on the low power structure of the node. While this is a good starting point it is not always helpful and can be misleading. For example, although an overall nodular pattern is characteristic of follicular lymphoma it can also be the dominant low-power feature of mantle cell and marginal zone lymphomas. We would nevertheless emphasize the importance of both low-power and high-power morphological examination based on good-quality sections. It is wise to arrive at a diagnosis, or differential diagnosis, based on morphology before ordering or embarking on the interpretation of immunohistochemical preparations. In the final analysis the morphology and immunohistochemistry should be compatible, and it is the concordance of these techniques that provides security of diagnosis.

We are aware of the time constraints facing pathologists and have aimed to make the basic information on entities easily accessible by presenting the clinical, morphological and immunohistochemical features of each disease together with illustrations in boxes. More detailed information is provided in the text.

Since we began writing this book we have seen a year-on-year growth of the proportion of lymph node biopsies received as needle biopsies. These are usually taken by radiologists using CT guidance. The most obvious value of this technique is in taking biopsies of deep-seated lesions and thus avoiding the need for surgery. Most pathologists would probably wish that superficial nodes were obtained by whole lymph node biopsy. However, as clinicians realize that a definitive diagnosis can be obtained on a high proportion of superficial nodes using needle biopsies, this type of biopsy is likely to become more common in view of its ease of application and low morbidity. We have therefore included in the book a chapter specifically on the interpretation of needle biopsies.

Acknowledgements

Firstly, we would like to thank Dr Bruce Addis and the previous editors for their work on the first two editions, without which there would not be a third edition. We thank our friends and colleagues in the South West Region for a constant supply of biopsy material. We are particularly indebted to Dr Denis Madders (Royal Bournemouth and Christchurch Hospitals) and Dr Kudair Hussein (Poole Hospital NHS Trust) for their help in the provision of illustrative cases. Dr Hilary Lawton (Peterborough Hospital) provided sections and photographs of unusual cases, and Professor John Chan (City of Hope, Duarte, California) provided material for the illustrations of Kimura disease, for which we are most grateful.

Dr Fabio Facchetti (Spedali Civili-University of Brescia) was most generous in the provision of Figures 3.40–3.42. Professor Ethel Cesarman (Weill Cornell Medical College) most generously provided the illustrations in Figure 8.11. We would like especially to express our warm gratitude to colleagues at Southampton General Hospital (Dr Vipul Fiora and Dr Sanjay Jogai) for their unremitting support in the identification and retrieval of representative cases. Finally, we wish to thank Barbara Norwitz and Henry Spilberg of CRC Press for their help and guidance.

Authors

Margaret Ashton-Key
Consultant Histopathologist
University Hospital Southampton NHS Foundation Trust
Honorary Senior Lecturer, University of Southampton, UK

Penny Wright
Consultant Histopathologist
Cambridge University Hospitals NHS Trust
Associate Lecturer, Cambridge University, UK

Dennis Wright
Emeritus Professor of Pathology
University of Southampton, UK

Abbreviations

AITL angioimmunoblastic T-cell lymphoma

ALCL anaplastic large cell lymphoma

ALK anaplastic lymphoma kinase

ALL acute lymphoblastic leukaemia

ALPS autoimmune lymphoproliferative syndrome

AML acute myeloid leukaemia

ANKL aggressive NK-cell leukaemia

ATLL adult T-cell leukaemia/lymphoma

B-CLL/SLL chronic lymphocytic leukaemia/small lymphocytic lymphoma

B-PLL B-cell prolymphocytic leukaemia

BSAP B-cell lineage activator protein

cHL classical Hodgkin lymphoma

CVID common variable immunodeficiency

DLBCL diffuse large B-cell lymphoma

EATL enteropathy-associated T-cell lymphoma

EBER Epstein–Barr virus–encoded RNA

EBNA Epstein-Barr nuclear antigen

EBUS endobronchial ultrasound

EBV Epstein–Barr virus

EMA epithelial membrane antigen

ESR erythrocyte sedimentation rate

EUS endoscopic ultrasound

FDC follicular dendritic cell

FISH fluorescence in-situ hybridization

FL follicular lymphoma

FNA fine needle aspiration

GCET1 germinal centre B cell expressed transcript-1

HCL hairy cell leukaemia

HCv hairy cell variant

H&E haematoxylin and eosin

HGAL human germinal centre-associated lymphoma gene

HHV-8 human herpesvirus 8

HIV human immunodeficiency virus

HMFGP human milk fat globule membrane

H/RS Hodgkin/Reed–Sternberg

HTLV human T-cell lymphotropic virus

HVCD hyaline vascular Castleman disease

ICOS inducible T-cell co-stimulator

IDRC interdigitating reticulum cell

Ig immunoglobulin

IgA immunoglobulin A

IgD immunoglobulin D

IgG immunoglobulin G

IgH immunoglobulin heavy chain

IgM immunoglobulin M

IRTA1 immunoglobulin superfamily receptor translocation-associated 1

ISH in-situ hybridisation

LBCL large B-cell lymphoma

LBL lymphoblastic lymphoma

LCA leukocyte common antigen

LDCHL lymphocyte depleted classical Hodgkin lymphoma

LDH lactate dehydrogenase

LEF1 lymphoid enhancer-binding factor-1

LGL large granular lymphocyte

L&H lymphocytic and histiocytic

LMP1 latent membrane protein

LPL lymphoplasmacytic lymphoma

LRCHL lymphocyte rich classical Hodgkin lymphoma

LYG lymphomatoid granulomatosis

MALT mucosa-associated lymphoid tissue

MCCHL mixed cellularity classical Hodgkin lymphoma

MCD multicentric Castleman disease

MCL mantle cell lymphoma

MUM1 multiple myeloma oncogene

MZL marginal zone lymphoma

NK natural killer

NLPHL nodular lymphocyte-predominant Hodgkin lymphoma

NOS not otherwise specified

NSCHL nodular sclerosis classical Hodgkin lymphoma

PAS periodic acid-Schiff

PCD plasma cell variant Castleman disease

PCR polymerase chain reaction

PDC plasmacytoid dendritic cell

PEL primary effusion lymphoma

PLL prolymphocytic leukaemia

PMBL primary mediastinal (thymic) large B-cell lymphoma

PNET primitive neuroectodermal tumour

PTCL peripheral T-cell lymphoma

PTGC progressive transformation of germinal centres

PTLD post-transplant lymphoproliferative disorder

SCID severe combined immunodeficiency

SLE systemic lupus erythematosus

TCR T-cell receptor

TdT terminal deoxynucleotidyl transferase

TFH T-follicular helper

THRLBCL T-cell/histiocyte–rich B-cell lymphoma

TIA-1 T-cell intracellular antigen 1

T-PLL T-cell prolymphocytic leukaemia

WHO World Health Organization

XLP X-linked lymphoproliferative disorder

ZAP-70 zeta chain–associated protein kinase 70

1

Handling of lymph node biopsies, diagnostic procedures and recognition of lymph node patterns

TAKING AND HANDLING OF LYMPH NODE BIOPSIES

Suboptimal techniques in the taking and handling of lymph node biopsies are probably the biggest obstacle to achieving a correct diagnosis. All concerned with this process should bear in mind that the objective of the biopsy is to achieve a timely and accurate diagnosis on which the subsequent management of the patient can be based. Feedback information at multidisciplinary team meetings is a valuable means of achieving and maintaining a high diagnostic standard of lymph node biopsies. In the absence of such meetings, personal contact is needed to ensure that any shortcomings in the biopsy technique and handling of the specimen are rectified.

Lymph nodes should be selected for biopsy on the likelihood that they contain the pathological process. They should be dissected out whole, if possible, and with the capsule intact. Fragmented nodes may be more difficult to diagnose than intact nodes, depending on the pathological process involved. Traction artefacts are usually most severe when the biopsy tissue is very fibrotic or has to be taken from a confined space, such as the anterior mediastinum.

Needle biopsies are now more frequently used for the diagnosis of lymph node pathology. When possible, open lymph node biopsies should be used for superficial, accessible lymph nodes; however, needle biopsies have a lower morbidity than open biopsies and are of particular value in sampling mediastinal, abdominal and retroperitoneal lymph nodes, avoiding the need for more invasive procedures, such as laparotomy. These biopsies are usually taken by radiologists using ultrasound or computed tomography (CT) guidance. If fixed quickly, needle biopsies give good morphological preservation, which, together with immunohistochemistry, allows the precise identification of most common lymphomas. The technique may be less successful in the identification of non-neoplastic proliferations. The handling and preparation of needle biopsies is discussed in more detail in Chapter 11.

Fine needle aspiration (FNA) biopsies have their greatest value in the separation of carcinoma from lymphoma and for the identification of recurrences or for staging. The role of this technique for the primary diagnosis of lymphoma is limited and presents many pitfalls unless in the hands of an expert cytopathologist. Alternatively, the lymph node aspirates obtained via ultrasound-guided endoscopic (endoscopic ultrasound [EUS]) or endobronchial (endobronchial ultrasound [EBUS]) sampling may be processed entirely as clot preparations, without making aspirate smears, thereby maximising the material available for immunohistochemistry. Additional information may be obtained if a separate aspirate sample is sent for flow cytometry. Ideally, the clot preparations obtained will contain lymph node microbiopsies, and serial sections may be stained with haematoxylin and eosin (H&E) whilst the intervening spare sections may be used for immunostaining or other special stains (Figure 1.1).

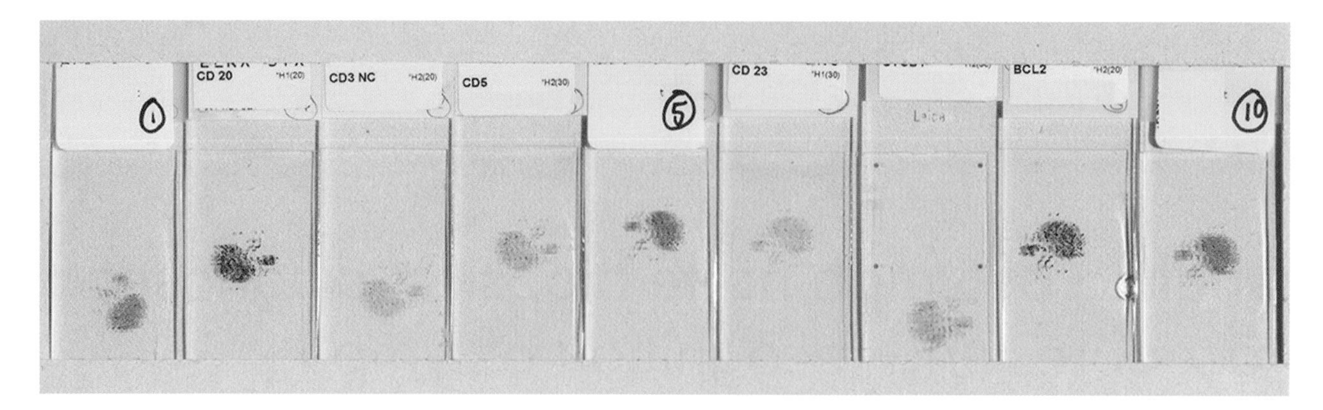

Figure 1.1 Series of sections from a fine needle aspiration cell block preparation demonstrating the use of 'spare sections' between haematoxylin and eosin (H&E) levels for immunohistochemistry.

Logistics dictate that many laboratories receive their lymph node biopsies in fixative. In such cases, the volume of the fixative should be at least ten times that of the specimen. Whole lymph nodes should be sliced as soon as possible to allow rapid penetration of the fixative.

Ideally, lymph node biopsies should be received fresh in the laboratory immediately after excision. This requires good communication between the pathologist and surgeon or radiologist, to ensure that there is minimum delay in the specimen reaching the laboratory. A slice taken from one end of the node can be gently touched onto a clean glass slide, and air-dried and stained by one of the rapid Romanowsky techniques to provide a rapid cytological assessment. Pathologists experienced with this technique may be able to give a provisional cytological diagnosis from this when appropriate. The technique is most useful, however, in determining the subsequent handling of the specimen in the laboratory. The slice of lymph node used to make the imprint preparation may be frozen for subsequent molecular investigation. It should not be used for histology, if this can be avoided, since the process of making imprint preparations often causes traction artefacts in the tissue (Box 1.1).

Fresh tissue may be sent for cytogenetic analysis and/or flow cytometry in appropriate cases. Frozen sections may be cut for morphology and immunohistochemistry when indicated. One or more slices of the node should be placed in fixative overnight or longer for histology and immunohistochemistry. Needle biopsies require a similar period of fixation. Fortunately, for diagnostic purposes a wide range of procedures, including immunohistochemistry, polymerase chain reaction (PCR) and fluorescence *in-situ* hybridization (FISH), can be performed on fixed tissue (Box 1.2).

BOX 1.1: Fresh whole lymph node biopsies

Slice using clean sharp blade; use slices as follows:

- Imprint cytology (tissue used to make imprints should not be used for histology) may be used for:
  - Rapid evaluation
  - Air-dried imprint slides stored for fluorescence *in-situ* hybridization (FISH) if required
- Histology and immunohistochemistry. Place slices in fixative; if using formalin fix for 12–24 hours
- Fresh tissue slices may be used for:
  - Cytogenetics
  - Molecular analysis
  - Cell culture
  - Microbiology

BOX 1.2: Fixed whole lymph node biopsies

- Cut into 5-mm slices with a sharp scalpel as soon as possible after biopsy
- Place in fixative, at least ten times the volume of the specimen
- Leave in fixative for 12–24 hours if formalin is the fixative
- Tissue for long-term storage should be blocked in paraffin after fixation, not left in fixative

PROCESSING, SECTIONING AND STAINING

Laboratories should maintain quality control of their reagents and equipment to ensure adequate processing, cutting, staining and immunohistochemistry. Cell morphology is important in haematopathology and can easily be obscured or distorted by poor fixation, processing and sectioning. Section thickness has a marked influence on cytological and histological appearances. The optimum thickness is 3–5 μm.

H&E is the stain most widely used in histopathology and is often the only one used in lymph node diagnosis. The Giemsa stain can add another dimension to haematopathology and is the stain of choice for this subspeciality in much of mainland Europe. The Giemsa stain highlights basophilia and eosinophilia, and this aids the identification of blast cells, plasma cells, eosinophils and mast cells.

When using this technique, care must be taken with the quality of the Giemsa stain used and the pH of the reagents; otherwise a section stained uniformly pale blue is obtained, which is of little diagnostic value.

The periodic acid–Schiff (PAS) stain is sometimes of value in haematopathology. It highlights intranuclear immunoglobulin M (IgM) inclusions (Dutcher bodies), basement membrane and ground substance, such as that seen around the blood vessels in angioimmunoblastic T-cell lymphoma. The reticulin stain can be of value in determining the overall structure of the lymph node, highlighting follicularity, sinus structure and blood vessels.

In some laboratories, all of the aforementioned stains are used as a 'lymph node set'. However, there is now a tendency to move directly from the H&E section to immunohistochemistry, when the additional use of one or more of these stains could be of greater diagnostic value. If only a small amount of biopsy material is available, as with most core biopsies, avoid cutting 'levels' routinely but keep spare unstained sections for immunohistochemistry: re-cutting the block inevitably wastes valuable tissue.

IMMUNOHISTOCHEMISTRY

With experience and good histological preparations it is possible to diagnose many of the common lymphomas on morphology alone. In most practices, however, a substantial number of cases cannot be categorized precisely without the aid of immunohistochemistry. Even in biopsies that are diagnosable with reasonable certainty on morphology alone, such as diffuse large B-cell lymphoma (DLBCL), immunohistochemistry will often provide additional prognostic information. It has therefore become standard practice in many laboratories to perform confirmatory immunohistochemistry on all lymphoma biopsies. The cost of this is trivial when set against the need to obtain an accurate diagnosis and the expense of treatment. Unfortunately, in the developing world, where obtaining and maintaining antibodies is often difficult and no costs are trivial, diagnostic immunohistochemistry is not widely practised.

In our experience, it is not uncommon to receive biopsies from pathologists who have made a reasonable morphological diagnosis but have been confused by the subsequent immunohistochemistry. To avoid this pitfall, pathologists should be aware of the staining characteristics of antibodies and the specificity of their reactivity. The laboratory should be subject to ongoing quality control. For most antibodies used in haematopathology, there will be internal controls within the tissues being investigated (e.g., reactive B- and T-cells, histiocytes). For antigens not commonly expressed within normal and reactive tissues, such as anaplastic lymphoma kinase 1 (ALK-1), external control tissues are necessary. Beware the section that is uniformly blue; it usually indicates technique failure.

Immunohistochemical techniques have improved considerably in the past decade, with the production of increased numbers of robust antibodies and the development of techniques for antigen retrieval.

Is it a lymphoma?

Large cell lymphomas may resemble other anaplastic neoplasms. The leukocyte common antigen (LCA, CD45) shows membrane expression in almost all lymphoid cells and has only rarely been reported in non-haematopoietic cells. However, there are pitfalls. LCA may be absent on precursor (lymphoblastic) B- or T-cell lymphomas. Anaplastic large cell lymphomas (ALCLs) are also often LCA negative, as are classic Hodgkin/Reed–Sternberg cells. ALCLs frequently express epithelial membrane antigen (EMA), which together with LCA negativity may suggest an epithelial neoplasm. EMA is expressed, however, on plasma cells and on the cells of a number of lymphomas. Most epithelial neoplasms (carcinomas) express low molecular weight cytokeratins. In the rare cases in which low molecular weight cytokeratins have been reported in lymphoma cells, this has usually been in the form of a paranuclear dot. The majority of malignant melanomas can be identified with antibodies to S100 protein and HMB45.

Is it a B-cell lymphoma?

IMMUNOGLOBULINS

B-lymphocytes are defined by their ability to synthesize immunoglobulins, which should therefore provide the most reliable means of identifying these cells. In practice, they are not used for this purpose in most laboratories. The main reason for this is that plasma immunoglobulins cause diffusion artefacts, particularly in poorly fixed specimens, that are often confusing and obscure specific staining. Cells that appear positive for immunoglobulins owing to passive uptake usually show smooth cytoplasmic staining that is most intense at the cell membrane. Within the node, these cells often occur in broad bands corresponding to the advancing front of the fixative as it diffuses into the tissue. Immunoglobulin in synthetic cells often appears granular, owing to its accumulation within the endoplasmic reticulum, or as larger inclusions. Synthesized immunoglobulin also frequently manifests as paranuclear (Golgi) staining and as strong staining around the nucleus corresponding to immunoglobulin within the perinuclear space. IgM is synthesized by a large proportion of DLBCLs and, because of its large molecular size and relatively low concentration in the plasma, shows less diffusion artefact than other immunoglobulins. Surface IgD can be recognized in well-fixed paraffin-embedded tissues. Reactive mantle cells are positive, as are most cases of B-cell chronic lymphocytic leukaemia/small lymphocytic lymphoma (B-CLL/SLL) and mantle cell lymphoma.

In addition to their use as B-cell lineage markers, immunoglobulins can be used to imply clonality. Clonal (neoplastic) populations are monotypic (i.e., they express only one light chain). It should be noted that not all monotypic populations show clonal immunoglobulin rearrangements (see Castleman disease and paediatric nodal marginal zone lymphoma). Monotypia may be obscured by a background population of reactive cells, which usually express kappa and lambda light chains in the ratio of two to one. In general, reactive cells show more intense staining for immunoglobulins than neoplastic cells.

Although immunohistochemistry can be used to identify light chain restriction in cells expressing cytoplasmic immunoglobulin, it is difficult to identify surface light chains except in the best fixed tissues. *In-situ* hybridization (ISH) for kappa and lambda mRNA offers an alternative and more reliable technique for demonstrating light chain synthesis, which avoids the problems of background staining.

CD20 (L26)

CD20 is a non-glycosylated phosphoprotein expressed on the membrane of B-cells. Although widely regarded as such, it is not a perfect B-cell marker because it is not expressed in the earliest stages of B-cell differentiation and is lost as the B-cell undergoes plasma cell change. It is absent therefore on many B-lymphoblastic lymphomas and plasmacytic tumours. Staining for CD20 should be on the cell membrane; cytoplasmic, nuclear and nucleolar staining are nonspecific. Because it is a surface membrane antigen, CD20 is often most strongly stained in biopsies that show some degree of shrinkage artefact and may appear less strongly stained in well-fixed tissues, such as needle biopsies. CD20 expression is of clinical importance because of the use of rituximab, an anti-CD20 antibody, in the treatment of lymphomas. Rarely, expression of CD20 may be lost in relapsed lymphomas after rituximab therapy.

CD20 has rarely been reported on T-cell lymphomas. It is also expressed on the epithelial cells of some thymic carcinomas (beware when diagnosing mediastinal large B-cell lymphoma).

CD79A

CD79 is a heterodimeric glycoprotein signal transduction molecule that associates with membrane immunoglobulin. Antibodies to the α-chain of the molecule (CD79a) provide an almost perfect B-cell lineage marker because CD79a is expressed throughout B-cell differentiation. It should be noted, however, that 50 per cent of T-lymphoblastic lymphomas express CD79a. Staining for CD79a is cytoplasmic and is strong on plasma cells. It is more strongly expressed on mantle cells than on germinal centre cells.

PAX5

PAX5 is a transcription factor that encodes the B-cell lineage specific activator protein (BSAP) that is expressed at earlier stages of B-cell differentiation but is usually not detected in plasma cells. It is a reliable marker of B-cell origin and is useful in the diagnosis of classical Hodgkin lymphoma. Rarely, aberrant expression may be seen in T-cell lymphomas, and PAX5 expression is also seen in small cell carcinoma and Merkel cell carcinoma.

OTHER MARKERS

Details of other markers of value in the diagnosis of B-cell lymphomas are as follows:

- **Annexin A1**—Is a member of the annexin family of calcium dependent phospholipid binding proteins. This is a highly sensitive marker for the diagnosis of hairy cell leukaemia (HCL). It is also abundantly expressed on T-lymphocytes and myeloid cells.
- **BCL-1 (cyclin D1)**—The cyclin D1 gene on chromosome 11q13 is translocated in mantle cell lymphoma, leading to overexpression of cyclin D1. Positive staining for cyclin D1 is seen in the nucleus; cytoplasmic staining is artefactual. This provides a valuable marker for the diagnosis of mantle cell lymphoma. Endothelial cells and histiocytes provide a positive internal control for the stain. Cyclin D1 expression may also be seen in HCL, some plasma cell neoplasms and intranodal palisaded myofibroblastoma.
- **BCL-2**—BCL-2 protein is coded by a gene on chromosome 18q21 that is involved in the 14:18 translocation characteristic of follicular lymphomas. It is a mitochondrial membrane protein that regulates apoptosis. In reactive tissues, small B- and T-cells are positive for BCL-2, whereas follicle centre B-cells are negative. Since 85 per cent of follicular lymphomas are BCL-2 positive, this provides a useful means of distinguishing between reactive and neoplastic follicles. BCL-2 expression in DLBCLs has been shown in some studies to be associated with a poor prognosis.
- **BCL-6**—This is a zinc finger transcriptional repressor coded for on chromosome 3q27. Specific staining is nuclear; cytoplasmic staining is artefactual. In normal and reactive lymphoid tissue, nuclear staining for BCL-6 is seen in follicle centres but not in prefollicular or postfollicular cells. Most follicular lymphomas are BCL-6 positive. Burkitt lymphoma cells are positive. A high proportion of DLBCLs express BCL-6 but the percentage of positive nuclei varies. Overexpression of BCL-6 may be caused by translocations involving 3q27 (present in up to 30 per cent of DLBCL) or mutations of the gene. BCL-6 positivity may be used to identify follicle centre cell origin or as a prognostic marker.
- **BRAF V600E**—This mutation is a specific characteristic of HCL, and a monoclonal antibody detecting the mutant protein is now available (anti-BRAF V600E [VE1]). This antibody is highly specific for HCL and does not stain other small B-cell lymphomas.
- **CD5**—This is a membrane glycoprotein expressed on T-cells. It is also expressed more weakly on the B-cells of B-CLL/SLL and mantle cell lymphoma. A small proportion of DLBCLs are positive for CD5.
- **CD10**—This recognizes a surface neutral endopeptidase (zinc-dependent metalloproteinase). Within the lymphoid system it is expressed on the cells of B-lymphoblastic lymphoma, some T-lymphoblastic lymphomas, and the

neoplastic cells of some angioimmunoblastic T-cell lymphomas. Follicle centre cells express CD10, as do most follicular lymphomas. Burkitt lymphoma cells are positive, as are a proportion of DLBCLs (germinal centre type). Outside the lymphoid system, CD10 is expressed on many stromal and epithelial cells, providing a positive internal control in many biopsies.

- **CD23**—This is a membrane glycoprotein expressed on activated B-cells. It also acts as a low-affinity receptor for IgE. In reactive tissues, CD23 is expressed on a variable proportion of mantle and follicle centre cells and on follicular dendritic cells (FDCs). It labels the cells of B-CLL/SLL and provides a useful marker for this disease. It does not label mantle cell lymphoma but, owing to its reactivity with FDCs, it highlights the dispersed pattern of these cells that is characteristic of this neoplasm.
- **CD38**—This is a 45-kDa transmembrane glycoprotein found on a number of different precursor cells, monocytes, activated T-cells and terminally differentiated B-cells. It is a useful marker of plasma cells and their neoplasms. It is also expressed in some cases of B-CLL/SLL and, like those expressing ZAP-70, this tends to be the group with an adverse prognosis.
- **CD45RA**—Antibodies to variants of the CD45 molecule (4KB5, MB1, MT2) have been used in the past as markers of B-cells but have been superseded by the more reliable and specific antibodies CD20 and CD79a. MT2 has been used as a marker for the distinction between reactive and neoplastic follicles. Although largely replaced by staining for BCL-2, this antibody might be worth considering in equivocal cases.
- **CDw75 (LN1)**—This is a sialylated carbohydrate determinant on the surface of B-cells in the germinal centre. It is not present on T-cells but is found on a variety of epithelial cells, including those in distal tubules, mammary glands, bronchus and prostate. Although there are more specific markers, CDw75 has been used for the identification of follicle centre cell lymphomas as it is not expressed on B-CLL/SLL or T-cell lymphomas. Staining is membranous with Golgi accentuation.
- **CD138**—This is a transmembrane heparan sulfate proteoglycan and a member of the syndecan proteoglycan family. Staining is membranous. It is expressed in late-stage differentiated B-cells, plasma cells and their neoplastic counterparts. Staining also occurs in primary effusion lymphoma and pyothorax-associated lymphoma. CD138 will also stain epidermis and a wide range of epithelial cells.
- **GCET1**—Germinal centre B-cell expressed transcript-1 protein expression is restricted to a subset of germinal centre B-cells and lymphomas showing germinal centre differentiation, follicular lymphoma, nodular LP Hodgkin lymphoma, Burkitt lymphoma and a subset of DLBCL. GCET1 is useful in the differentiation of FL and marginal zone lymphoma.
- **HGAL (GCET2)**—Human germinal centre-associated lymphoma gene protein can be detected in both reactive and neoplastic follicles and is a sensitive germinal centre marker. There is high expression in follicular lymphomas and low expression in other small B-cell lymphomas.
- **LEF1**—Lymphoid enhancer-binding factor-1 is a nuclear protein expressed in pre-B- and T-cells. LEF1 is overexpressed in CLL, and strong nuclear staining of LEF1 is observed in virtually all cases of CLL including CD5-negative cases. Other low-grade B-cell lymphomas are LEF1 negative.
- **MUM1/IRF4**—The *MUM1* (multiple myeloma oncogene)/*IRF4* (interferon regulatory factor 4) gene was identified as a myeloma-associated proto-oncogene that is activated at the transcriptional level as a result of t(6;14) (p25;q32) chromosomal translocation. Its expression is restricted to cells of B-cell lineage and activated T-cells. MUM1/IRF4 is used with CD10 and BCL-6, mainly for the separation of germinal centre and non–germinal centre types of DLBCL. It is also a useful marker of plasma cells and plasma cell neoplasms.
- **SOX11**—This is a neural transcription factor that is overexpressed in both conventional mantle cell lymphoma and cyclin D1–negative mantle cell lymphoma. Nuclear expression of SOX11 is highly specific for mantle cell lymphoma.
- **ZAP-70**—Zeta chain–associated protein kinase 70 (ZAP-70) is a member of the tyrosine kinase family that is normally expressed in T-cells and NK cells and known to be important in their signalling processes. It is absent in normal peripheral B-cells. It is expressed in a subgroup of B-CLL/SLL, corresponding to those cases with a poorer prognosis in which the *IgVH* genes are unmutated. It is absent in the group with a better prognosis and mutated genes.

Is it a T-cell lymphoma?

Many B-cell lymphomas contain large numbers of T-cells; indeed, these may be the majority population, as in some follicular lymphomas and in T-cell/histiocyte-rich B-cell lymphomas. The majority of these reactive T-cells will be small lymphocytes.

T-CELL RECEPTOR

The defining antigen of the T-cell is the T-cell receptor (TCR) molecule (αβ or γδ). Until recently the only TCR antibodies that could be used with paraffin-embedded tissues were to the TCR β-chain, βF1. Monoclonal antibodies detecting the constant region of the TCR γ-chain and TCR δ-chain in paraffin sections are now commercially available, allowing for positive identification of γδ T cells. With the use of these antibodies, the majority of T-cell lymphomas can be assigned to one or the other lineage (αβ or γδ). However, these antibodies are not in widespread use, and a subset of cases is either TCR silent or dual TCR positive.

CD3

The CD3 molecule is a complex of four distinct glycoprotein chains that associate with the TCR. It is only when the CD3/TCR complex is fully assembled that it is inserted into the cell membrane. CD3 may therefore be detected either within the cytoplasm or at the cell surface. Theoretically, it provides the most reliable T-cell lineage marker, although it is lost or very weakly expressed in 20–25 per cent of T-cell lymphomas.

CD5

CD5 is a 67-kDa glycoprotein that is expressed on the surface of all mature T-cells and a small number of B-cells. It provides an excellent marker of T-cells and is expressed in most T-cell lymphomas. The much weaker reaction of B-CLL/SLL and mantle cell lymphoma with CD5 is unlikely to cause diagnostic confusion because these tumours also express CD20 and CD79a.

OTHER MARKERS OF T-CELL LYMPHOMAS

- **ALK-1**—ALK-1 recognizes a tyrosine kinase that is overexpressed in 85 per cent of T/Null ALCLs as a result of t(2;5) in 90 per cent of cases and of variant translocations in the remainder. Those cases with t(2;5) translocation show cytoplasmic, nuclear and nucleolar staining with ALK-1, as a result of the activity of the native nucleophosmin. The variant translocations give cytoplasmic and/or membrane staining only. A small number of DLBCLs express ALK. These tumours usually have plasmablastic features and lack expression of CD20, CD79a and CD30. Rare cases in this group show t(2;5), others show t(2;17) involving the clathrin gene. Reactivity with ALK-1 in the latter group shows characteristic granular cytoplasmic staining. Some inflammatory myoblastic tumours show translocations of the *ALK* gene and express cytoplasmic ALK.
- **CD1a**—This reacts with cortical thymocytes but is of most value in the diagnosis of Langerhans cell histiocytosis.
- **CD2**—The CD2 molecule is one of the earliest T-cell markers, being present on over 95 per cent of thymocytes. It is a glycosylated transmembrane receptor that is responsible for spontaneous E-rosetting of human T-cells with sheep red cells. Like CD3, it is a useful marker of T-cells and T-cell lymphomas.
- **CD4**—This is a marker of helper/inducer T-cells and their neoplasms. It is also present on histiocytes, which may make interpretation of sections difficult.
- **CD7**—This is expressed by T-cells and most T-cell neoplasms. Loss of CD7 expression is frequently seen in T-cell lymphomas. However, it may also occasionally be lost in reactive conditions. Aberrant expression may be seen on myeloid neoplasms.
- **CD8**—This is a marker of suppressor/cytotoxic T-cells and neoplasms derived from them.
- **CD25**—This is an interleukin-2 α-chain receptor expressed by a subset of CD4+ regulatory T-cells. It was first described in association with human T-cell lymphotropic virus type 1 (HTLV-1)–related lymphomas (adult T-cell lymphoma/leukaemia), and it was later observed that HCL also shows a high frequency of CD25 expression. The majority of cases of ALCL in children are also positive, and this is unrelated to HTLV-1.
- **CD43 (MT1)**—Antibodies to CD43 recognize a membrane mucin known as sialophorin. Although once widely used as a T-cell marker, CD43 has broad reactivity to many cell types, including some B-cell lymphomas. It is a useful marker of undifferentiated granulocytic neoplasms.
- **CD45RO**—This was once widely used as a T-cell marker but has a broad reactivity. It has been largely superseded by antibodies to CD3 and CD5.
- **CXCL13**—This is a chemokine that is specifically expressed in follicular T-helper cells. Like PD-1, it is expressed in angioimmunoblastic T-cells and in the rosetting T-cells of NLPHL. The antigen is localized to the cytoplasm with paranuclear dot enhancement.
- **Cytotoxic granule-associated proteins (TIA-1, perforin and granzyme B)**—Cytotoxic T-cells and natural killer (NK) cells are characterized by the presence of cytotoxic granules that are released in response to target cell recognition, leading to apoptosis. These antigens are revealed as cytoplasmic granules. They are expressed by cytotoxic T-cells and some T-cell lymphomas (e.g., ALCL and enteropathy-associated T-cell lymphomas).
- **PD-1**—Programmed cell death-1 (PD-1 or CXCR5) is a member of the CD28 co-stimulatory receptor family that is expressed by germinal centre-associated T-cells in reactive tissue. PD-1+ T-cells are found in angioimmunoblastic T-cell lymphoma and also form the rosetting cells around the multilobated LP cells of nodular lymphocyte-predominant Hodgkin lymphoma (NLPHL).

Is it an NK- or a T/NK-cell lymphoma?

CD56

CD56 is a membrane glycoprotein and a member of the immunoglobulin superfamily. It was originally identified in brain and designated neuronal cell adhesion molecule (NCAM). CD56 is a marker for NK- and T/NK-cell neoplasms. It is expressed in a proportion of cases of acute myeloid leukaemia and is strongly expressed on most neuroectodermal tumours.

CD57

The CD57 antigen is a glycoprotein expressed on a variable proportion of peripheral blood lymphocytes, not all of which have NK-cell activity. Approximately half of these cells are NK cells, the remainder being CD8+ T-cells. Some T- and NK-cell neoplasms express CD57, but less than 10 per cent of CD56+ nasal type T/NK-cell neoplasms express this antigen.

Many of the CD4 T-cells found in germinal centres express CD57. The cells that rosette the LP cells of NLPHL express CD4, CD57, PD-1 and CXCL13, a feature that might be of value in differentiating these tumours from other types of Hodgkin lymphoma and T-cell/histiocyte-rich B-cell lymphoma.

CD57 is also expressed on a wide range of nonhaematopoietic neoplasms.

Is it a precursor B- or T-cell lymphoma?

TERMINAL DEOXYNUCLEOTIDYL TRANSFERASE

Terminal deoxynucleotidyl transferase (TdT) is an intranuclear DNA polymerase that catalyses the addition of deoxynucleotidyl residues to DNA. It is in part responsible for creating diversity of antibodies or TCRs in the early stages of B- or T-cell development. It therefore provides a marker for precursor B- or T-cell lymphomas. It is also expressed in cases of myeloid leukaemia in lymphoid blast crisis. In normal tissues, TdT positivity is seen in cortical thymocytes and in 1–2 per cent of bone marrow lymphocytes (2–7 per cent in neonates).

Nuclear staining indicates positivity; cytoplasmic staining is artefactual and of no diagnostic significance.

Is it Hodgkin lymphoma?

CD30

CD30 is a member of the tumour necrosis factor/nerve growth factor receptor family, its expression being associated with activation. Immunochemically, it is seen as membrane and/or paranuclear staining. Diffuse cytoplasmic staining, often associated with poor fixation, is of no significance. Scattered small parafollicular blast cells are labelled in reactive tissues. Strong uniform positivity is seen in most ALCLs. The majority of Hodgkin/Reed–Sternberg (H/RS) cells in classical types of Hodgkin lymphoma are positive. A variable proportion of cells in some large B- and T-cell lymphomas express CD30.

CD15

Antibodies to CD15 recognize a specific sugar sequence known as X hapten. It is expressed in the later stages of myeloid differentiation, and strong staining of polymorphs usually provides a good internal control. H/RS cells of classical Hodgkin lymphoma express CD15 on the cell membrane and/or as a granular paranuclear aggregate in 70–80 per cent of cases. Weaker staining may be seen in a small proportion of B- and T-cell lymphomas, including ALCL. Many epithelial cells and carcinomas express CD15.

Is it a histiocytic or dendritic cell proliferation?

CD21

This antibody recognizes the complement receptor C3d that acts as the B-cell receptor for Epstein–Barr virus (EBV). It is expressed by a range of B-cells but, in paraffin-embedded tissues, the strong reactivity is with FDCs, for which it is an excellent marker.

CD23

In addition to its use in the diagnosis of B-CLL/SLL, CD23 is a marker of FDCs.

CD35

CD35 is an epitope of the receptor for the C3b fragment of human complement. It is expressed on FDCs and, to a lesser extent, on mantle cells. Like CD21 and CD23, it is used mainly for the demonstration of FDC networks and FDC tumours.

LYSOZYME

Lysozyme is an enzyme (muramidase) produced at many mucosal surfaces. It is also a good marker for normal and neoplastic cells of the myelomonocytic series, including mature histiocytes. In poorly fixed tissues it often shows confusing diffusion artefacts.

CD68

CD68 is a glycosylated transmembrane protein involved in lysosomal trafficking and is expressed in all cells containing lysosomes. There are several good monoclonal antibodies to CD68 that react with different epitopes on the molecule and give different reactivities. Thus the antibody KP1 reacts with benign and neoplastic histiocytes as well as myeloid precursors, granulocytes and most acute myeloid leukaemias. Antibody PGM1 reacts with benign and neoplastic monocytes and histiocytes, but not with granulocytic cells or their precursors.

CD123

CD123 is the interleukin-3 receptor α-chain and is a reliable marker of plasmacytoid dendritic cells. It is useful in the diagnosis of blastic plasmacytoid dendritic cell neoplasms and for detecting reactive populations of plasmacytoid dendritic cells.

CD163

CD163 is an acute phase-regulated transmembrane protein that mediates the endocytosis of haptoglobin–haemoglobin complexes. Like CD68, it is expressed on monocytes and tissue macrophages as well as myeloid leukaemia cells with monocytic differentiation.

S100 PROTEINS

S100 proteins are calcium-binding proteins originally identified in brain tissue. The designation S100 was given because these proteins are soluble in 100 per cent neutral ammonium sulfate. Their main value in haematopathology is as a marker for Langerhans cells and interdigitating reticulum cells.

They also label many neural tumours and the majority of malignant melanomas.

LANGERIN (CD207)

Langerin is a membrane-associated lectin expressed exclusively by Langerhans cells. It is localized not only on the cell surface but also intracellularly in association with Birbeck granules.

PODOPLANIN (D2-40)

Podoplanin is a small mucin-like membrane protein expressed in diverse cells, including renal podocytes, mesothelial cells, myoepithelial cells, and endothelial cells. It has sensitivity that is equal to or greater than that of CD21, CD23 and CD35 for FDCs. Staining is membranous and cytoplasmic.

CLUSTERIN

Clusterin is a glycoprotein expressed on a wide variety of cell types. Among the haematopoietic neoplasms, it has been demonstrated in ALCL and in about 50 per cent of primary cutaneous ALCLs, with dot-like Golgi staining. It is also a marker for FDCs.

Other markers

PROLIFERATION MARKER: Ki67

Ki67 is a large non-histone nucleoprotein expressed through all phases of the cell cycle except G0. Expression begins at the end of G1 and reaches a maximum in the mitotic phase of the cycle. Thus the intensity of the nuclear staining is often variable. The proliferation fraction, as determined by Ki67 labelling, may have prognostic significance and is often diagnostically helpful, for example in the distinction among small B-cell lymphomas (1–15 per cent), DLBCLs (40–95 per cent) and Burkitt lymphoma (100 per cent).

TRANSCRIPTION FACTORS—PAX5, OCT2, BOB1, PU.1

All these transcription factors, which show nuclear localization, are expressed in the majority of B-cell lymphomas and NLPHL, but are downregulated in RS and Hodgkin cells of classical Hodgkin lymphoma. PAX5 is expressed in a significant proportion of classical Hodgkin lymphomas. There is inconsistency in the reported incidence of expression of Oct-2, BOB.1 and PU.1. Whereas the neoplastic cells in NLPHL consistently express BOB1, it is also weakly expressed by a minority of H/RS cells in classical Hodgkin lymphoma. Oct2 is strongly expressed in NLPHL, but it also stains H/RS cells in many cases. PU.1 may prove to be more diagnostically useful because it is more consistently negative in classical Hodgkin lymphoma but positive in NLPHL.

ADDITIONAL MARKERS FOR GRANULOCYTES

- **CD117 (c-KIT)**—This is a tyrosine kinase receptor found in a variety of normal cell types and a number of tumours, including connective tissue tumours, carcinomas and some melanomas. In haematopathology, it is a useful marker of early myeloid cells and mast cells.
- **CD99 (MIC2)**—The transmembrane glycoprotein CD99 is a product of the *MIC2* gene. It is expressed widely in normal tissues and has been used as a marker for Ewing sarcoma–primitive neuroectodermal tumour (PNET). CD99 is positive in T-cell acute lymphoblastic leukaemias/lymphomas (T-ALLs) and most B-ALLs. It may also be positive in acute myeloid leukaemias, some B-cell lymphomas including DLBCLs, and is reported to be frequently positive in ALCL and adult T-cell leukaemia/lymphoma (ATLL).
- **CD34**—This is used mainly as a marker of endothelial cells. It is also expressed by pluripotential haematopoietic stem cells and is positive in both T- and B-ALL and in about 40 per cent of cases of acute myeloid leukaemia.

Viral agents

EPSTEIN–BARR VIRUS

The presence of latent EBV infection provides an important aid to diagnosis in several forms of lymphoma. Cells harbouring latent EBV carry multiple copies of the viral episome and express a set of viral gene products, the so-called latent proteins, consisting of six nuclear antigens (Epstein-Barr nuclear antigen [EBNAs]) and three latent membrane proteins (LMPs). In addition to the latent proteins, they also show expression of the small non-coding EBV-encoded RNAs (EBERs): the function of these transcripts is not clear, but they are consistently expressed in all forms of latent EBV infection. Three patterns of latency are described. Latency pattern I is seen in Burkitt lymphoma, where EBER transcription is associated with selective expression of EBNA1. In latency pattern II, seen in EBV-positive classical Hodgkin lymphoma and some T-cell lymphomas, EBNA1 is expressed together with LMP1 and EBERs. Lymphomas developing in transplant recipients and patients with other forms of immunodeficiency, which appear to be primarily driven by EBV, show a latency III pattern of expression with EBNA2 and LMP1. For diagnostic purposes, LMP1 and EBNAs can be demonstrated by immunohistochemistry and EBERs by ISH. In acute infectious mononucleosis, EBERs can be demonstrated in the infected B-cells by ISH, and scanty EBER-positive lymphocytes can often be identified in healthy carriers (Table 1.1).

KAPOSI SARCOMA–ASSOCIATED HERPESVIRUS OR HUMAN HERPESVIRUS 8

Kaposi sarcoma–associated herpesvirus (KSHV) is associated with Kaposi sarcoma, primary effusion lymphoma and some cases of multicentric Castleman disease. The latency-associated nuclear antigen (LANA) of KSHV has a number of actions, including an ability to bind with p53 and interfere with apoptosis. Antibodies are available to LANA-1 and give nuclear staining in positive cells.

Table 1.1 Latency patterns of Epstein-Barr virus (EBV)

	Latency pattern I EBER EBNA1	Latency pattern II EBER EBNA1 LMP1	Latency pattern III EBER EBNA2 LMP1	EBER
Burkitt lymphoma	×			
EBV+ cHL and some T-cell lymphomas		×		
PTLD lymphomas			×	
Acute IM (and carriers)				×

cHL, classical Hodgkin lymphoma; EBER, Epstein–Barr virus–encoded RNA; EBNA, Epstein-Barr nuclear antigen; EBV, Epstein–Barr virus; LMP, latent membrane protein; PTLD, post-transplant lymphoproliferative disease; IM, infectious mononucleosis.

MOLECULAR TECHNIQUES

Increasing use is being made of other techniques to confirm or refine lymphoma diagnosis.

Mention has already been made of ISH for demonstrating immunoglobulin light chains and EBERs.

Fluorescent *in-situ* hybridization

A number of lymphomas are associated with specific chromosomal translocations. Conventional cytogenetics has several disadvantages for the detection of chromosomal abnormalities, the main one being the requirement for fresh tissue. FISH, which can be applied to routinely processed and embedded tissue, offers a more convenient and reliable method of identifying translocations using labelled probes to known break points. For example, a mixture of locus-specific labelled DNA probes containing sequences homologous to the *IgH* and *bcl-2* genes is used to detect the t(14;18) typical of follicular lymphoma. Other diagnostic translocations include t(11;14) in mantle cell lymphoma and t(2;5) in ALCL. In Burkitt lymphoma, demonstration of a solitary *MYC* translocation, usually t(8;14), makes an important contribution to the diagnosis.

Polymerase chain reaction

The majority of lymphomas are diagnosed using immunohistochemistry with appropriate choice of antibodies. Even with reliable techniques and educated interpretation of the results, there remains a small number of cases where the diagnosis is uncertain. This can usually be resolved by demonstrating clonal rearrangement of the immunoglobulin heavy chain gene or the *TCR* genes. However, careful interpretation of the results is required, and correlation with clinical findings, since clonal populations may arise, for example, in viral infections. Because the technique relies on extraction of DNA from the tissue, it is subject to technical limitations imposed by the amount of tissue available and differences in fixation and processing.

DIAGNOSING THE 'UNDIAGNOSABLE' BIOPSY

Biopsies may be deemed undiagnosable for a number of reasons: they are too small, there is a severe crush or traction artefact or there is extensive necrosis. In the past, such biopsies were categorized as inadequate and a repeat biopsy requested—a step that is not always clinically possible or desirable. With modern antigen-retrieval techniques and immunohistochemistry, it is often possible to obtain useful information from such biopsies, and even to arrive at a conclusive diagnosis.

It is worth reiterating that the biopsies in which the diagnosis is most difficult to make are those subjected to poor fixation (poor-quality fixative, small volume of fixative, slow penetration of fixative because of tissue size, inadequate fixation time). That is to say, it is the handling of the biopsy after surgery that is at fault, rather than the taking of the biopsy. In such biopsies, not only is the cytomorphology degraded, but so are many immunohistochemical reactions. Two antibodies that are particularly useful in biopsies showing traction artefact are CD20 (L26) and Ki67. CD20 is a robust membrane antigen that will often be found to outline large B-cells in biopsies that in H&E-stained sections appear totally crushed. It may also identify B-cells in infarcted tissue; however, the overinterpretation of background staining in necrotic tissue must be avoided. Ki67, a nuclear antigen, commonly shows a clear-cut labelling or proliferation fraction in biopsies showing traction or crush artefact. The labelling fraction is useful in distinguishing between slowly proliferating (low-grade) and rapidly proliferating (high-grade) lymphomas.

WHERE TO BEGIN? RECOGNIZING LYMPH NODE PATTERNS

An experienced haematopathologist is usually able to assign a lymph node biopsy to one or more possible diagnostic categories (reactive, neoplastic, small cell, large cell, mixed) on the initial reading of the histology. Confirmation and refinement of the diagnosis may then be made by more detailed morphological assessment, immunohistochemistry, molecular genetic analysis, and so on. It is wasteful of resources, often difficult and sometimes misleading to attempt to make a diagnosis without having first placed the biopsy into one or more possible morphological subgroups. We would counsel against the blind application of large antibody 'panels' in the hope that the diagnosis will reveal itself. Ultimately, there should be no discrepancy between the morphology and the ancillary investigations. Pathologists lacking in experience in haematopathology may find this initial categorization of biopsies difficult, which can lead to misdirection of further investigations.

Attempts have been made to construct an algorithm that will lead the inexperienced to the correct diagnosis. We have found that the breadth and complexity of haematopathology makes it difficult to construct and use such algorithms. The following sections outline some of the features to look for in lymph node biopsies in order to place them in a diagnostic category.

SINUS ARCHITECTURE

Is the sinus architecture intact, or partially or completely destroyed?

The sinus architecture is often best visualized in reticulin-stained preparations. An intact sinus structure is usually seen in reactive lymph nodes, whereas complete or partial destruction of the sinus architecture is usual in most lymphomas. Exceptions are leukaemias and lymphomas that have leukaemic manifestations (myeloid, B-CLL/SLL, lymphoblastic), which may infiltrate nodes, leaving much of the sinus structure intact.

Are the sinuses very prominent? What cells do they contain?

Sinus histiocytosis may be seen in a number of reactive and inflammatory lymphadenopathies, particularly those involving the mesenteric lymph nodes. Lipid-filled histiocytes may be seen in Whipple disease. A sinus pattern is characteristic of many cases of Langerhans cell histiocytosis and of Rosai–Dorfman disease.

A prominent sinus pattern is seen with metastatic neoplasms, ALCL and some cases of DLBCL.

CAPSULE

Is the capsule thickened; does the lymphoproliferation extend into the perinodal tissues?

Thickening of the capsule is common in reactive/inflammatory processes. It is characteristic of the nodular sclerosing subtype of Hodgkin lymphoma. Extension beyond the capsule is often seen in lymphomas (e.g., follicular lymphoma, angioimmunoblastic T-cell lymphoma), but is less commonly seen in reactive/inflammatory processes.

REACTIVE FOLLICLES

Are reactive follicles present?

Reactive follicles are, of course, characteristic of most reactive/inflammatory lymph nodes. They may, however, be seen in a number of lymphomas (e.g., marginal zone lymphoma, interfollicular Hodgkin lymphoma) and in lymph nodes showing early or partial infiltration by lymphoma. Thus, residual reactive germinal centres may be seen in lymphoblastic lymphoma, B-CLL/SLL and mantle cell lymphoma.

OVERALL GROWTH PATTERN

Is the growth pattern follicular/nodular or diffuse?

Follicular lymphomas recapitulate much of the structure of reactive lymphoid follicles and have a follicular growth pattern. Mantle cell and marginal zone lymphomas surround and then colonize reactive follicles, frequently giving the node an overall nodular structure. Prominent proliferation centres in B-CLL/SLL may make the growth pattern appear nodular on low-power inspection. Hodgkin lymphomas of the nodular lymphocyte-predominant, lymphocyte-rich classical and nodular sclerosis subtypes have a nodular growth pattern.

In addition to follicular hyperplasia, a nodular growth pattern is characteristic of Castleman disease.

PARACORTEX

Is the paracortex expanded? If so, by what cells?

Paracortical expansion is characteristic of some reactive lymphadenopathies and is associated with viral infections. It is seen in its most extreme form in infectious mononucleosis when the paracortex is expanded by B- and T-blasts. In dermatopathic lymphadenopathy the paracortex is expanded by pale-staining interdigitating reticulum cells.

MARGINAL ZONE

Are there prominent sinusoidal, parasinusoidal or perifollicular monocytoid B-cells (cells with oval nuclei and clear cytoplasm)?

Focal monocytoid B-cell proliferation is characteristic of lymphadenopathy in human immunodeficiency virus (HIV) infection and in toxoplasmosis. Admixed neutrophils are characteristically seen, whereas T-cells are infrequent in these foci. In contrast to normal marginal zone B-cells, monocytoid B-cells are negative for BCL-2.

NECROSIS

Does the lymph node show areas of necrosis?

Caseous necrosis is characteristic of tuberculosis. Serpiginous areas of necrosis may be seen in cat scratch disease and in lymphogranuloma venereum. 'Normal' or neoplastic lymph nodes may undergo infarction. Areas of necrosis are relatively common in large cell lymphomas but rare in small cell lymphomas. When present in small cell lymphomas, necrosis usually indicates additional pathology such as virus infection. Angioinvasive lymphoproliferations, such as extranodal NK/T-cell lymphoma, nasal-type and lymphomatoid granulomatosis, usually show varying degrees of necrosis.

APOPTOSIS

Do many cells show apoptosis?

Apoptosis is a characteristic finding in reactive germinal centres. It is commonly seen in malignant lymphomas with a high growth fraction and is particularly prominent in Burkitt lymphoma. The apoptotic debris may be seen between viable cells or within macrophages. Apoptosis is a striking feature of Kikuchi–Fujimoto disease.

Apoptotic granulocytes may be seen in inflammatory conditions. If there is any doubt as to the nature of these bodies, they may be identified by immunohistochemistry using a granulocytic marker such as CD15.

GRANULOMAS

Does the lymph node contain epithelioid cell/giant cell granulomas? Are these well defined? Do they show necrosis?

Foreign material and infection may induce granulomatous inflammation in lymph nodes. Tuberculosis (necrotizing confluent granulomas) and sarcoidosis (non-necrotizing discrete granulomas) are among the most common causes of granulomatous lymphadenitis.

Epithelioid cell clusters, sometimes with giant cells, are frequently seen in Hodgkin lymphoma and some T-cell lymphomas (Lennert lymphoma). Large numbers of epithelioid cells in sheets or clusters may be seen in other lymphomas. They are often prominent in marginal zone lymphoma, T-cell histiocyte-rich B-cell lymphoma and ALCL (lymphohistiocytic variant).

2

Normal/reactive lymph nodes: Structure and cells

Lymph nodes from different anatomical sites show variation in their structure. Cervical lymph nodes exhibit the characteristic pattern of follicles, paracortex, medulla and sinuses. Axillary lymph nodes in their resting state appear as a rim of lymphoid tissue around a core of fat. In malignant lymphomas and other lymphoproliferations, this fat is colonized and may disappear completely. Mesenteric nodes have more prominent sinuses and usually less conspicuous follicles and paracortex. Pelvic lymph nodes often show prominent sclerosis.

LYMPHOID FOLLICLES

These structures generate T-dependent antibody responses. They are the site at which the development of antibody diversity and isotype switching occurs. Primary follicles are composed of small B-cells, bearing surface immunoglobulin M (IgM) and IgD, and follicular dendritic cells (FDCs). If a secondary follicle is sectioned at one pole, so as not to include the germinal centre, it will appear as a primary follicle. Secondary follicles have a germinal centre composed of blast cells (centroblasts) and their progeny (centrocytes). These cells show polarity, with the blast cells forming the dark zone (because of their deep cytoplasmic basophilia in Giemsa-stained preparations) and the centrocytes the light zone. The cells of the dark zone show numerous mitotic figures and have a high proliferation fraction; many may show features of apoptosis. Follicular polarity is usually better seen in lymphoid follicles at mucosal surfaces, such as the tonsil, than in lymph nodes.

Germinal centre B-cells do not express BCL-2 and are thus susceptible to apoptosis. Only those B-cells selected for good antibody affinity are allowed to re-express BCL-2 and survive. A high proportion of follicle centre B-cells undergo apoptosis and are ingested by macrophages within the centre (tingible body macrophages).

The germinal centre is surrounded by the mantle zone composed of small B-lymphocytes of the same phenotype as those seen in primary follicles. Marginal zone B-cells with clear cytoplasm may be seen outside the mantle zone. These are often seen in mesenteric nodes, but are usually inconspicuous at other sites. Lymphoid follicles contain a network of FDCs. FDCs capture and retain immune complexes on their surface for presentation to B- and T-cells. The nuclei of FDCs are characteristic, often appearing binucleate or multinucleate. The dendritic processes are not identifiable in routinely stained sections but are often highlighted in sections stained for IgM, which labels immune complexes on the surface of the processes. Staining for CD21 and CD23 also highlights these processes. Involuting germinal centres often contain interstitial eosinophilic proteinaceous material. Plasma cells are sometimes seen in the germinal centres of reactive lymph nodes. Lymphoid follicles contain a variable number of small T-cells, many of which express CD3, CD4, PD-1 and CD57.

PARACORTEX

The paracortex, deep to the follicles, is composed largely of T-cells with scattered B-cells. Depending on the reactive state of the node, there may be variable numbers of B- and T-blasts present. Characteristic high endothelial venules are seen within the paracortex. In these vessels, the endothelial cells appear cuboidal. Selectins expressed on these cells direct the traffic of lymphocytes from the blood into the lymph node. These small lymphocytes are often seen between the endothelial cells, or between them and the basement membrane.

Non-phagocytic antigen-processing cells, known as interdigitating reticulum cells (IDRCs), are present in the paracortex. These cells have numerous filamentous cytoplasmic processes when seen in cell suspensions. It is the complex interdigitating of these processes between adjacent cells, as seen in electron micrographs, that gives them their name. IDRCs have many similarities to Langerhans cells, but do not contain Birbeck granules. They are positive for S100 protein and HLA-DR. They have complex, deeply cleft nuclei with small nucleoli. Scattered phagocytic histiocytes are also found in the paracortex.

Cells previously known as plasmacytoid T-cells are found in the paracortex in variable numbers. These were originally thought to be T-cells because of their reactivity with CD4, but this antigen is also expressed by cells of the monocyte lineage to which these cells belong. They express CD68 and are now designated as plasmacytoid monocytes. Plasmacytoid monocytes may occur as single cells, small clusters or larger aggregates that may mimic follicle centres. The cells are of uniform size with rounded, often eccentric, nuclei that lack the characteristic clumped chromatin of plasma cells. In Giemsa-stained preparations the cytoplasm is basophilic and electron microscopy shows stacks of rough endoplasmic reticulum. Apoptotic cells are characteristically seen within clusters of plasmacytoid monocytes, although the proliferative activity of these cells is low. Their accumulation and survival is cytokine dependent and they are thought to be precursors of antigen-presenting dendritic cells.

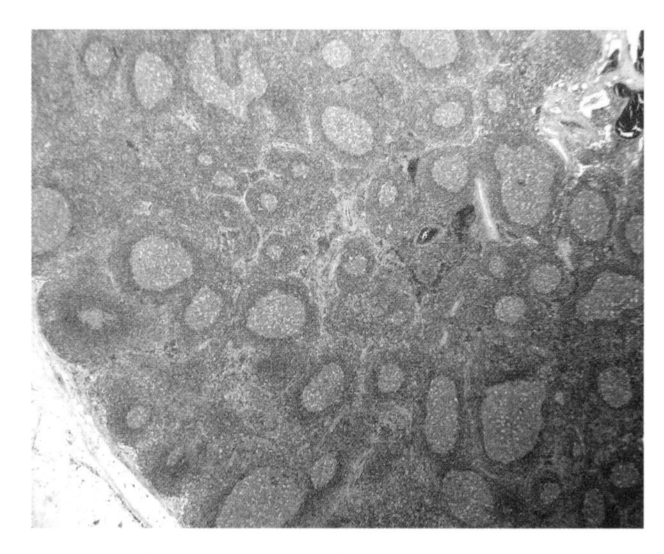

Figure 2.1 Low-power view of a reactive lymph node. Note the irregularity of size and shape of the reactive follicles, which have well-defined mantles. The subcapsular sinus and some medullary sinuses are recognizable.

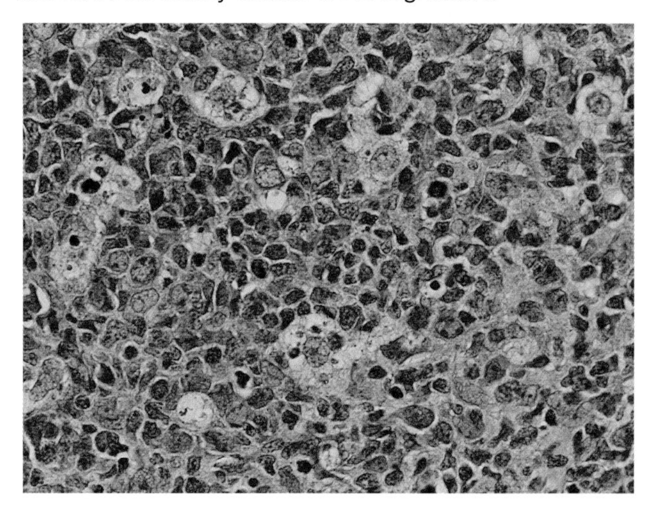

Figure 2.2 High-power view of a reactive germinal centre containing tingible body macrophages and showing zonation of centrocytes and centroblasts.

MEDULLARY CORDS

Medullary cords occur towards the hilum of the lymph node and contain a variable mixture of small lymphocytes, blast cells and plasma cells. Plasma cells are particularly prominent in chronic inflammatory diseases, such as rheumatoid disease.

SINUSES

Afferent lymphatics enter the lymph node along its convex surface and an efferent vessel leaves from the concave hilum of the node. Lymph enters the subcapsular sinus and then percolates through the network of medullary sinuses. The sinuses are defined by a scaffolding of type IV collagen to which are attached the sinus lining cells. These cells have dendritic processes and attach to one another through desmosomes. They are not phagocytic and appear to act as antigen-presenting cells. In addition to the sinus lining cells, the sinuses contain variable numbers of mononuclear cells, including phagocytic histiocytes (see Figures 2.1–2.19).

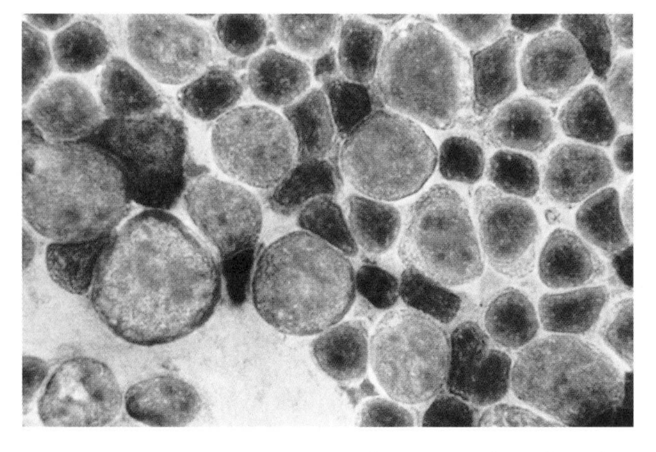

Figure 2.3 Imprint preparation of a reactive lymph node showing centroblasts with basophilic cytoplasm and visible nucleoli. The smaller lymphoid cells are a mixture of centrocytes and T-cells.

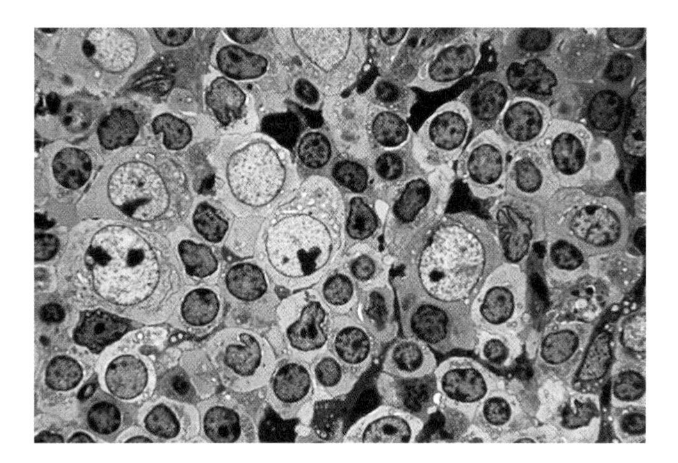

Figure 2.4 Plastic-embedded section of a reactive follicle showing centroblasts with visible nucleoli together with smaller centrocytes, a few of which show cleft nuclei.

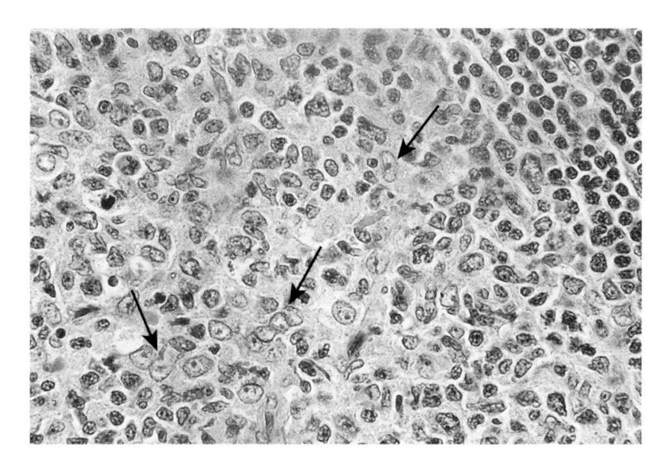

Figure 2.5 Regressing reactive germinal centre consisting mainly of centrocytes. The nuclei of several dendritic reticulum cells are present; mononuclear, binuclear and multinucleated cells are highlighted with arrows. These have oval nuclei with a well-defined nuclear membrane and a single eosinophilic nucleolus. These nuclei appear singly, in pairs or as small clusters.

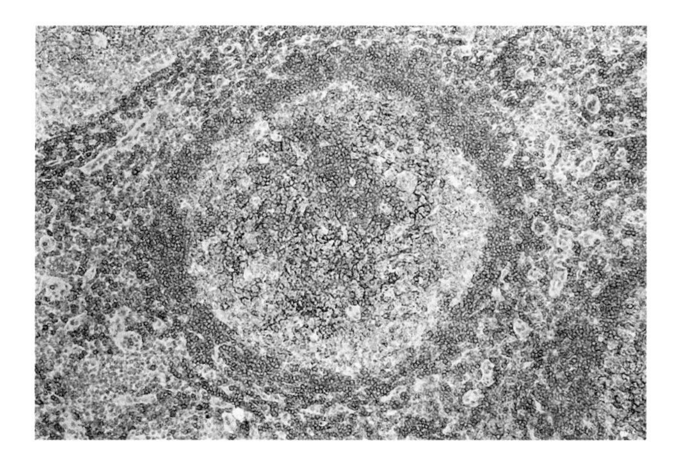

Figure 2.6 Reactive germinal follicle stained for immunoglobulin M (IgM). Note the positive staining of mantle cells and deposition of IgM, in the form of immune complexes, on follicular dendritic cells.

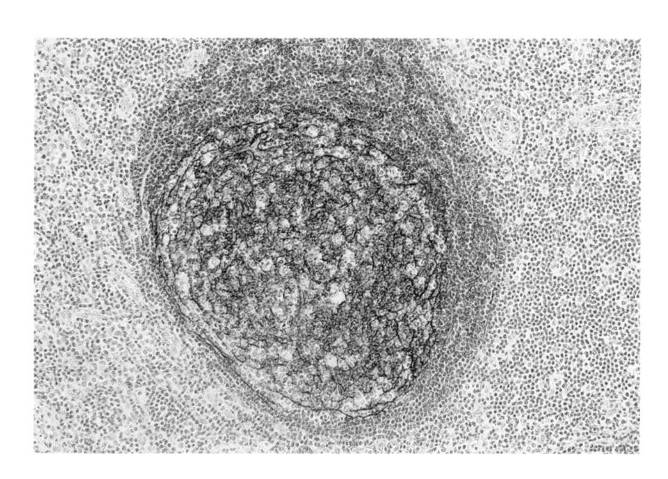

Figure 2.7 Reactive germinal follicle stained for CD21 to show follicular dendritic cells.

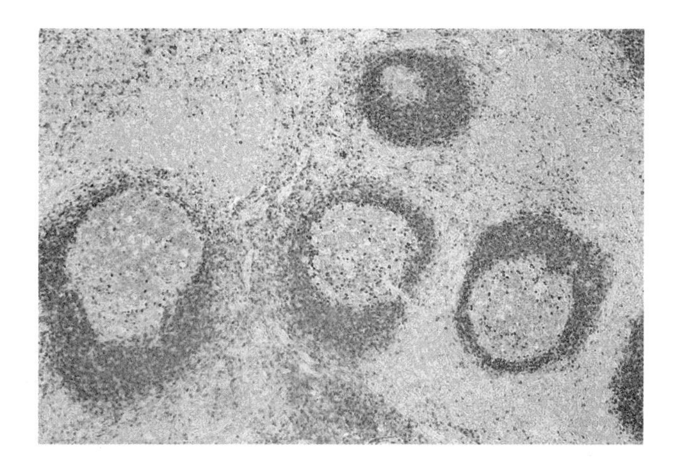

Figure 2.8 Reactive germinal follicles stained for CD79a. Note the well-defined, strongly stained mantle zones.

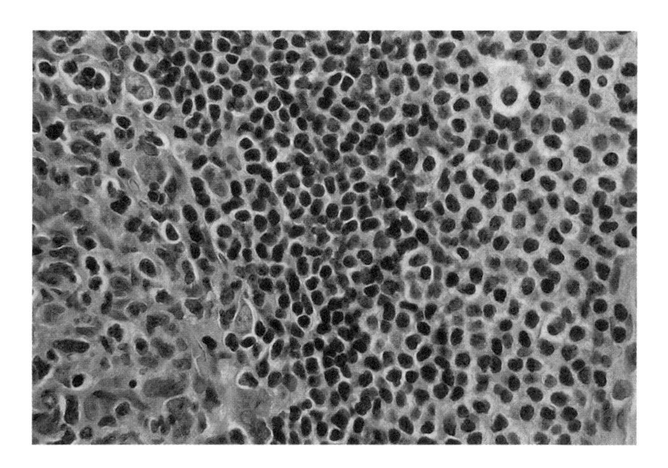

Figure 2.9 High-power view of a reactive germinal follicle showing the germinal centre cells (left), mantle cells (centre) and marginal zone cells with increased clear cytoplasm (right).

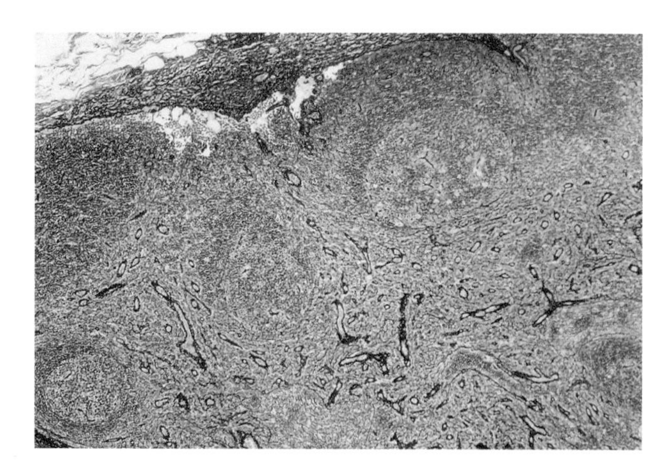

Figure 2.10 Reactive lymph node stained by the Gordon and Sweet method to show reticulin. Note the vascularity of the paracortex.

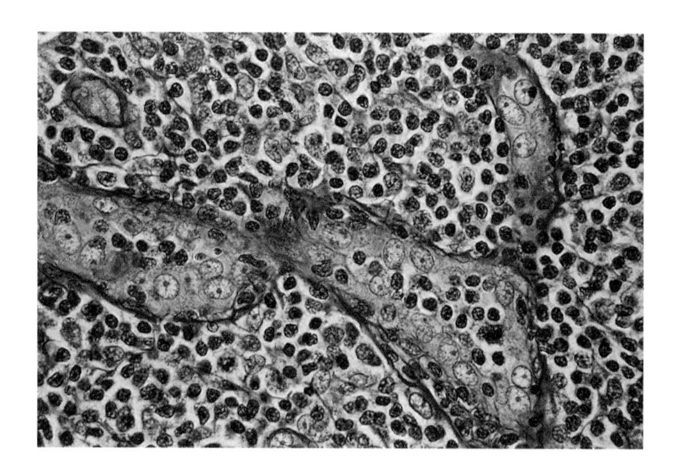

Figure 2.11 Paracortex of a reactive lymph node stained by the periodic acid–Schiff (PAS) technique. PAS-positive staining outlines a high endothelial venule containing many small lymphocytes in the lumen.

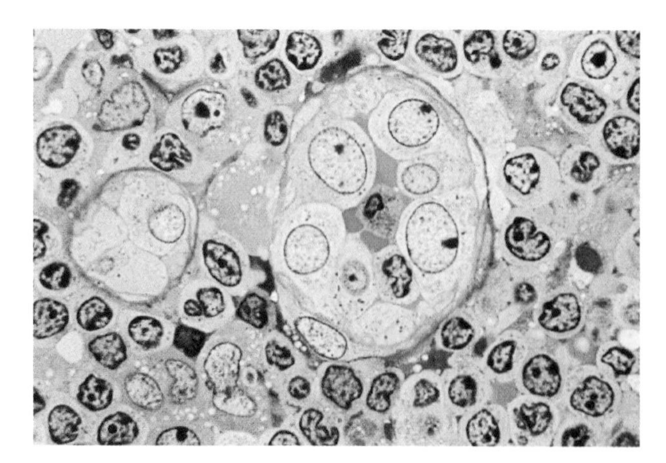

Figure 2.12 Plastic-embedded section from the paracortex of a reactive lymph node showing a high endothelial venule. Note the lymphocyte that appears to be passing between endothelial cells.

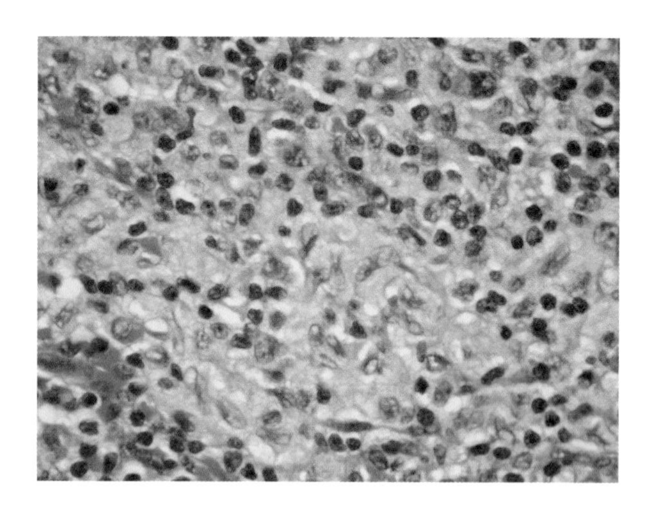

Figure 2.13 Section of paracortex showing many interdigitating reticulum cells. The twisted grooved nuclei of these cells have an appearance said to resemble a 'wrung-out dishcloth'. They have abundant, ill-defined, pale pink cytoplasm.

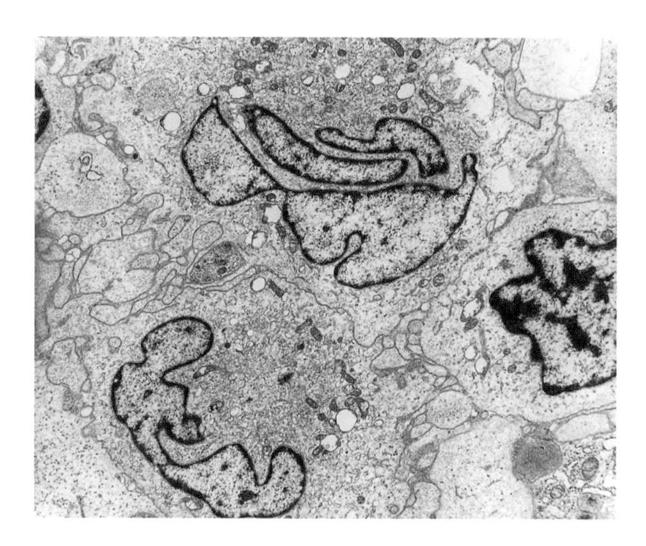

Figure 2.14 Electron micrograph of interdigitating reticulum cells. Note the complexity of the nuclear shape and the interdigitating cell membranes that give this cell its name.

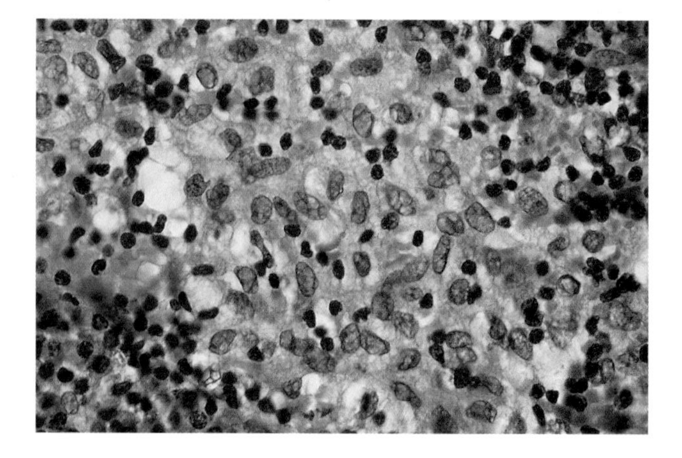

Figure 2.15 Sinus lining cells showing oval or bean-shaped nuclei with inconspicuous nucleoli and abundant pale-staining cytoplasm.

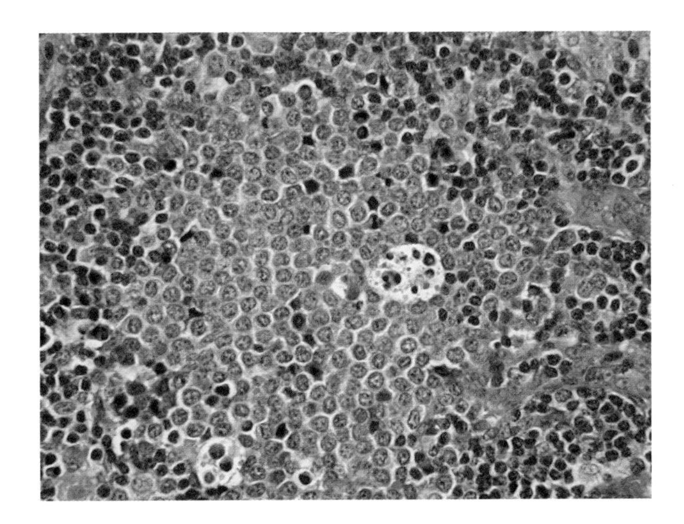

Figure 2.16 High-power view of an island of plasmacytoid monocytes. Note the regular rounded nuclei of these cells and their amphophilic cytoplasm. Many apoptotic cells are present together with two 'tingible body' macrophages.

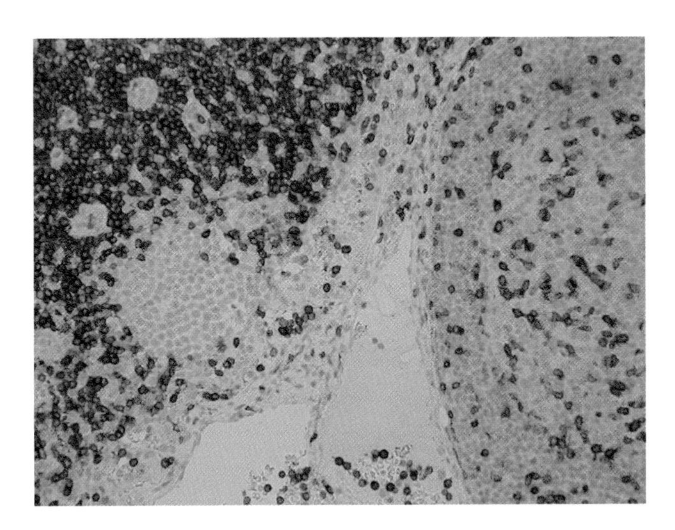

Figure 2.17 Section of reactive lymph node stained for CD3. Note scattered T-cells in the reactive follicle (right) and more closely packed T-cells in the paracortex (left). The island of plasmacytoid monocytes is unstained.

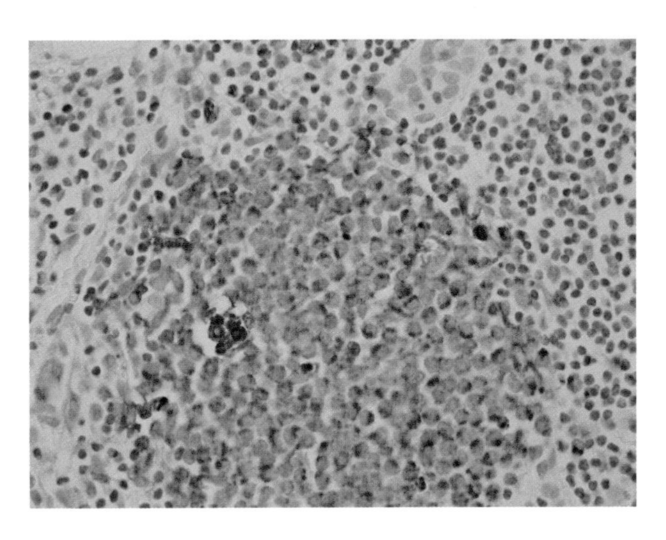

Figure 2.18 Island of plasmacytoid monocytes stained for CD68. Note strong granular staining of the plasmacytoid monocytes and strong staining of a 'tingible body' macrophage.

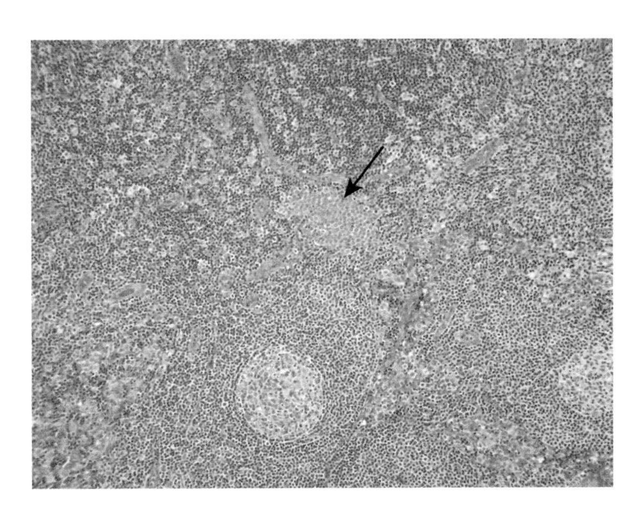

Figure 2.19 Section of reactive lymph node showing two follicles and an island of plasmacytoid monocytes (arrow).

3

Reactive and infective lymphadenopathy

We have categorized this disparate group of lymphadenopathies on the basis of their most prominent histological feature (Box 3.1). This is to some extent arbitrary, since some have overlapping features and these may vary at different stages of the disease. Biopsy is most likely to be performed in patients with persistent lymphadenopathy, often to exclude malignancy. Well-taken, well-fixed, whole lymph node biopsies are desirable in such cases because a number of reactive lymphadenopathies may closely mimic malignant lymphoma.

FOLLICULAR HYPERPLASIA

NON-SPECIFIC FOLLICULAR HYPERPLASIA

Follicular hyperplasia is probably the most common pattern of lymph node reaction seen and is characterized by enlarged follicles with prominent germinal centres. It is often accompanied by the presence of plasma cells in the medullary cords and throughout the interfollicular parenchyma. The causes of reactive follicular hyperplasia generally include antigen challenges that stimulate a B-cell response. Ancillary diagnostic procedures identify many of the aetiological agents but in the absence of demonstrable infecting microorganisms and any specific histological feature that may point to the aetiology, this reaction is termed 'non-specific hyperplasia'. Such non-specific reactions are more common in children and young adults, and are often seen in nodes draining infected sites, such as tonsils, skin or intestinal tract. In reactive lymphadenopathy, the follicles retain a distinct mantle of small lymphocytes, and the germinal centres may show polarization of the centroblasts and centrocytes. They are generally distributed predominantly in the cortex of the node and are often irregular in shape and size. Reactive germinal centres show a large number of mitotic figures and numerous apoptotic bodies that are frequently phagocytosed by 'tingible body macrophages'.

RHEUMATOID LYMPHADENOPATHY

The lymphadenopathy in rheumatoid arthritis is not limited to nodes draining affected joints but may often be generalized as part of this systemic disease. There is marked follicular hyperplasia. The enlarged germinal centres may contain amorphous periodic acid–Schiff (PAS)–positive hyaline deposits, and infrequently sarcoid-like granulomas may accompany the follicular hyperplasia. Large numbers of plasma cells, often with Russell bodies, infiltrate the medullary cords and may also be present within germinal centres.

Gold lymphadenopathy is a complication of longstanding intramuscular injections of colloidal gold for the

BOX 3.1: Patterns of reactive and infective hyperplasia

FOLLICULAR HYPERPLASIA

Non-specific follicular hyperplasia
Rheumatoid lymphadenopathy
Syphilis
Toxoplasmosis
Kimura disease
Measles
Cytomegalovirus lymphadenitis
Human immunodeficiency virus (HIV) lymphadenitis
Progressive transformation of germinal centres
Castleman disease

PARACORTICAL EXPANSION

Dermatopathic lymphadenopathy
Inflammatory pseudotumour of lymph nodes
Kikuchi–Fujimoto disease
Systemic lupus erythematosus
Viral lymphadenitis, including infectious mononucleosis
Drug-induced lymphadenopathy
Autoimmune lymphoproliferative syndrome (ALPS)

SINUS EXPANSION

Rosai–Dorfman disease (sinus histiocytosis with massive lymphadenopathy)
Langerhans cell histiocytosis
Lipogranulomatous reaction
Silicone lymphadenopathy and storage diseases

GRANULOMATOUS LYMPHADENOPATHY

Suppurative

Cat scratch, lymphogranuloma venereum, tularaemia, *Yersinia, Listeria, Corynebacterium*
Epithelioid granulomas

Necrotizing

Tuberculosis, atypical mycobacteria, leprosy, fungi

Non-necrotizing

Sarcoidosis, berylliosis, protozoa and metazoa, lymph nodes draining carcinoma
Reaction to foreign materials

treatment of rheumatoid arthritis. The lymph nodes show changes similar to those of rheumatoid lymphadenopathy. In addition, there is sinus histiocytosis with histiocytes and giant cells containing black pigment. This pigment gives a characteristic orange-red birefringence in polarized light (Box 3.2 and Figures 3.1 to 3.4).

BOX 3.2: Rheumatoid lymphadenopathy

- Lymphadenopathy may be widespread
- Follicular hyperplasia
- Follicles may contain plasma cells
- Mantle zone retained
- Plasma cell infiltration of medullary cords

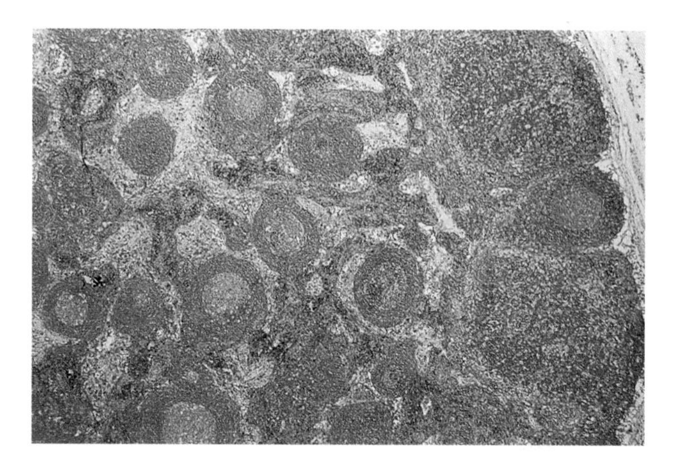

Figure 3.1 Rheumatoid lymphadenopathy: lymph node stained with methyl green pyronin. The node shows follicular hyperplasia with numerous red (pyroninophilic) plasma cells in the medullary cords.

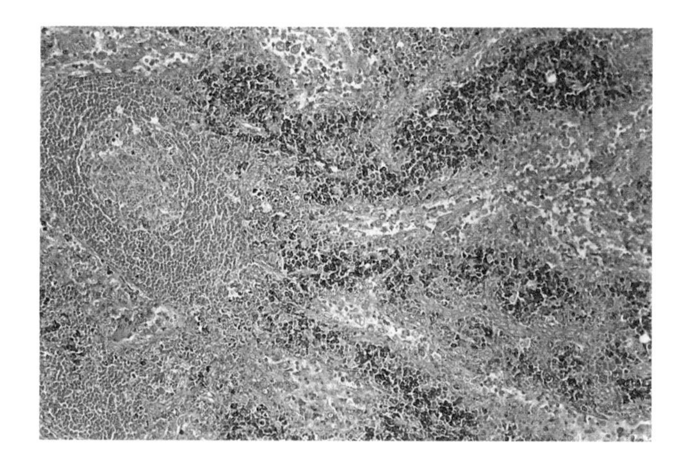

Figure 3.2 Rheumatoid lymphadenopathy: lymph node stained by Giemsa stain. The deep basophilia of the plasma cell cytoplasm highlights these cells in the medullary cords.

SYPHILIS

In primary syphilis, the site of entry of *Treponema pallidum* is characterized by a chancre that heals after 2–4 weeks but the treponemes spread from the chancre to regional lymph nodes, which become enlarged, hard and painless. Secondary syphilis begins 6–8 weeks after infection, and manifests with generalized lymphadenopathy with localized or generalized skin and

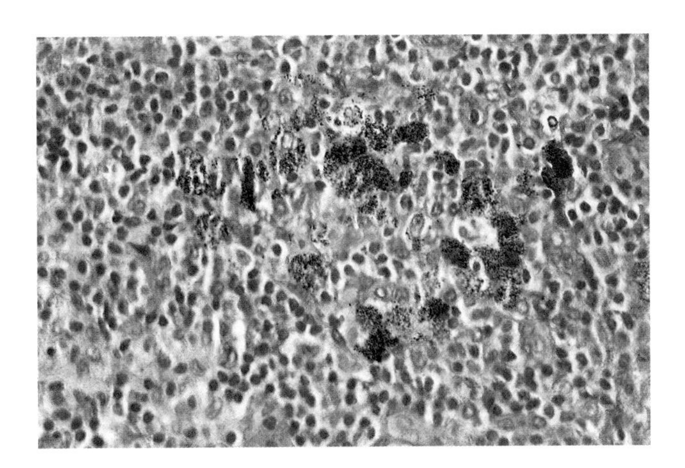

Figure 3.3 'Gold lymphadenopathy'. Axillary lymph node from a patient with rheumatoid disease treated with colloidal gold. Black pigment is seen within histiocytes.

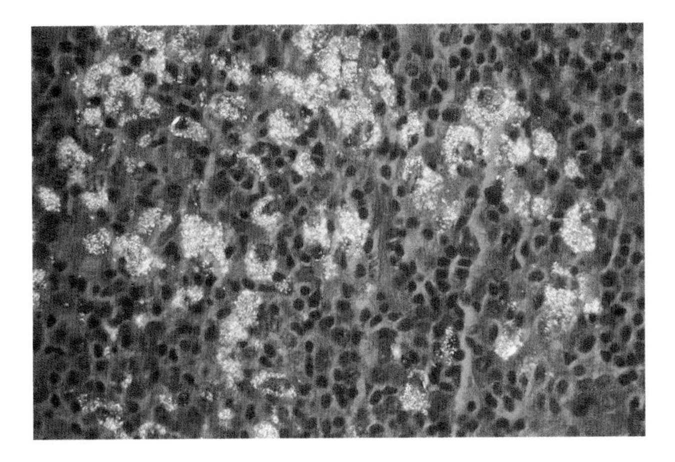

Figure 3.4 'Gold lymphadenopathy' viewed in polarized light showing characteristic orange–red birefringence.

mucosal eruptions. After a latency period of as long as 2 years with occasional relapses, tertiary syphilis develops with the formation of gummata, ulcers and nodules in various organs, including the cardiovascular and central nervous systems. The lymphadenopathy that accompanies all three stages of the disease is a result of persistence of the organisms with continuous antigenic stimulation of B- and T-cells. The former is represented by marked follicular hyperplasia that extends into the medulla. This is accompanied by expansion of the paracortex and marked fibrosis of the capsule, the fibrosis sometimes penetrating into the node. Plasmacytosis is prominent within the medullary cords. Small, non-caseating epithelioid granulomas are sometimes present with single so-called 'naked' giant cells in the paracortex. There is arteritis and phlebitis of the numerous vessels that form in the capsule and pericapsular tissues, and these are characterized by a prominent cuff of plasma cells and lymphocytes.

Silver stains reveal the spirochaetes in the walls of high endothelial venules and capsular vessels in all three stages of the disease. Immunohistochemistry using antibodies to *T. pallidum* is a more reliable method of identifying these spirochaetes than silver stains (see Inflammatory pseudotumour of lymph nodes, later) (Box 3.3 and Figures 3.5 to 3.7).

BOX 3.3: Syphilitic lymphadenopathy

- Generalized lymphadenopathy a feature of secondary syphilis
- Follicular hyperplasia
- Plasma cells expand medullary cords
- Epithelial cell clusters in paracortex
- Naked giant cells
- Capsular fibrosis
- Arteritis and phlebitis of capsular vessels
- Perivascular lymphoplasmacytic cuffing
- Spirochaetes demonstrated by silver stains

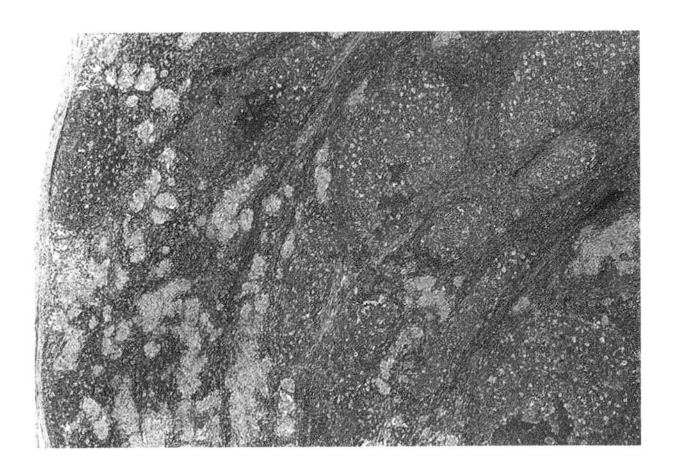

Figure 3.5 Secondary syphilis. Lymph node showing marked follicular hyperplasia with small epithelioid granulomas in the paracortex.

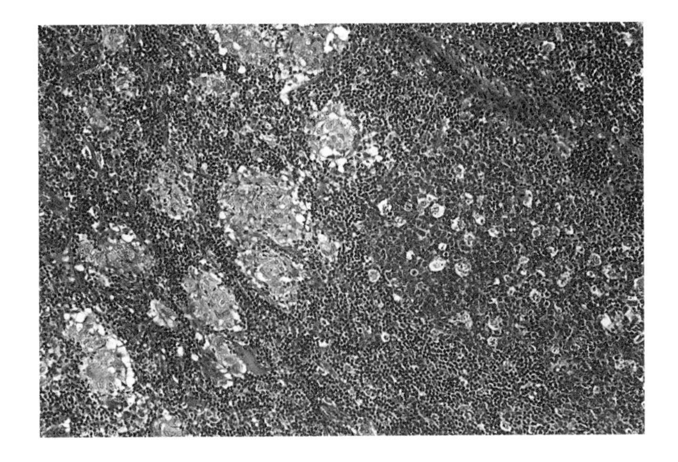

Figure 3.6 Secondary syphilis. Lymph node showing epithelioid cell clusters in the paracortex adjacent to a reactive follicle.

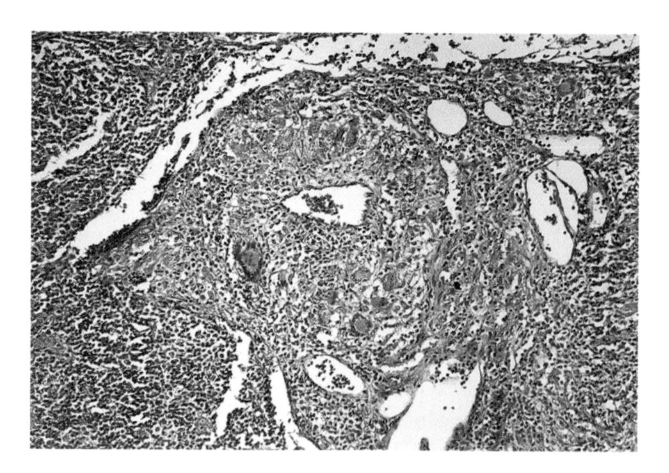

Figure 3.7 Syphilitic lymph node showing marked venulitis with giant cells.

TOXOPLASMA LYMPHADENOPATHY

Lymphadenitis is the most frequent manifestation of symptomatic toxoplasmosis. *Toxoplasma gondii*, a coccidian parasite, is one of the most prevalent and geographically widespread protozoan infections of humans. The histological changes of toxoplasmic lymphadenitis are a mixed pattern of prominent follicular hyperplasia and monocytoid B-cell hyperplasia, with small to large clusters of epithelioid histiocytes in the paracortex and sometimes extending into the germinal centres. Monocytoid B-cells do not express BCL-2 protein, a feature that can help highlight them. Very rarely, parasitic cysts are found, usually within the cortical sinus of the node. These cysts do not appear to evoke a tissue response, and the low rate of detection of *Toxoplasma* genomes by polymerase chain reaction (PCR) has raised the suspicion that the lymphadenitis may be a reaction to protozoan antigens rather that to direct contact with the organisms (Box 3.4 and Figures 3.8 to 3.10).

BOX 3.4: Toxoplasmic lymphadenitis

- Prominent follicular hyperplasia
- Monocytoid B-cell hyperplasia
- Epithelioid cell clusters that often impinge on germinal centres
- *Toxoplasma* pseudocysts filled with merozoites are exceptional

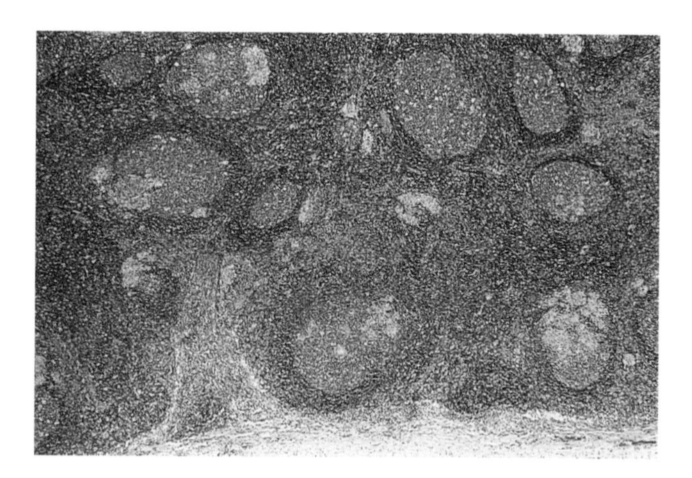

Figure 3.8 Toxoplasmosis. Lymph node showing marked follicular hyperplasia and numerous small epithelioid cell clusters.

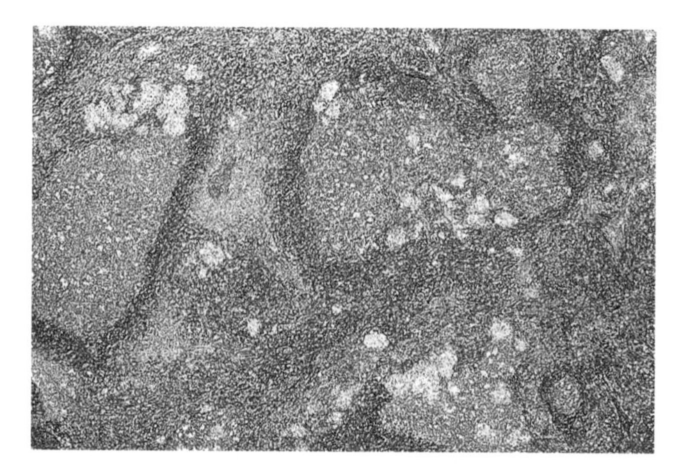

Figure 3.9 Toxoplasmosis. The epithelioid cell clusters appear to encroach on and enter the germinal centres. A collection of monocytoid B-cells is seen between the two germinal centres.

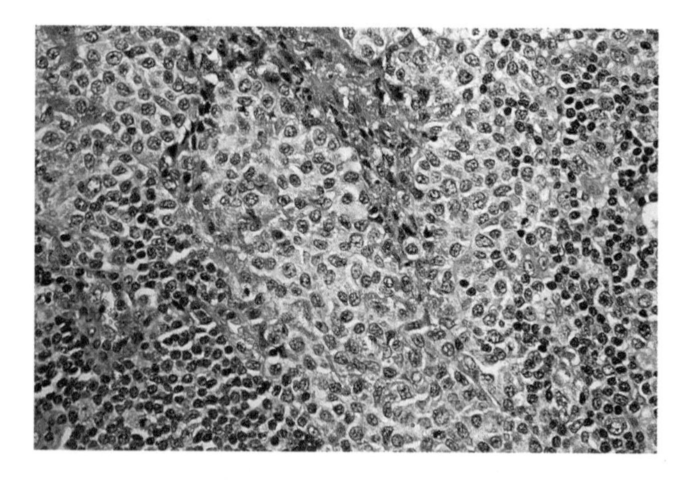

Figure 3.10 Toxoplasmosis. Prominent monocytoid B-cells distributed in and around a sinus.

KIMURA DISEASE

Kimura disease is a self-limiting condition that is prevalent among but not exclusive to Asian populations, with a striking male predominance. If untreated, this idiopathic condition usually remains static and may show regression. The nodes involved measure up to a few centimetres in diameter, and multiple enlarged nodes may become matted. The histological changes are characterized by follicular hyperplasia, paracortical expansion with prominent high endothelial venules and marked eosinophil infiltration. There is often a proteinaceous precipitate in the germinal centres, and immunostains reveal immunoglobulin E (IgE) deposited on the follicular dendritic cell network. The germinal centres may show foci of necrosis or folliculolysis. There is an accompanying eosinophil infiltration of the germinal centres that can form eosinophil micro-abscesses. Eosinophils infiltrate the sinuses and paracortex, where micro-abscesses may also form. The expanded paracortex contains plasma cells, small lymphocytes, mast cells and occasional Warthin–Finkeldey-type giant cells. Patchy fibrosis may occur around venules (Box 3.5 and Figures 3.11 to 3.13).

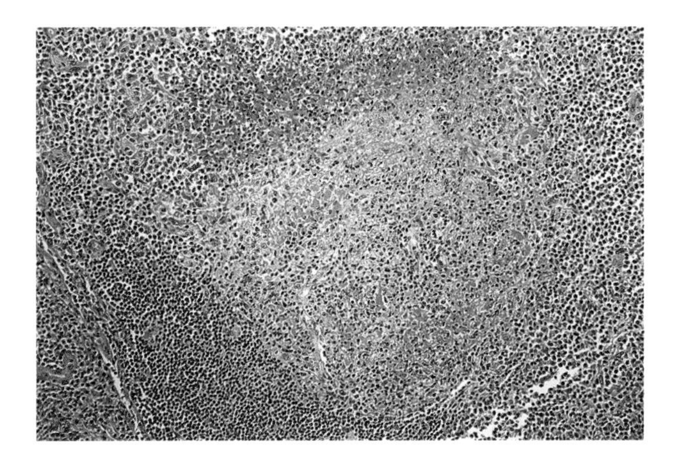

Figure 3.12 Kimura disease. High-power view of the follicle containing an eosinophil micro-abscess.

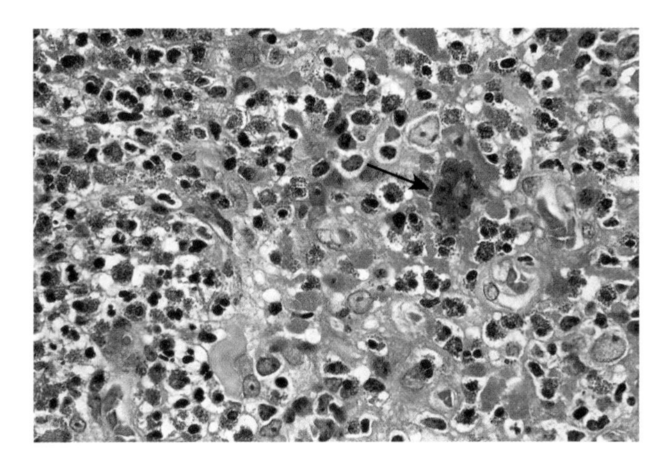

Figure 3.13 Kimura disease. Follicle centre largely replaced by eosinophils and amorphous eosinophilic material. Also shown, by the arrow, is a follicular dendritic cell polykaryon.

BOX 3.5: Kimura disease

- Common in Asians, male predominance
- Enlarged lymph nodes in neck and periauricular region, may be matted
- Occasional involvement of salivary glands
- Blood eosinophilia
- Florid follicular and germinal centre hyperplasia with proteinaceous precipitate
- Immunoglobulin E (IgE) deposition on follicular dendritic cells
- Paracortical expansion by plasma cells, small lymphocytes and mast cells
- Marked eosinophil infiltration of germinal centres, paracortex and medulla with micro-abscess formation
- Warthin–Finkeldey-type polykaryocytes

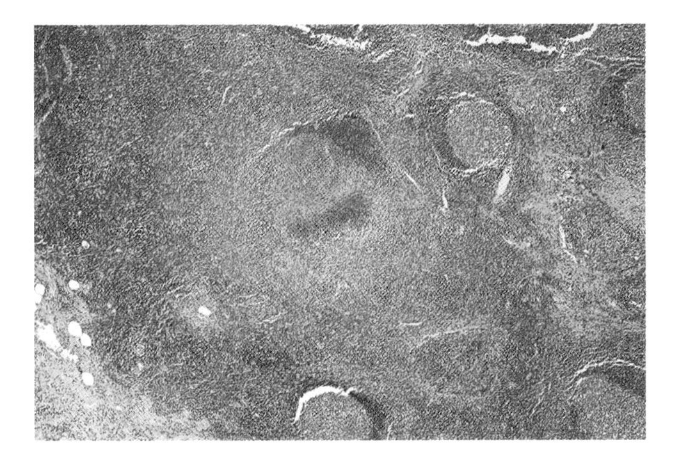

Figure 3.11 Kimura disease. Lymph node shows follicular hyperplasia. One follicle shows necrosis and an eosinophil micro-abscess.

HUMAN IMMUNODEFICIENCY VIRUS/ ACQUIRED IMMUNE DEFICIENCY SYNDROME LYMPHADENITIS

Persistent generalized lymphadenopathy accompanied by weight loss, fever, night sweats, malaise, diarrhoea and hypergammaglobulinaemia may be a presenting manifestation of human immunodeficiency virus/acquired immune deficiency syndrome (HIV/AIDS) (Box 3.6 and Figures 3.14 and 3.15).

HIV shows strong tropism for lymphoid tissues, especially CD4+ T-cells, monocytes and dendritic cells. In the acute phase of the infection, HIV infects mononuclear cells in the peripheral blood, which migrate to lymphoid organs causing acute reactive lymphadenitis. Macrophages and dendritic cells in lymph nodes form the reservoirs for the virus, and circulating T-cells disseminate the virus. The result of continued infection causes cytopathic destruction of CD4+ T-cells and dendritic cells in germinal centres so that the latter involute. Opportunistic infections may supervene.

BOX 3.6: Human immunodeficiency virus lymphadenitis

APPROPRIATE CLINICAL SETTING

- Fever, weight loss, diarrhoea, hypergammaglobulinaemia
- Persons at risk for infection
- Palpable lymph nodes at two or more sites
- Decreased peripheral CD4+ T-cells
- Reversed CD4+/CD8+ T-cell ratio
- Positive human immunodeficiency virus (HIV) antigen or antibody test

FLORID FOLLICULAR HYPERPLASIA (ACUTE STAGE)

- Hyperplastic, irregular follicles
- Mantle may be attenuated or disrupted
- Monocytoid B-cell hyperplasia
- Warthin–Finkeldey giant cells

FOLLICULAR INVOLUTION (SUBACUTE AND CHRONIC STAGE)

- Follicular effacement
- Germinal centre involution with accumulation of hyaline material
- Lymphocyte depletion in paracortex
- Plasma cell accumulation in paracortex
- Vascular proliferation in paracortex

LYMPHOCYTE DEPLETION (BURNOUT)

- Atrophic or absent follicles
- Hyalinized germinal centres with prominent thick vessels and periodic acid–Schiff (PAS)-positive deposits
- Lymphocyte depletion in paracortex
- Extensive vascular proliferation and fibrosis

Lymph node histology progresses with evolution of the disease through the following stages:

- Florid follicular hyperplasia.
- Mixed follicular hyperplasia and follicular involution.
- Follicular involution.
- Lymphocyte depletion.

Lymph nodes at the stage of florid follicular hyperplasia show very prominent, often irregular (geographic) follicles. The mantle cells are often attenuated and in places may be disrupted. Naked germinal centres may be seen. There is often marginal zone B-cell hyperplasia. Multinucleated giant cells of the Warthin–Finkeldey type are scattered randomly in the parenchyma.

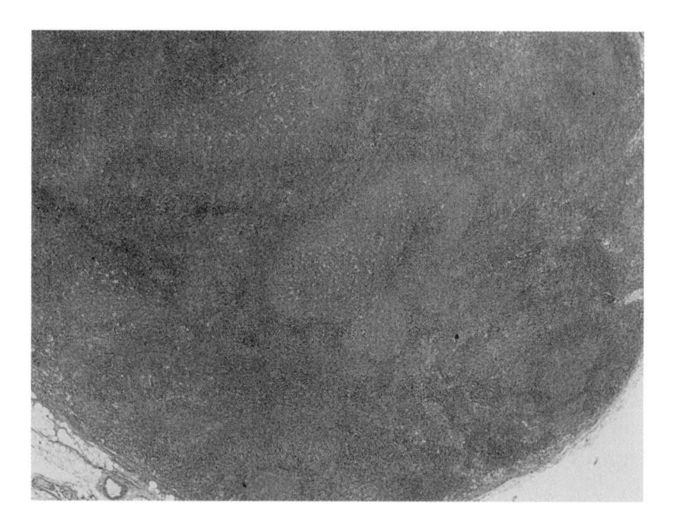

Figure 3.14 Human immunodeficiency virus lymphadenopathy. Lymph node showing follicular hyperplasia with large irregular germinal centres. Islands of pale-staining monocytoid B-cells can be seen in the centre of the field.

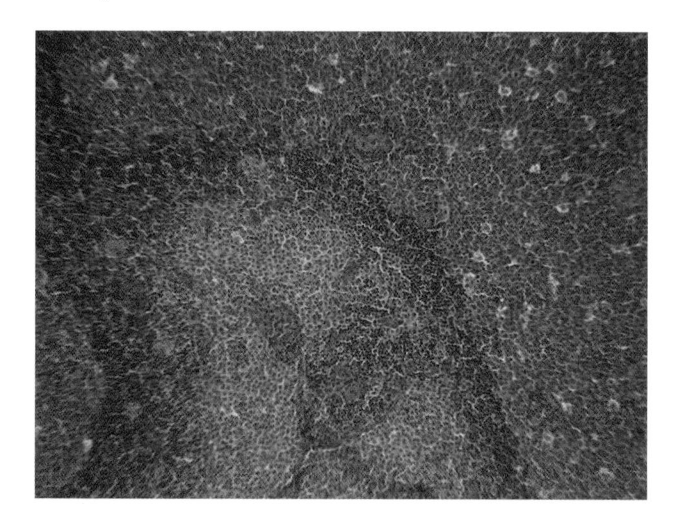

Figure 3.15 Human immunodeficiency virus lymphadenopathy. High-power view showing part of a germinal centre and an island of monocytoid B-cells.

Follicular lysis follows infiltration of the germinal centres by small lymphocytes, resulting in disruption of the follicles. PAS-positive material accumulates in the follicles, making them less cellular. Follicular lysis is accompanied by progressive plasma cell accumulation.

Lymphocyte depletion represents the burnt-out stage of HIV/AIDS lymphadenitis with atrophic follicles, lymphocyte depletion and extensive diffuse vascular proliferation. The follicles are small and depleted of lymphoid cells and contain thick collagen-ensheathed vessels surrounded by deposits of PAS-positive material. The follicular atrophy may progress to complete hyalinization, and interfollicular and paracortical zones show lymphocyte depletion and extensive vascularization. There is a prominence of plasma cells and diffuse fibrosis so that the overall appearance is that of an exhausted, burnt-out node.

The presence of HIV in infected nodes can be demonstrated with immunostaining to various HIV antigens, such as the core protein p24. Staining for this antigen is localized to the follicular dendritic cells similar to the distribution of the virus detected by electron microscopy. HIV RNA can also be demonstrated by *in-situ* hybridization and PCR.

PROGRESSIVE TRANSFORMATION OF GERMINAL CENTRES

Progressive transformation of germinal centres (PTGC) is most commonly seen in adolescent and young adult males, presenting as solitary painless lymphadenopathy, sometimes of long duration. The cervical nodes are most commonly involved, axillary and inguinal nodes much less frequently. The disease runs a benign course but it may be recurrent. Rare cases show a synchronous or metachronous association with nodular lymphocyte-predominant Hodgkin lymphoma (NLPHL) (Box 3.7 and Figures 3.16 to 3.18).

Lymph nodes are usually considerably enlarged and may have been present for many months. They show one or several expanded follicles interspersed between the reactive follicles. The expansion of the follicles and disruption of the germinal centres is caused by the influx of mantle cells. This is seen most clearly with immunohistochemistry. Mantle cells stain strongly with antibodies to IgD and CD79a, and can be seen breaking up the germinal centres into small clusters of cells. Antibodies to follicular dendritic cells (CD21, CD23, CD35) show an expanded dendritic cell network. Loose clusters of epithelioid histiocytes may be seen in PTGC; in some cases these surround the expanded follicle.

The main differential diagnosis of PTGC is with NLPHL. In the former, the expanded follicles are scattered among reactive follicles; in the latter, any residual reactive follicles are compressed towards the periphery of the node.

BOX 3.7: Progressive transformation of germinal centres

- Most common in young males
- May be recurrent
- Single, asymptomatic, enlarged node
- Expanded follicles in a background of follicular hyperplasia
- Thick mantle around expanded follicles
- Mantle cells infiltrate germinal centres, which break up into small groups of centroblasts

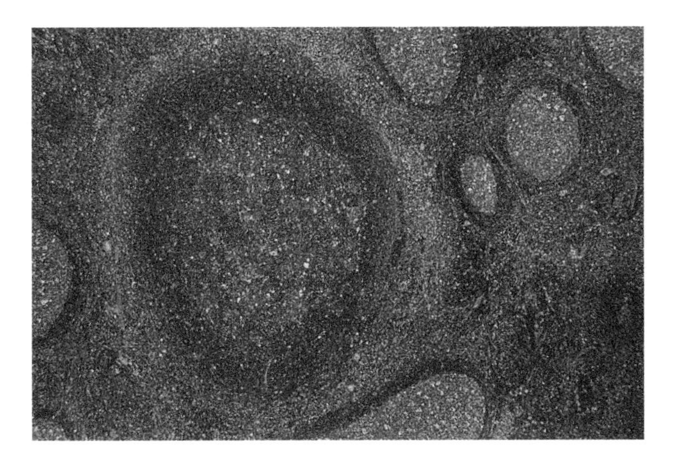

Figure 3.16 Progressive transformation of germinal centres. Section showing a 'transformed' expanded germinal centre with adjacent reactive lymphoid follicles.

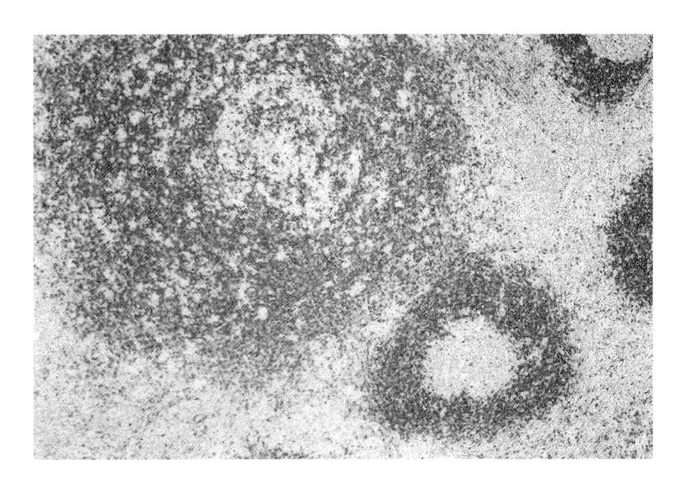

Figure 3.17 Progressive transformation of germinal centres. Section stained for CD79a showing strong positive staining of mantle cells. The 'transformed' follicle is expanded and broken up by mantle cells. Staining for immunoglobulin D would show a similar appearance.

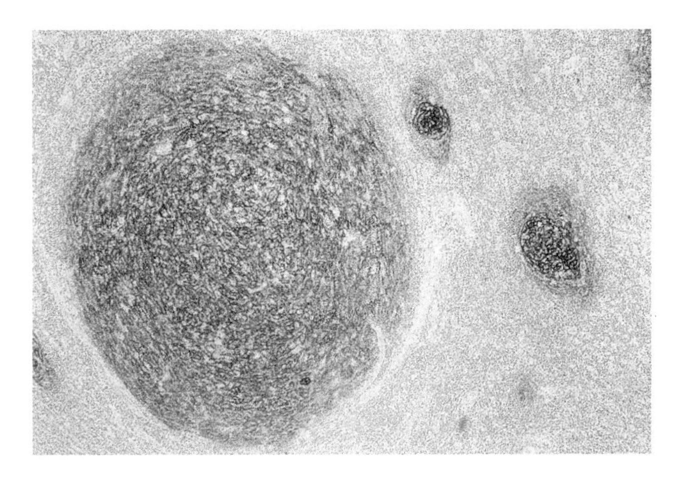

Figure 3.18 Progressive transformation of germinal centres. Section stained for CD21 showing follicular dendritic cells. The 'transformed' follicle shows an expanded network of follicular dendritic cells in comparison with the tight networks seen in the adjacent reactive follicles.

Unlike the individual L and H cells ('popcorn cells') seen within nodules of NLPHL, the cells at the centre of the nodules of PTGC consist of clusters of centroblasts.

CASTLEMAN DISEASE

Castleman disease is conveniently divided into three types: hyaline vascular, plasma cell variant—localized, and multicentric Castleman disease. There may appear to be some overlap among these categories, possibly because the follicles in the plasma cell variant become more hyaline with time.

Hyaline vascular Castleman disease

Hyaline vascular Castleman disease (HVCD) occurs most commonly in young adults and presents most frequently in the mediastinum. Peripheral lymph nodes may be involved, as also may be various extranodal sites. The lesion is usually solitary and not accompanied by systemic symptoms.

Histologically, HVCD shows characteristic follicles with a broad mantle of small lymphocytes that often assume a concentric onion-skin pattern. The centres of the follicles are hypocellular, consisting mainly of endothelial cells and dendritic reticulum cells, some of which may show nuclear atypia. Occasional follicles, depending on the plane of section, show penetration through the mantle by vessels ensheathed with collagen, giving a lollipop appearance.

The interfollicular tissues are composed predominantly of a network of vessels on a thick collagenous scaffold, a feature well demonstrated in sections stained for reticulin. Scattered or clustered small lymphocytes, plasma cells and plasmacytoid monocytes are seen among these vessels. Sinus structures are not usually identifiable (Box 3.8 and Figures 3.19 to 3.20).

BOX 3.8: Castleman disease

HYALINE VASCULAR CASTLEMAN DISEASE

- Most common in young adults
- Commonly presents in mediastinum
- Hyaline follicles consist mainly of endothelial cells and dendritic reticulum cells
- Mantle zone shows onion-skin layering
- Interfollicular areas vascular
- Sinus structure not usually visible

CASTLEMAN DISEASE PLASMA CELL VARIANT—LOCALIZED

- Wide age range
- Systemic symptoms
- Anaemia, hyperglobulinaemia, elevated erythrocyte sedimentation rate (ESR), increased plasma cells in bone marrow
- Abdominal nodes most commonly involved
- Follicular hyperplasia with narrow mantle zones
- Dense interfollicular infiltrate of plasma cells obscures underlying architecture
- One third of cases show light chain restriction

MULTICENTRIC CASTLEMAN DISEASE

- Older age groups
- May be primary or associated with other diseases such as human immunodeficiency virus (HIV) and acquired immune deficiency syndrome (AIDS)
- Overproduction of interleukin-6 (IL-6) from endogenous sources or from human herpesvirus 8 (HHV-8)
- Histology similar to localized disease
- Sixty per cent of cases associated with POEMS syndrome (polyneuropathy, organomegaly, endocrine abnormalities, monoclonal gammopathy, skin rashes)

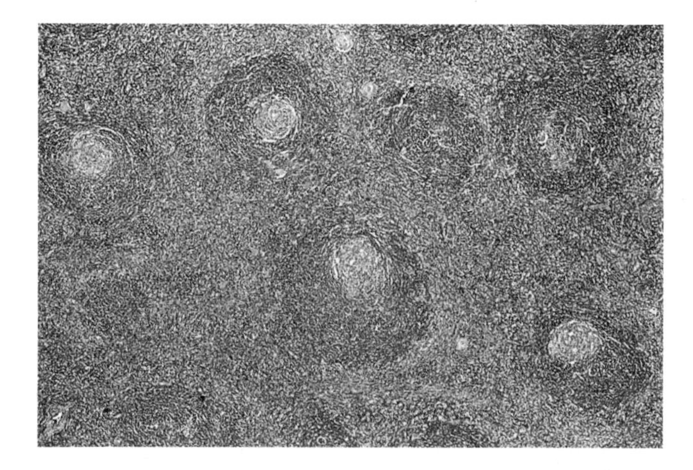

Figure 3.19 Castleman disease, hyaline vascular type. Note hyalinized follicles and lack of sinus structures.

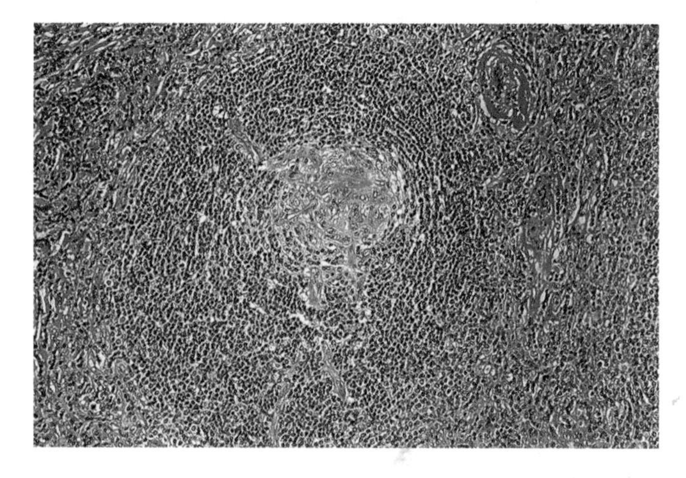

Figure 3.20 Castleman disease, hyaline vascular type. Follicle showing hyalinization of the germinal centre with penetrating vessels, onion-skin layering of the mantle cells and vascularized interfollicular tissue.

Typical HVCD is so characteristic that it is easily recognizable. Difficulties may arise when few or many hyaline vascular follicles are found in what otherwise appears to be a reactive lymph node, such as might be seen in the late stages of HIV infection. These cases, in contrast with HVCD, usually show some residual sinus structure, contain more plasma cells and lack an interfollicular vascular network.

HVCD is a benign proliferation that is usually cured by local resection. It has a rare association with follicular dendritic cell sarcoma (see Chapter 9).

Castleman disease plasma cell variant—localized

Castleman disease plasma cell variant—localized (PCD) has a wide age spectrum. It presents most commonly with abdominal lymphadenopathy involving one or a group of nodes. Mediastinal and peripheral lymphadenopathy are much less common than in HVCD. Patients typically have systemic symptoms and abnormal laboratory test results: anaemia, raised polyclonal gamma globulin, elevated erythrocyte sedimentation rate (ESR) and increased plasma cells in the bone marrow. These symptoms disappear and the laboratory test results revert to normal after surgical removal of the affected nodes. The systemic effects appear to be mediated by interleukin-6 (IL-6) secreted by the affected nodes.

Histologically, PCD shows varying degrees of follicular hyperplasia with a narrow mantle zone surrounded by sheets of plasma cells. The density of the plasma cells often obscures the underlying sinus structure, although this is usually identifiable in areas. In the majority of cases of PCD the plasma cells show polytypic immunoglobulin light chain expression; over one third show light chain restriction (usually lambda light chain). In most situations in haematopathology, light chain restriction is interpreted as evidence of monoclonality and hence of neoplasia. In PCD the monotypic plasma cells, on molecular examination, are found to be polyclonal. This monotypia may reflect an abnormal IL-6–induced proliferative drive (Box 3.8 and Figures 3.21 to 3.25).

The main histological differential diagnosis of PCD is with reactive lymphadenopathies showing marked plasmacytosis, as seen in rheumatoid disease and syphilis. In these nodes the underlying sinus structure is usually more apparent and the interfollicular infiltrate less uniformly plasmacytic.

Multicentric Castleman disease

Multicentric Castleman disease (MCD) occurs in an older age group than PCD or HVCD. It may appear as a primary disease or in association with HIV infection, Kaposi sarcoma, plasma cell neoplasms, malignant lymphomas and autoimmune disease. In common with PCD, it is associated with systemic symptoms and

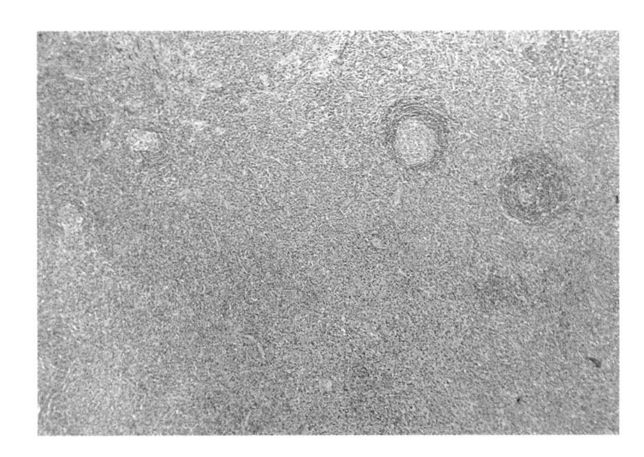

Figure 3.21 Castleman disease, plasma cell variant. Reactive follicles in a background of plasma cells. No sinus structure visible.

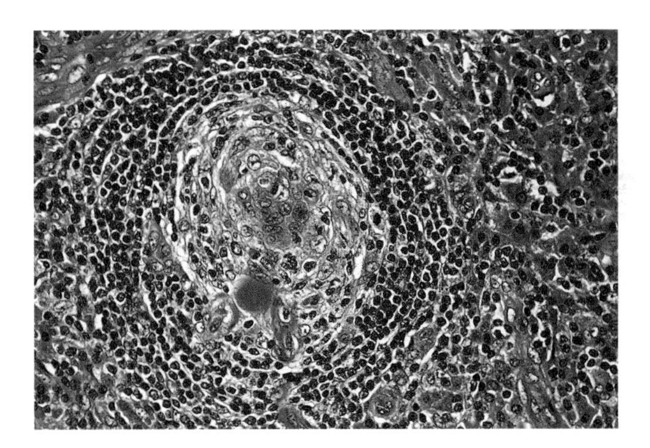

Figure 3.22 Castleman disease, plasma cell variant. Hyalinized follicle composed mainly of dendritic and endothelial cells surrounded by plasma cells.

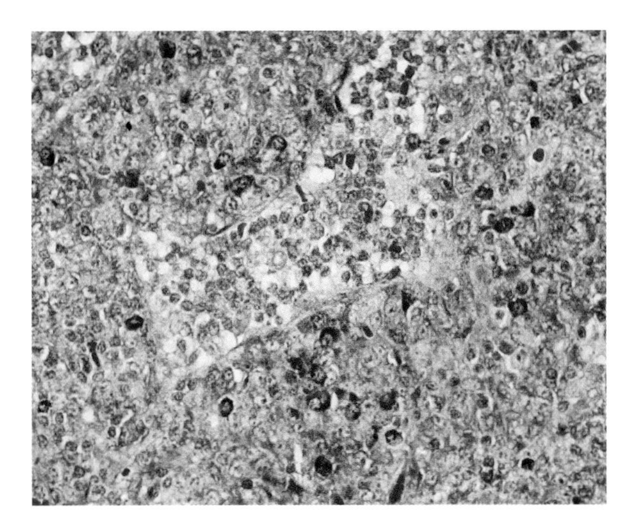

Figure 3.23 Giemsa-stained section of Castleman disease plasma cell variant. This case showed kappa light chain restriction. Note the scattered plasma cells that have more normal features (cytoplasmic basophilia, pale paranuclear hof, clumped nuclear chromatin) than the majority plasmacytoid cells.

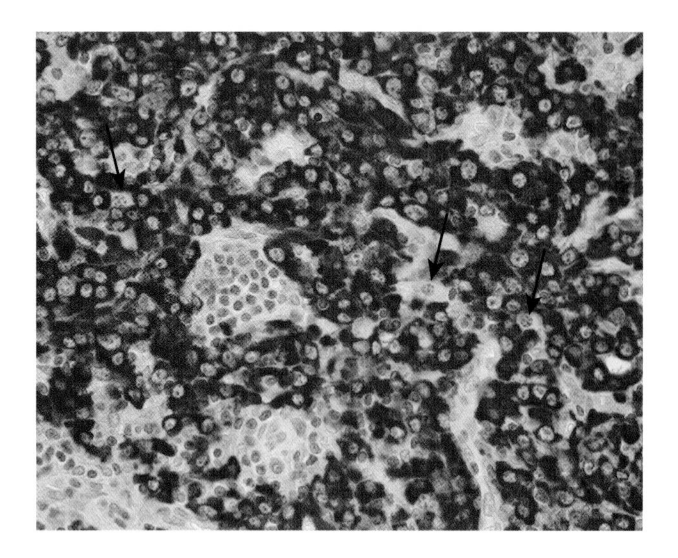

Figure 3.24 Castleman disease, plasma cell variant, showing monotypic kappa light chain staining. Arrows show unstained lambda-expressing plasma cells that have more mature nuclear features than the kappa-positive cells.

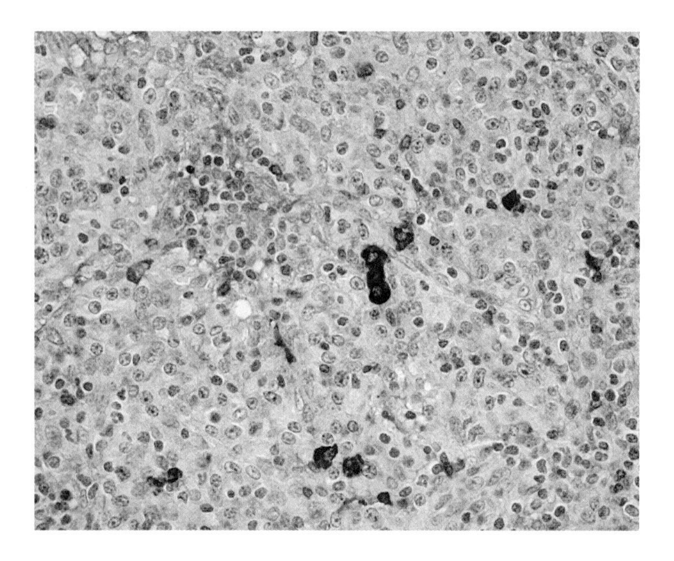

Figure 3.25 Same biopsy as shown in Figure 3.24 stained for lambda light chain. Note that the few lambda-positive plasma cells show much darker staining than the majority kappa-positive cells in Figure 3.24.

abnormal laboratory findings. The underlying pathogenesis appears to be the overproduction of IL-6 either from endogenous sources or from the viral homologue of IL-6 produced by human herpesvirus 8 (HHV-8). It has been suggested that MCD should be called IL-6 lymphadenopathy (Box 3.8).

The histopathology of MCD is similar to that of PCD. It may progress to a burnt-out phase with abundant hyaline vascular germinal centres, when it might be interpreted as HVCD or mixed HVCD/PCD.

The POEMS syndrome is associated with MCD in approximately 60 per cent of cases. The features of this syndrome (polyneuropathy, organomegaly, endocrine abnormalities, monoclonal gammopathy, skin rashes) are thought to result from the production of autoantibodies and from cytokine abnormalities.

Patients with HHV-8–associated Castleman disease, particularly those who are also HIV positive, may develop a plasmacytoid diffuse large B-cell lymphoma (Du *et al.* 2001, see Chapter 5). Patients with MCD, with or without lymphoma, have a poor prognosis.

PARACORTICAL EXPANSION

VIRAL LYMPHADENOPATHY

In using lymph node architecture to categorize reactive and infective lymphadenopathies, the viral lymphadenopathies present the problem that they may be associated with both follicular and paracortical expansion. We have therefore divided them on the basis of their most prominent feature, including HIV/AIDS under 'follicular hyperplasia' and infectious mononucleosis and other viral lymphadenopathies under 'paracortical expansion'.

Measles lymphadenopathy

The measles virus, a paramyxovirus, produces a marked systemic lymphoid response as the virus multiplies in macrophages and lymphocytes in the unimmunized patient. After vaccination with live attenuated virus, there may be associated lymphadenitis. The infection is characterized by a marked proliferation of immunoblasts in the paracortex accompanied by follicular hyperplasia. Nodal architecture may be partially obliterated by diffuse sheets of immunoblasts with a relative depletion of lymphocytes, so that a mottled appearance is produced. Scattered among these cells and within follicles are Warthin–Finkeldey cells, which also appear in various hyperplastic lymphoid tissues during the prodromal stage of the infection. These large, syncytial polykaryocytes are 25–150 μm in diameter with as many as 50 nuclei; they result from the fusion of various cell types mediated by the measles virus (Box 3.9 and Figures 3.26 and 3.27).

BOX 3.9: Measles lymphadenitis

- Measles infection or recent vaccination
- Nodal and extranodal lymphoid tissue involved
- Follicular hyperplasia and paracortical expansion
- Mottled appearance caused by marked proliferation of immunoblasts and relative depletion of lymphocytes in the paracortex
- Warthin–Finkeldey giant cells

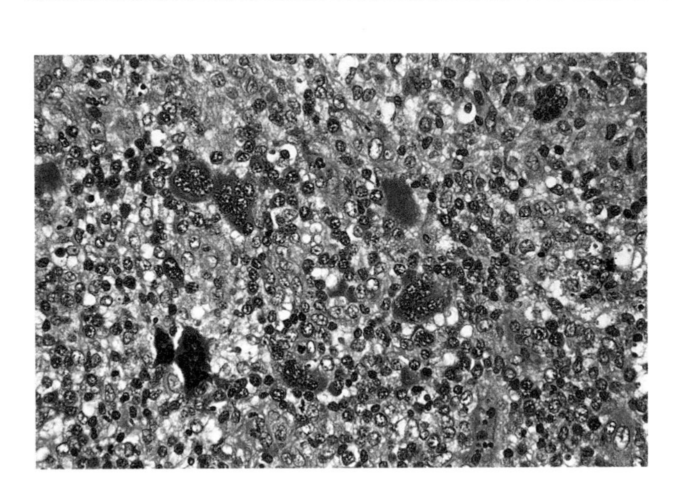

Figure 3.26 Measles lymphadenopathy. Warthin–Finkeldey giant cells in paracortex of lymph node after measles vaccination.

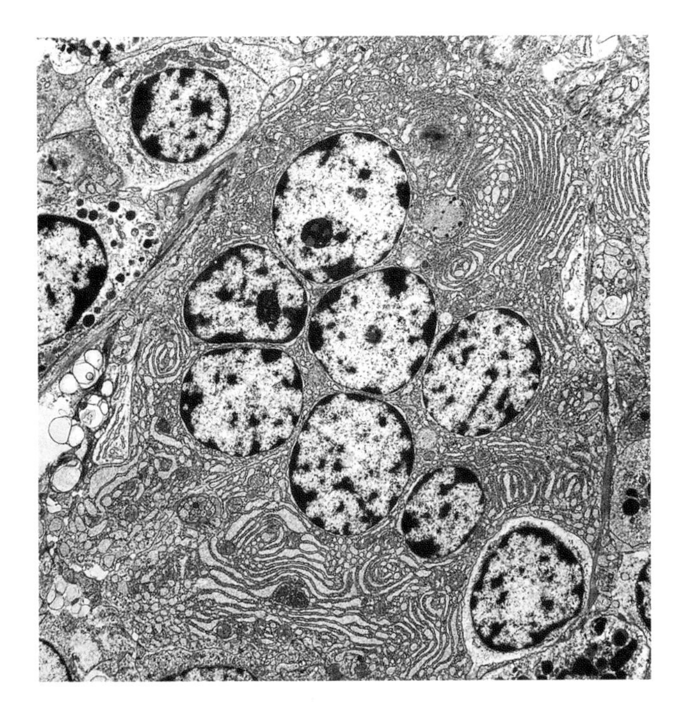

Figure 3.27 Electron micrograph of Warthin–Finkeldey giant cell formed by the fusion of plasma cells in a case of measles lymphadenopathy.

Cytomegalovirus lymphadenopathy

Cytomegalovirus (CMV) is ubiquitous and infects a large percentage of the human population. For most people the infection causes little or no morbidity and remains latent. Infection in neonates and reactivation or primary infection in patients with immunodeficiency can cause severe systemic disease. Primary infection, particularly in adolescents and young adults, can result in an acute flu-like illness with lymphadenopathy.

Histologically, the nodes show follicular hyperplasia and varying degrees of monocytoid B-cell proliferation. Polymorphs are usually seen within the monocytoid B-cell areas, and there may be small foci of necrosis. The typical 'owl eye' nuclear inclusions may be found in endothelial cells, histiocytes or unidentifiable cells. A frequently quoted single-case report concluded that CMV specifically infected T-cells that had lost HLA-DR and IL-2R expression. It was proposed that the downregulation of these two antigens might be the mechanism by which CMV evades host defences (Younes *et al.* 1991). In practice it is often difficult to identify the exact nature of the infected cells. In addition to the intranuclear inclusions, granular cytoplasmic inclusions may be seen. In many cases the viral inclusions are very scanty and can be found only after a long search. Identification of the virus may be made by *in-situ* hybridization to detect viral DNA or by immunohistochemistry to identify viral antigens (Figures 3.28 to 3.30).

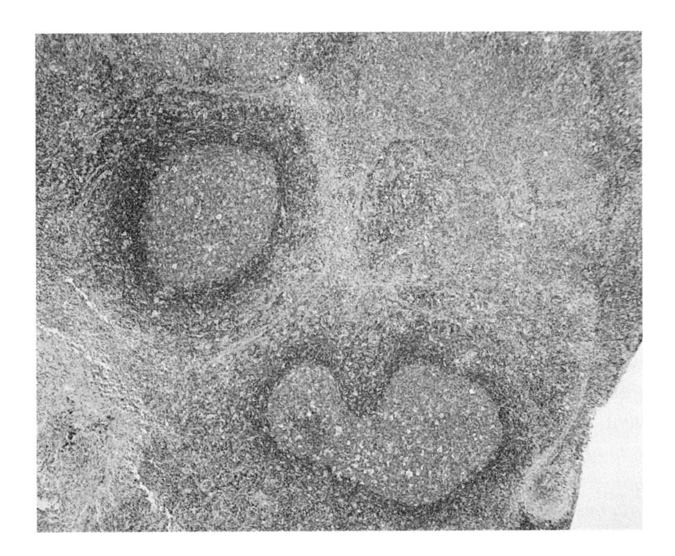

Figure 3.28 Low-power view of cytomegalovirus (CMV)-infected lymph node showing follicular hyperplasia and an expanded zone of monocytoid B-cells (upper right).

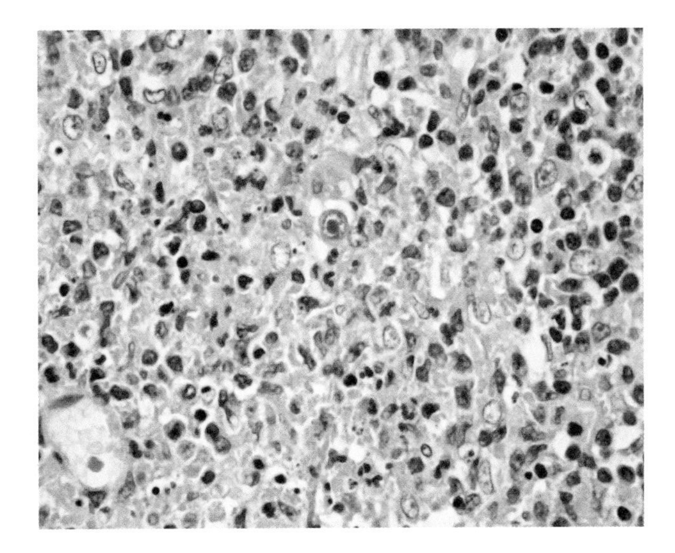

Figure 3.29 High-power view of cytomegalovirus (CMV)-infected lymph node. A cell at the centre shows a characteristic CMV inclusion. Many of the surrounding monocytoid B-cells show apoptosis and there are scattered polymorph leukocytes.

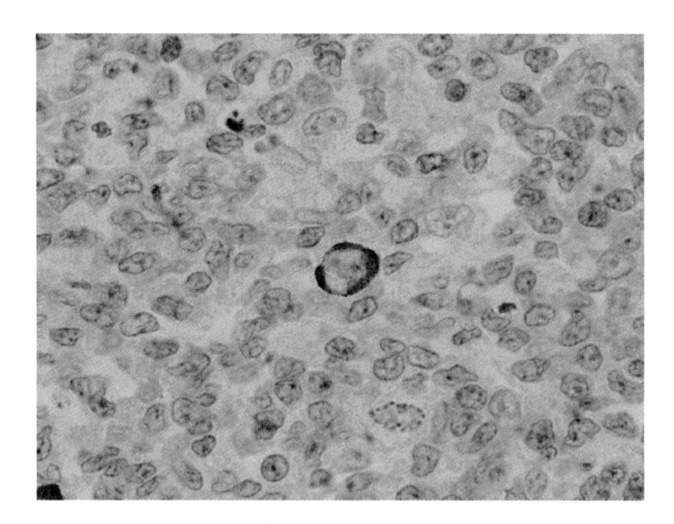

Figure 3.30 Immunohistochemical preparation using a polyclonal rabbit antiserum to cytomegalovirus (CMV). The CMV antigens stained are predominantly intracytoplasmic.

The differential diagnosis will be with other inflammatory/reactive lymphadenopathies that show follicular hyperplasia and monocytoid B-cell proliferation. A number of publications list Hodgkin lymphoma in the differential diagnosis on the grounds that the eosinophilic nuclear inclusions mimic the large eosinophilic nucleoli of Hodgkin/Reed–Sternberg (H/RS) cells, together with the fact that the 'owl eye' cells of CMV express CD15 (Rushin *et al.* 1990). However, the architecture of CMV-infected nodes is that of a reactive/inflammatory process and not compatible with any subtype of Hodgkin lymphoma.

Infectious mononucleosis lymphadenopathy

The Epstein–Barr virus (EBV) infects susceptible lymphoid cells in the oropharynx and persists as a latent virus throughout life. Infection in infants is usually symptomless or trivial. When older patients are infected, the disease may be severe and may simulate lymphoma. In the acute infection, the virus replicates in perifollicular B-cells, stimulating a vigorous humoral and cellular immune response.

Involved lymph nodes are enlarged but not matted. They may show varying degrees of follicular hyperplasia but the most striking feature is paracortical expansion. Large numbers of blast cells, many of immunoblast morphology, are seen within the paracortex. Immunohistochemistry shows that these are of both B- and T-cell phenotype. Occasionally, atypical cells and cells resembling H/RS cells are seen. However, this is a feature more frequently encountered in tonsils than in lymph nodes. These cells, in common with EBV-infected H/RS cells, express latent membrane protein 1 (LMP1) and Epstein–Barr virus-encoded RNA 1 (EBER1). They also express CD30 but not CD15 (Isaacson *et al.* 1992, Reynolds *et al.* 1995) (Box 3.10 and Figures 3.31 to 3.34).

BOX 3.10: Infectious mononucleosis lymphadenitis

- Teenagers and young adults affected
- Paracortical expansion
- Numerous B- and T-immunoblasts present in paracortex
- Atypical Hodgkin/Reed–Sternberg (H/RS)–like cells may be seen but are more common in the tonsil than in lymph nodes

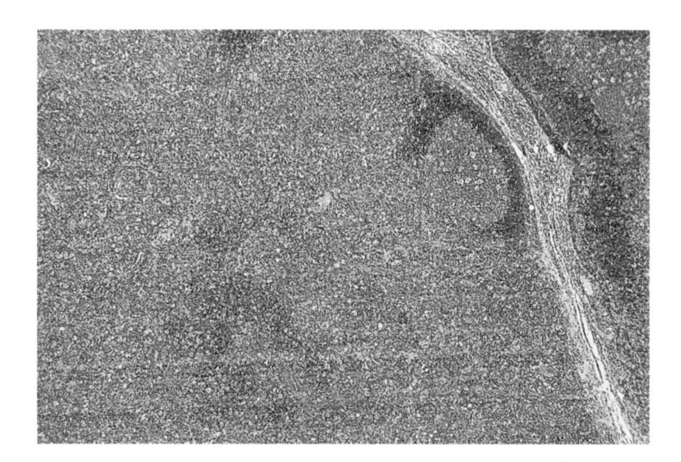

Figure 3.31 Infectious mononucleosis lymphadenopathy. Low-power view showing paracortical blast cell hyperplasia partially surrounding a reactive follicle.

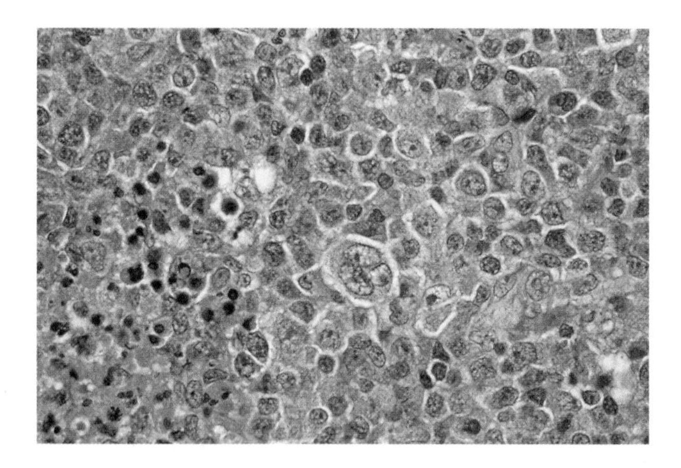

Figure 3.32 Infectious mononucleosis lymphadenopathy. Paracortex showing blast hyperplasia; note multinucleated cell and apoptotic nuclei.

Other viral lymphadenopathies

Other common viruses that cause lymphadenitis include herpes simplex, varicella–herpes zoster and vaccinia; however, such cases are rarely biopsied. Although there may be some element of follicular hyperplasia, such nodes usually show a predominant pattern of paracortical hyperplasia.

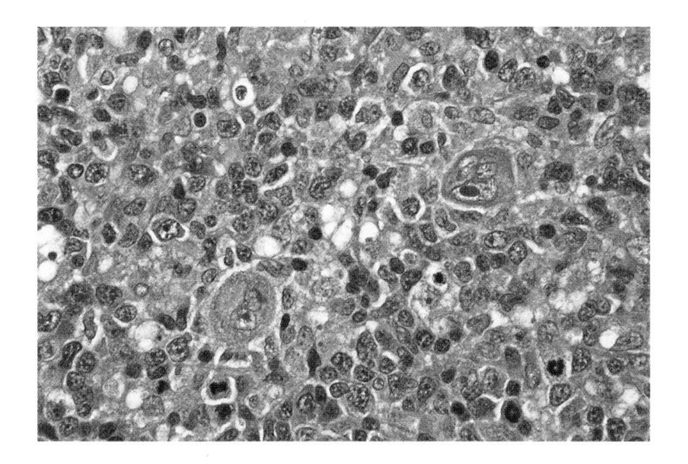

Figure 3.33 Infectious mononucleosis lymphadenopathy showing cells resembling Reed–Sternberg cells in the paracortex.

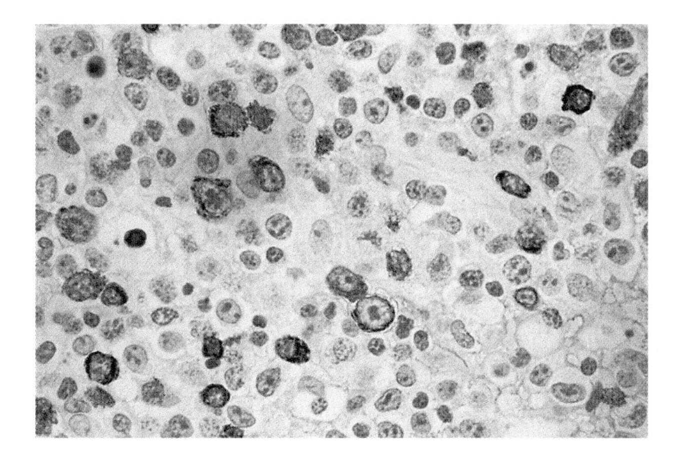

Figure 3.34 Infectious mononucleosis lymphadenopathy. Immunoperoxidase technique showing IgM+ B-cell blasts in the paracortex.

Monocytoid B-cell hyperplasia may be seen at some stages of the infection, and areas of necrosis may occur. Characteristic intranuclear and/or cytoplasmic inclusions may be seen. Immunostaining with specific antibodies is available for herpes simplex and herpes zoster.

DERMATOPATHIC LYMPHADENOPATHY

Dermatopathic lymphadenopathy shows paracortical hyperplasia resulting from the accumulation of interdigitating reticulum cells (IDRCs), Langerhans cells, histiocytes containing lipid and melanin, and paracortical T-cells. Dermatopathic lymphadenopathy represents the reaction of a superficial lymph node to the drainage of skin antigens and melanin from various chronic dermatoses. In patients with cerebriform cutaneous T-cell lymphomas (mycosis fungoides and Sézary syndrome), it may be difficult or impossible to determine whether lymph nodes showing dermatopathic lymphadenopathy contain neoplastic T-cells. In such cases, it is necessary to look for T-cell receptor clonality in order to confirm neoplastic infiltration.

Grossly, the enlarged lymph node may show a distinct rim of pigment immediately beneath the capsule. The nodal architecture is preserved and there is marked paracortical expansion comprising irregular pale staining nodules of IDRCs and Langerhans cells, histiocytes and intermingled T-cells. Scattered phagocytic histiocytes are found intermingled with the IDRCs. These may have foamy cytoplasm and contain ingested melanin and lipid. IDRCs and Langerhans cells express S100 protein, which provides a good immunohistochemical marker for dermatopathic lymphadenopathy (Box 3.11 and Figures 3.35 to 3.39).

BOX 3.11: Dermatopathic lymphadenopathy

- Superficial lymph nodes draining chronic dermatoses
- Cut surface of node may show subcapsular rim of pigment
- Paracortical expansion with aggregates of interdigitating reticulum cells and Langerhans cells
- Scattered macrophages containing lipid and melanin

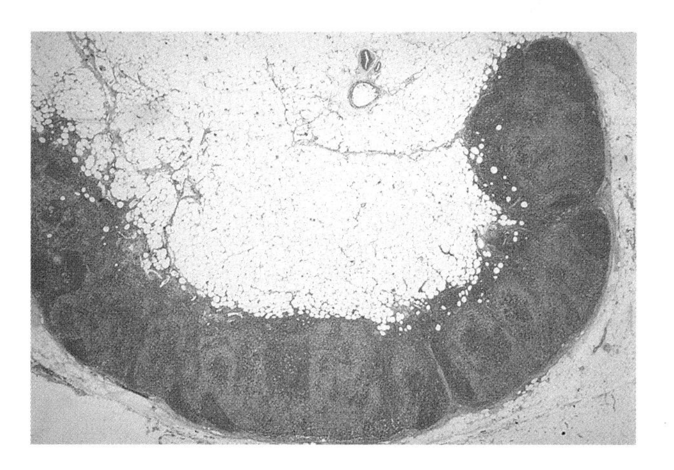

Figure 3.35 Dermatopathic lymphadenopathy. Axillary lymph node showing expanded pale-staining paracortex.

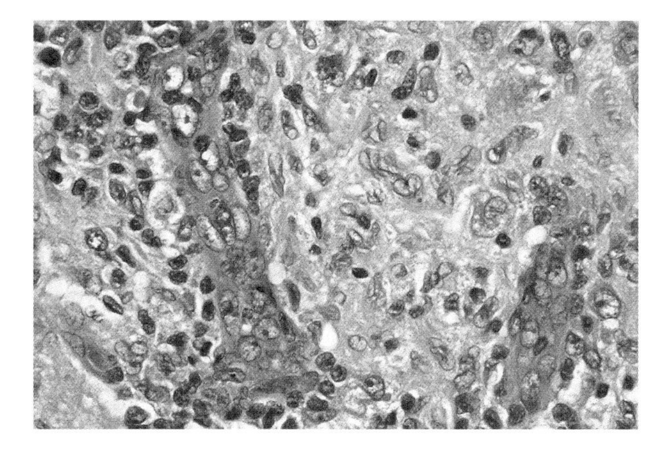

Figure 3.36 Dermatopathic lymphadenopathy. High-power view showing characteristic interdigitating reticulum cells with elongated grooved and twisted nuclei and abundant pale cytoplasm.

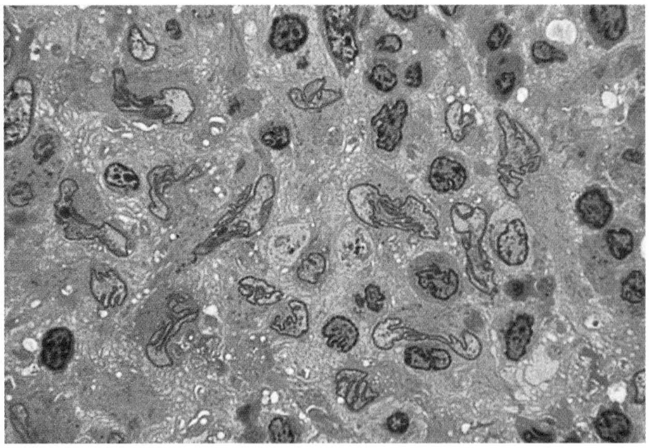

Figure 3.37 Dermatopathic lymphadenopathy. Plastic-embedded section showing the characteristic morphology of interdigitating reticulum cells.

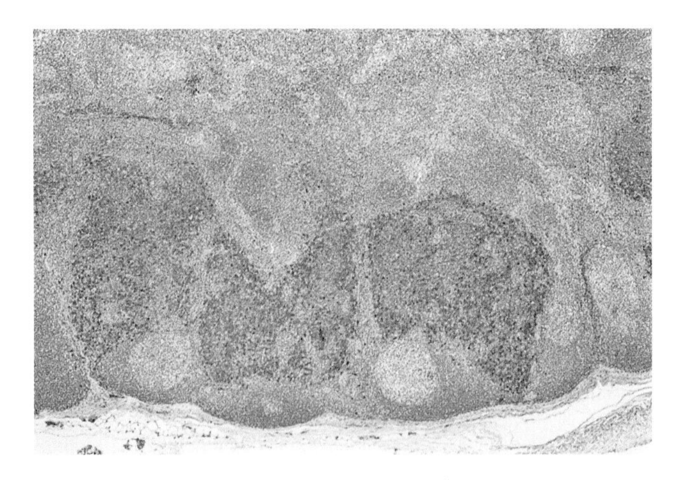

Figure 3.38 Dermatopathic lymphadenopathy. Section stained for S100 protein showing the expanded network of interdigitating reticulum cells in the paracortex.

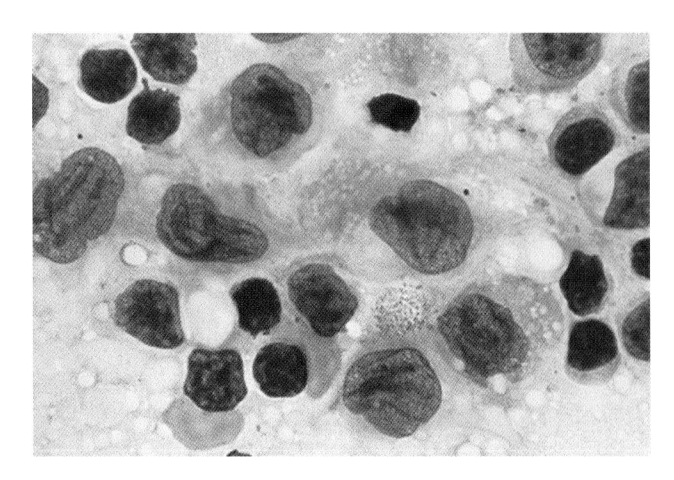

Figure 3.39 Dermatopathic lymphadenopathy, imprint cytology. The interdigitating reticulum cells have grooved nuclei and abundant pale-staining cytoplasm.

INFLAMMATORY PSEUDOTUMOUR OF LYMPH NODES

Inflammatory pseudotumour of lymph nodes was first described by Perrone *et al.* (1988) in a report of seven cases. In this series and a larger series (Moran *et al.* 1997) there was an age range of 8 to 81 years. Lymphadenopathy is usually confined to a single lymph node group. In addition to lymphadenopathy, more than half the patients have systemic symptoms, including weight loss and fever, sometimes associated with a raised ESR, anaemia and hypergammaglobulinaemia. Morphologically, lymph nodes show a vascular spindle cell infiltrate, often with a storiform pattern, typically involving the hilum and medulla of the node and sometimes extending through the capsule into the perinodal tissues. There is a variable infiltrate of lymphocytes, plasma cells and neutrophils. Moran *et al.* (1997) recognized three stages of the lesion (stage I localized, stage II widespread throughout the medulla of the node and stage III fibrotic). Angioinvasion may be seen, with areas of infarction (Box 3.12 and Figures 3.40 to 3.42).

Immunohistochemistry shows that many of the spindle cells have the immunophenotype of myofibroblasts, expressing vimentin and actin. Many cells also express CD68 and other histiocyte markers. Lymphocytes may be of T- or B-cell lineage.

The pseudotumour reaction may be caused by cytokine production in response to infection. A number of infective agents have been detected in cases of inflammatory pseudotumour. Arber *et al.* (1995) detected EBV DNA in two of ten inflammatory pseudotumours of lymph node and five of eight extranodal cases involving spleen or liver. In the nodal cases EBERs were detected in lymphocytes, whereas in the extranodal proliferations they were present in the spindle cells. In a more recent publication using a polyclonal antibody to *T. pallidum*, Facchetti *et al.* (2009) detected *T. pallidum* spirochaetes in large numbers in four of nine

BOX 3.12: Inflammatory pseudotumour of lymph nodes

- Wide age range
- Usually affects single nodes or nodal groups
- Systemic symptoms common
- May have raised erythrocyte sedimentation rate (ESR), anaemia and hyperglobulinaemia
- Spindle cell proliferation consisting of histiocytes, myofibroblasts and blood vessels often showing a storiform pattern
- Variable infiltrate of lymphocytes, plasma cells and neutrophils
- Proliferation may be focal or spread throughout the medulla, interfollicular areas and sometimes perinodal tissues
- Varying degrees of fibrosis may be seen

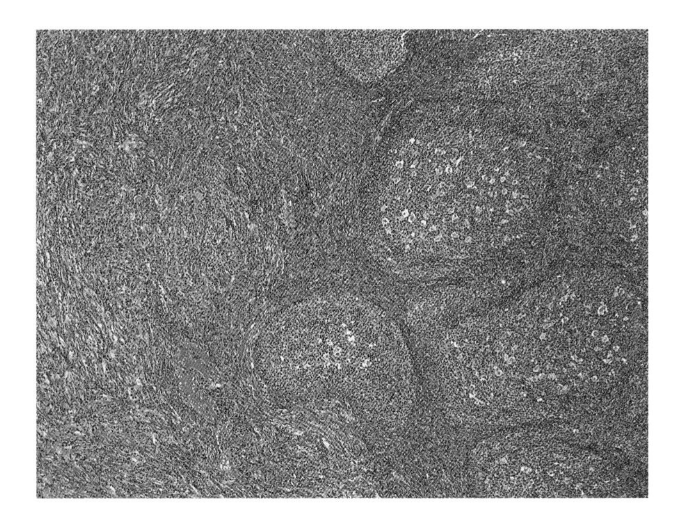

Figure 3.40 Inflammatory pseudotumour of lymph node showing follicular hyperplasia to the right and the spindle cell pseudotumour proliferation to the left.

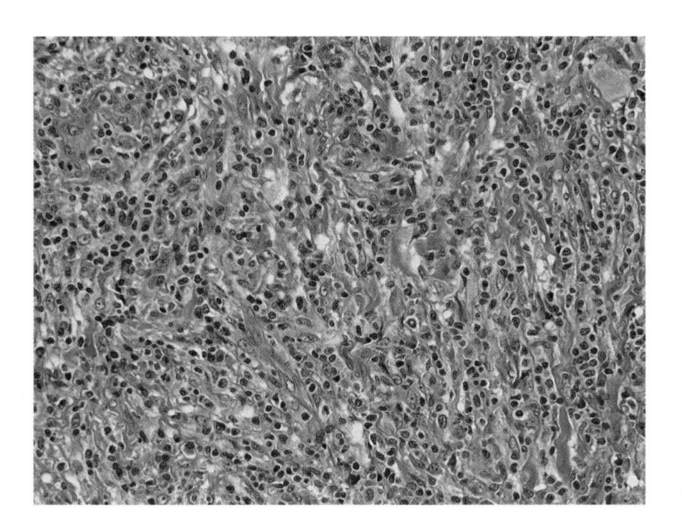

Figure 3.41 Higher power view of inflammatory pseudotumour showing a spindle cell proliferation of myofibroblasts, histiocytes and blood vessels infiltrated by lymphocytes and plasma cells.

cases of nodal inflammatory pseudotumour, with two further positive cases added in proof. The spirochaetes were detected mainly in the cytoplasm of histiocytes. The presence of follicular hyperplasia distinguished the *T. pallidum*–positive from the *T. pallidum*–negative cases.

The main differential diagnosis of nodal inflammatory pseudotumour is with follicular dendritic cell sarcoma. The distinction can be made using immunohistochemistry, with CD21, CD23 and CD35 being expressed on dendritic cell sarcomas. Cheuk *et al.* (2001) reported 11 extranodal cases labelled as inflammatory pseudotumour-like follicular dendritic cell sarcoma. These had a more indolent behaviour than typical follicular dendritic cell sarcoma. All were positive for EBER, which was detected in the spindle cells. The differential diagnosis of nodal inflammatory pseudotumour also includes mycobacterial pseudotumour, typically

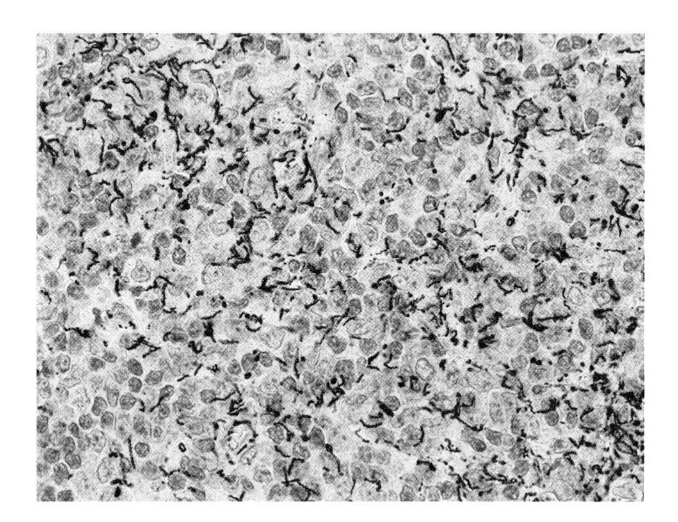

Figure 3.42 Immunoperoxidase stain of the same lymph node using a polyclonal rabbit antiserum to *Treponema pallidum*. Numerous spirochaetes are seen throughout the lymph node.

seen in HIV-positive patients. Stains for acid-fast bacilli will identify these cases (Logani *et al.* 1999).

Inflammatory pseudotumour of lymph nodes should not rest as a final diagnosis but prompt a search for possible aetiological agents.

IgG4-RELATED DISEASE

IgG4-related disease (IgG4-RD) is an increasingly recognised fibroinflammatory condition that commonly exhibits multisystem involvement, with localised or diffuse patterns of tissue involvement (Deshpande *et al.* 2012). It is classically characterised by lymphoplasmacytic inflammation, storiform fibrosis and obliterative venulitis although this is less frequently seen in lymph nodes than in involvement of other sites (Ferry 2013). Five major patterns of lymph node involvement are described: follicular hyperplasia, PTGC, interfollicular expansion by plasma cells and immunoblasts, Castleman disease–like and the more typical inflammatory pseudotumour appearance (Grimm *et al.* 2012). Rarely granulomas may also be present (Bateman *et al.* 2015). All of these are seen in association with prominent IgG4+ plasma cells together with an IgG4+/IgG+ ratio of over 40 per cent on immunohistochemistry (Figures 3.43 to 3.46).

KIKUCHI DISEASE

Kikuchi disease (also known as Kikuchi–Fujimoto disease, histiocytic necrotizing lymphadenitis) is a self-limiting disease occurring predominantly in adolescent and young adult females. It is more prevalent in Asia than in the rest of the world. The most common presentation is with one or more enlarged cervical lymph nodes that are frequently painful and may be associated with fever and systemic symptoms. Other superficial lymph nodes are much less

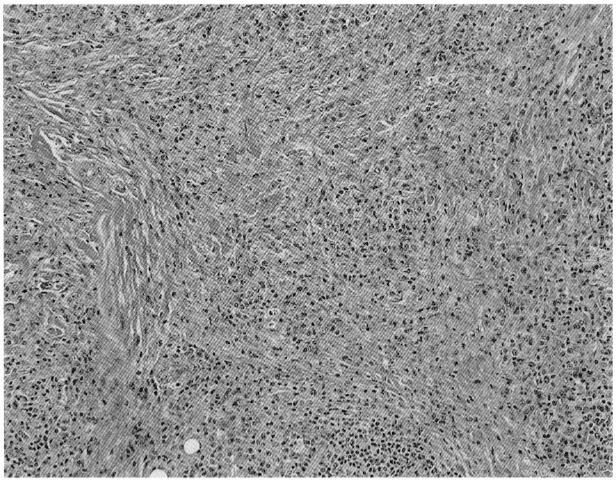

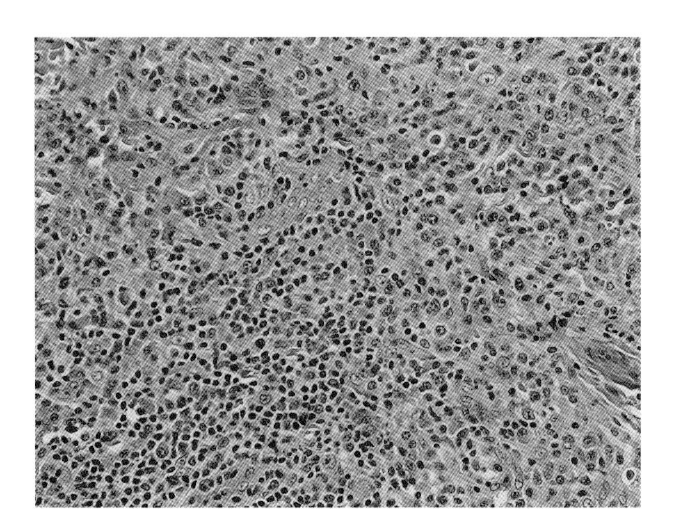

Figure 3.43 Immunoglobulin G4 (IgG4)–related disease in lymph node; low-power view showing storiform fibrosis.

High-power view image of immunoglobulin G4 disease

Figure 3.44 High-power view of immunoglobulin G4 (IgG4) disease demonstrating the plasmacytic infiltrate.

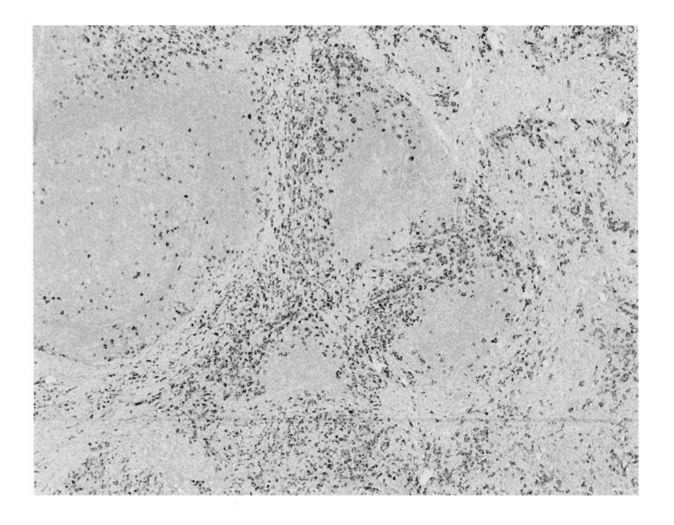

Figure 3.45 Immunoglobulin G4 (IgG4) disease immunoperoxidase stained for IgG highlighting the IgG-positive plasma cells.

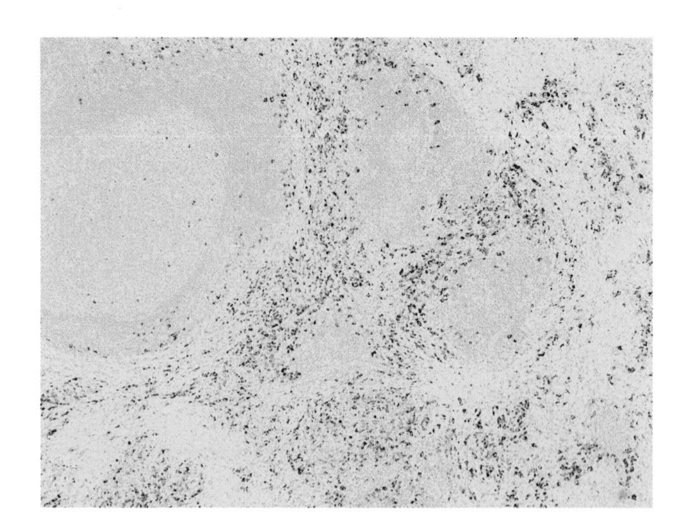

Figure 3.46 This is the same field as in Figure 3.45 immunoperoxidase stained with an antibody specific for immunoglobulin G4 (IgG4). The ratio of IgG4- to IgG-stained plasma cells is greater than 40 per cent.

frequently affected. The aetiology is unknown (Box 3.13 and Figures 3.47 to 3.51).

The histology of Kikuchi disease varies as the disease progresses. In the early stages there is variable follicular hyperplasia with expansion of the paracortex by small lymphocytes, B- and T-cell blasts, plasmacytoid monocytes and histiocytes. Apoptosis is usually prominent among these cells. As the disease progresses, areas of necrosis appear. Neutrophils are not associated with these areas of necrosis, presumably because the cells are undergoing apoptosis. As the apoptosis and necrosis progress, the number of histiocytes increases until they become the predominant cell type. Many of these histiocytes contain ingested cell debris and characteristically have crescentic nuclei.

Immunohistochemical markers for CD68 will identify histiocytes and plasmacytoid monocytes. The histiocytes in Kikuchi disease often co-express the macrophage-restricted epitope of CD68 (detected with the PG-M1 antibody) and myeloperoxidase. This co-expression is not seen in other histiocytic proliferations (Pileri *et al.* 2001). Both B- and T-cell antibodies will often identify B- and T-cell blasts among the non-necrotic cells. The majority of the small lymphocytes are CD8+ T-cells.

Kikuchi disease must be differentiated from non-Hodgkin lymphoma. In the proliferative phase of the disease the appearance of sheets of blast cells, as also is the case in infectious mononucleosis, may appear alarming. The presence of an underlying normal nodal structure and the morphological and immunohistochemical heterogeneity of the cells distinguishes Kikuchi disease from lymphoma. Kikuchi disease may be confused with other diseases that cause lymph node necrosis in which fragmented neutrophil nuclei may give an appearance suggesting preceding apoptosis. Markers for granulocytes, such as CD15, will identify these neutrophils, cells that are almost invariably absent from Kikuchi disease.

BOX 3.13: Kikuchi disease

- Most common in Asia
- Most common in adolescent and young adult females
- Cervical lymphadenopathy, often painful; other nodes less commonly affected
- Systemic symptoms and fever common
- Paracortical expansion with small lymphocytes, blast cells, plasmacytoid monocytes and histiocytes
- Plasma cells uncommon
- Widespread apoptosis and areas of necrosis without neutrophils
- Phagocytic histiocytes characteristically have crescentic nuclei

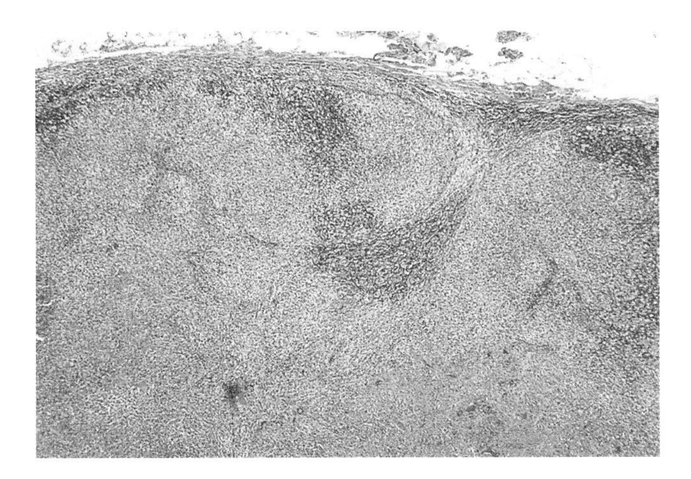

Figure 3.47 Kikuchi disease. Lymph node showing partial loss of architecture with widespread apoptosis and necrosis.

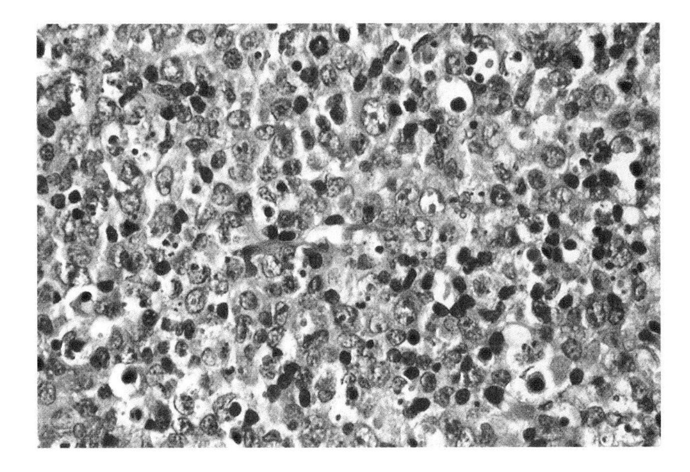

Figure 3.48 Kikuchi disease. Blast cells with numerous apoptotic bodies in the background.

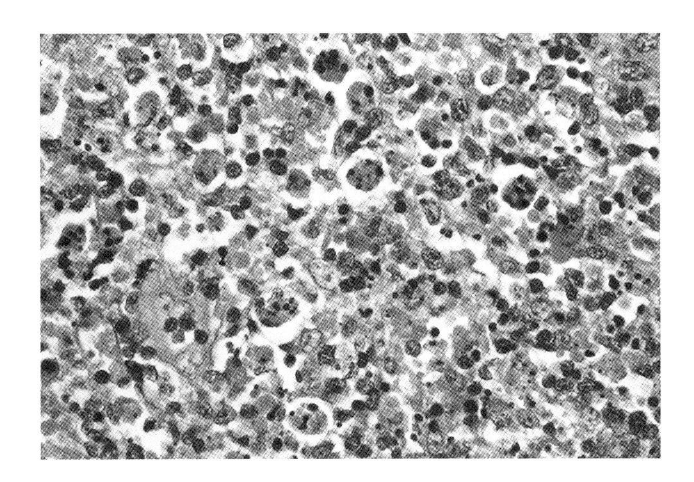

Figure 3.49 Kikuchi disease. An area of almost complete apoptosis and necrosis.

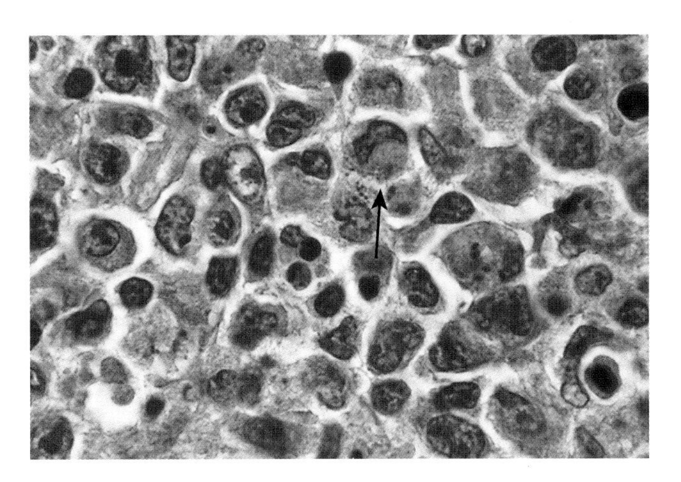

Figure 3.50 Kikuchi disease. High-power view showing blast cells and histiocytes with characteristic crescentic nuclei (arrow).

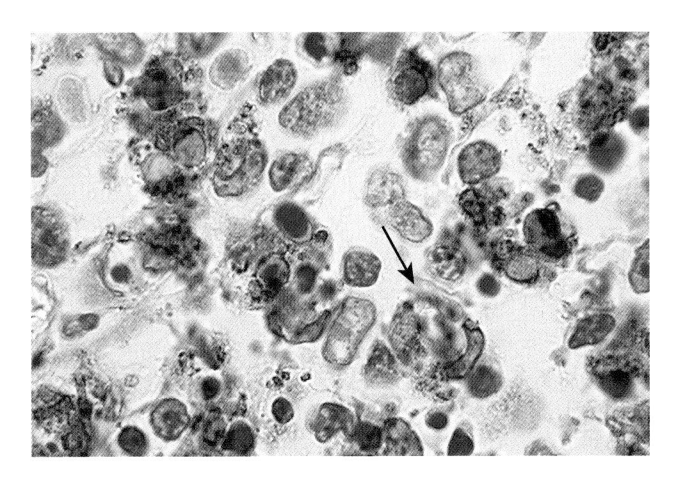

Figure 3.51 Kikuchi disease. Section stained for CD68, showing histiocytes with crescentic nuclei (arrow) and phagocytosed apoptotic debris.

SYSTEMIC LUPUS ERYTHEMATOSUS LYMPHADENOPATHY

Lymphadenopathy may occur in patients with systemic lupus erythematosus (SLE), but in practice biopsy of such nodes is rare, probably because other manifestations of the disease have already established the diagnosis. The histological findings in lymph nodes from patients with SLE have features in common with Kikuchi disease, and it is probably wise to bring this to the attention of the clinicians when reporting Kikuchi disease. Features that differentiate the two conditions are the presence of haematoxylin bodies (aggregates of DNA and anti-DNA antibodies), vasculitis, DNA deposition on blood vessels and plasma cell infiltrates in SLE (Figures 3.52 and 3.53).

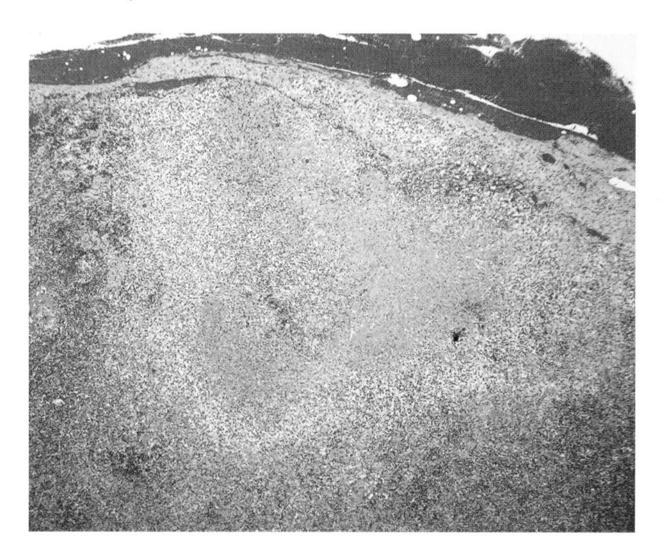

Figure 3.52 Systemic lupus erythematosus of lymph node showing a wedge-shaped area of subcapsular necrosis. The area of necrosis is less extensive than that usually seen in Kikuchi disease.

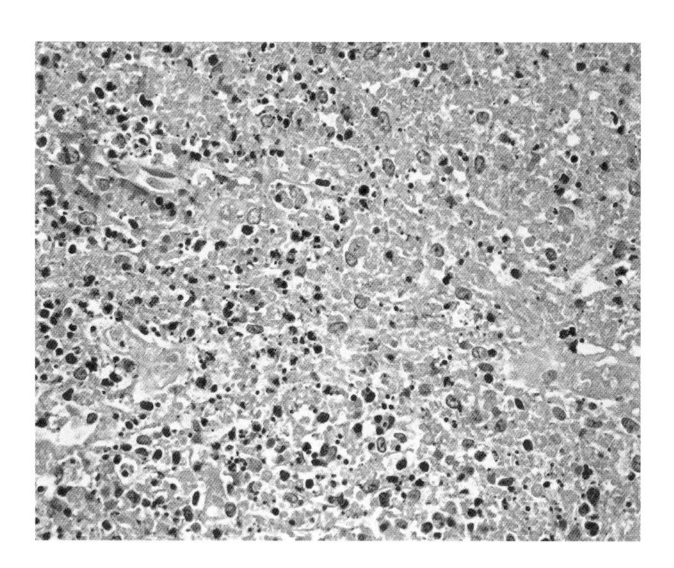

Figure 3.53 Higher power view of systemic lupus erythematosus of lymph node showing apoptosis and necrosis without polymorph infiltration.

DRUG-INDUCED LYMPHADENOPATHY

Drug-induced lymphadenopathy is most commonly seen in patients exhibiting hypersensitivity to anticonvulsant drugs (carbamazepine, phenytoin [Dilantin]), but it may also be seen in association with hypersensitivity to other drugs. Patients typically show fever, skin rashes and generalized lymphadenopathy with eosinophilia (Box 3.14 and Figures 3.54 to 3.56).

BOX 3.14: Drug-induced lymphadenopathy

- History of exposure to drug, particularly anticonvulsant therapy
- Fever, skin rashes and lymphadenopathy
- Peripheral blood eosinophilia
- Paracortical expansion with mixed infiltrate including blast cells and eosinophils
- Small areas of necrosis may be present, often associated with eosinophils
- Regresses following drug withdrawal

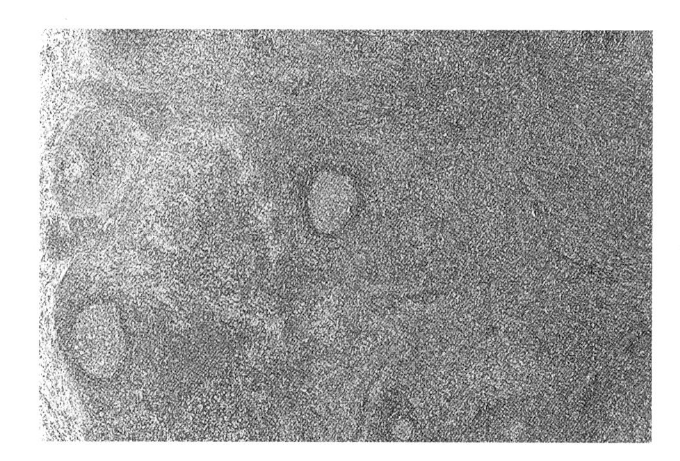

Figure 3.54 Drug-induced lymphadenopathy, showing reactive follicles and expanded paracortex.

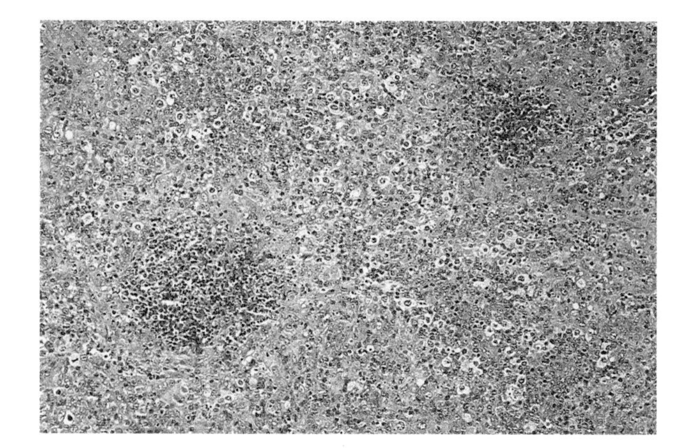

Figure 3.55 Drug-induced lymphadenopathy. Paracortex showing blast cell hyperplasia and eosinophil micro-abscesses.

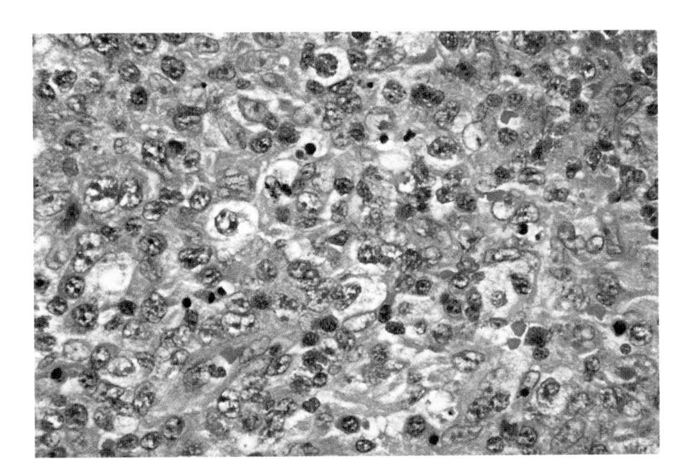

Figure 3.56 Drug-induced lymphadenopathy. High-power view of paracortex showing blast cells and eosinophils.

Lymph nodes show a predominantly paracortical expansion with a population of B- and T-cell blasts, small lymphocytes, histiocytes, neutrophils and eosinophils. The blast cells may have prominent nucleoli but they do not exhibit atypia.

AUTOIMMUNE LYMPHOPROLIFERATIVE SYNDROME

Autoimmune lymphoproliferative syndrome (ALPS) is an uncommon syndrome caused by germline or somatic mutations in the Fas-mediated apoptotic pathway. The resultant dysregulation of lymphocyte homeostasis leads to the development of autoimmune disease, typically causing cytopenias but also affecting solid organs. Patients present with persistent lymphadenopathy with or without hepatosplenomegaly. The disease usually manifests in childhood and may ameliorate with increasing age. Adult relatives may have mutations in the Fas pathway genes but have little evidence of disease (Sneller *et al.* 1997, Teachey *et al.* 2009). A high percentage of double negative T-cells (CD4– CD8–) is found in the peripheral blood and in the tissues. Lymph nodes show marked paracortical hyperplasia caused by expansion by T-cells showing a high proliferation index. These cells express CD3 but are negative for CD4 and CD8 as well as CD45RO. Many of the T-cells express the cytotoxic granule markers TIA-1 and perforin. Most lymph nodes show florid follicular hyperplasia, sometimes with PTGC or with regressive changes (Lim *et al.* 1998). These morphological features may be influenced by preceding treatments for autoimmune disease. Polyclonal plasmacytosis is frequently present (Box 3.15 and Figures 3.57 to 3.63).

The marked paracortical T-cell expansion may lead to an erroneous diagnosis of lymphoma. This is well illustrated by a case in which a 33-year-old man was diagnosed as having a T-cell lymphoma. His lymphadenopathy persisted despite treatment with chemotherapy and radiotherapy. Three years later his 9-year-old son was found to have ALPS.

BOX 3.15: Autoimmune lymphoproliferative syndrome (ALPS)

- Caused by mutations of genes in the Fas apoptosis pathway; relatives may carry mutations without overt evidence of disease
- Disease usually presents in childhood and may ameliorate with age
- Patients present with lymphadenopathy often accompanied by hepatosplenomegaly
- Patients usually have autoimmune disease at presentation; autoimmune cytopenias are the most common
- Increased levels of double negative T-cells (CD4– CD8–) are found in the blood and tissues
- Lymph nodes show marked paracortical expansion with T-cells/blasts and polytypic plasma cells
- Germinal centres usually prominent, may show progressive transformation
- Increased incidence of T- and B-cell lymphomas and Hodgkin lymphoma in relatives

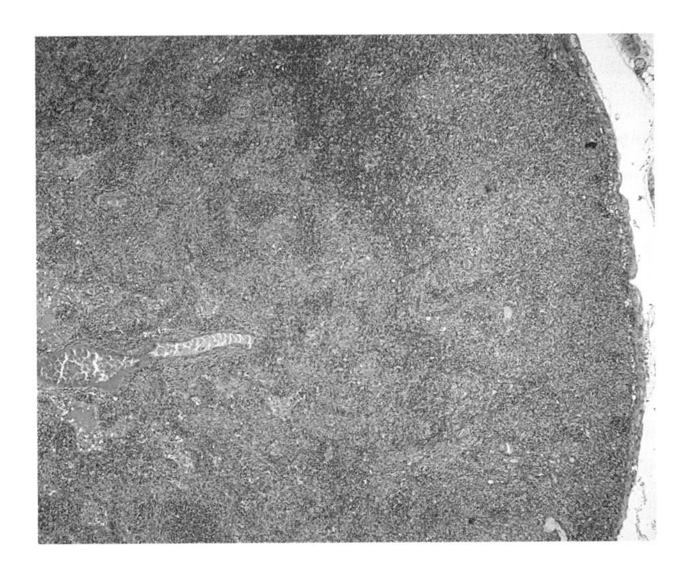

Figure 3.57 Low-power view of a node from a patient with autoimmune lymphoproliferative syndrome (ALPS) showing marked parafollicular hyperplasia.

This prompted a review of the father's lymph node biopsy which was reinterpreted as ALPS. He was found to have clinical and laboratory features of ALPS, but not the autoimmune features that had caused his son to present (Van der Werff ten Bosch *et al.* 1999). Despite this cautionary report it should be noted that kindred of patients with ALPS who have germline Fas mutations have an increased risk of developing non-Hodgkin and Hodgkin lymphoma (14 times greater than expected for T- and B-cell lymphomas and 51 times greater for Hodgkin lymphoma) (Straus *et al.* 2001).

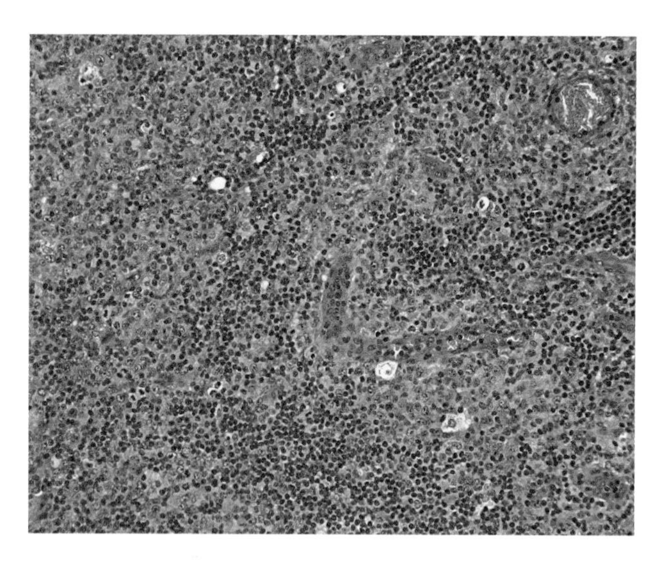

Figure 3.58 Higher power view of autoimmune lymphoproliferative syndrome (ALPS) lymph node showing expansion of the paracortex by medium-sized blast cells and small lymphocytes.

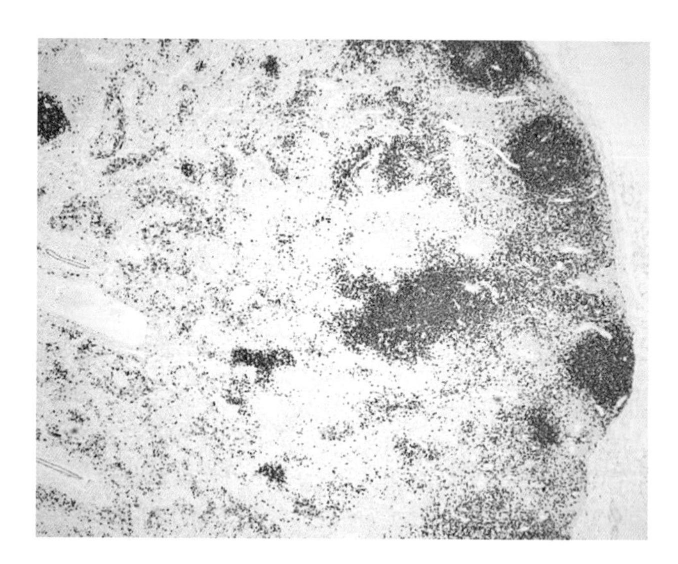

Figure 3.60 Autoimmune lymphoproliferative syndrome (ALPS) lymph node stained for CD79a. In addition to the primary follicles, large numbers of plasma cells are seen in the paracortex.

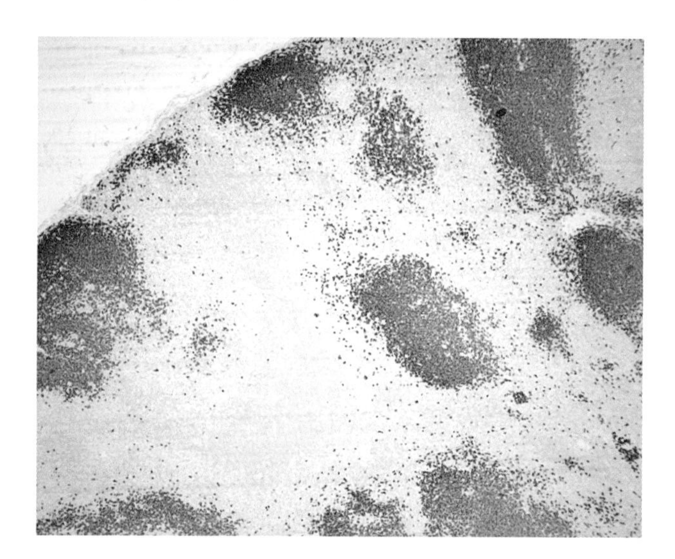

Figure 3.59 Immunoperoxidase stain of autoimmune lymphoproliferative syndrome (ALPS) lymph node for CD20 showing several primary follicles.

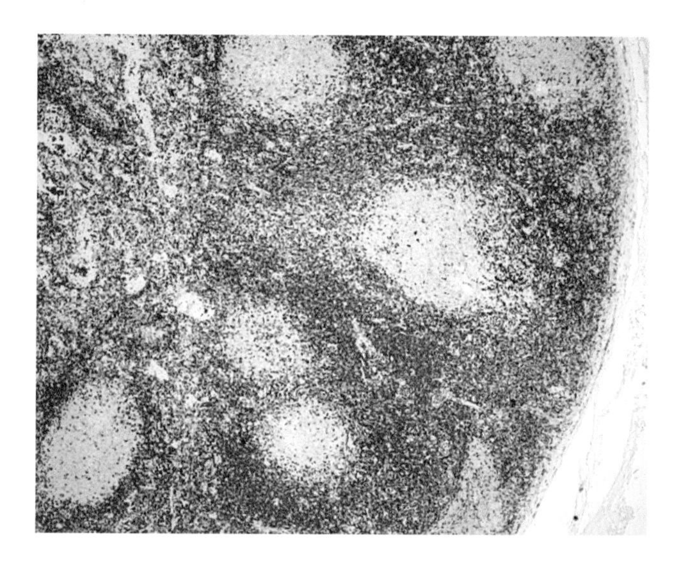

Figure 3.61 Autoimmune lymphoproliferative syndrome (ALPS) lymph node immunostained for CD5 showing that the paracortical expansion is mainly a result of T-cells.

SINUS EXPANSION

SINUS HYPERPLASIA (SINUS HISTIOCYTOSIS)

Prominent sinuses containing many histiocytes are a feature seen in reactive lymph nodes, particularly from the mediastinum and the mesentery. This is often seen in lymph nodes draining carcinomas. The histiocytes do not show atypical features, and other inflammatory cells may accompany them. Occasionally the histiocytes contain oval bodies of variable size that have a green coloration in haematoxylin and eosin (H&E)-stained sections. These inclusions, known as Hamazaki–Wesenberg bodies, are composed of ceroid and are acid fast in Ziehl–Neelsen preparations. They probably result from the increased turnover of lipid membranes and should not be confused with organisms.

ROSAI–DORFMAN DISEASE (SINUS HISTIOCYTOSIS WITH MASSIVE LYMPHADENOPATHY)

This disease was first described in English by Rosai and Dorfman as sinus histiocytosis with massive lymphadenopathy (Rosai and Dorfman 1969). Since the disease may be extranodal as well as nodal, the use of the eponymous title is more appropriate. It may occur at all ages but is most

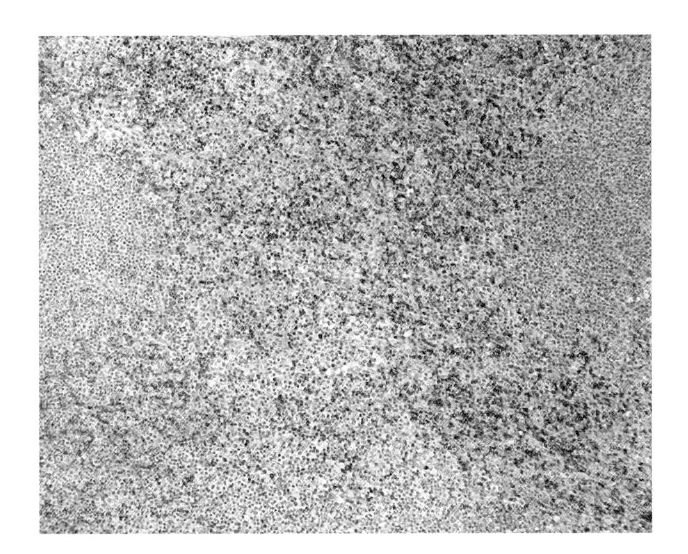

Figure 3.62 Autoimmune lymphoproliferative syndrome (ALPS) lymph node immunostained for CD4. There are relatively few CD4 cells present, presumably because of the increased number of CD4− CD8− T-cells.

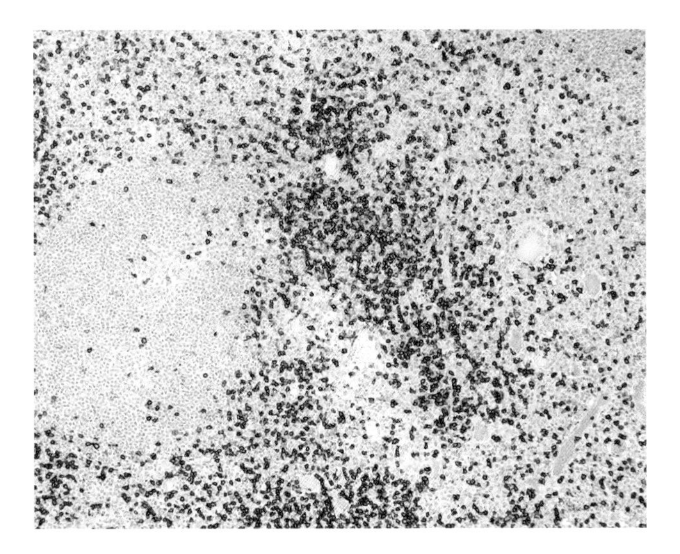

Figure 3.63 Autoimmune lymphoproliferative syndrome (ALPS) lymph node immunostained for CD8. The number of CD8 T-cells is not unusual but it highlights the paucity of CD4 T-cells.

common in childhood and young adult life. There is a male predominance of 4:3.

Patients usually present with cervical lymphadenopathy, other lymph nodes being much less frequently involved. The lymphadenopathy is often massive and may be of long duration. Systemic symptoms of fever, night sweats, weight loss, malaise and arthralgia may be present. Laboratory tests often show hypochromic microcytic anaemia, hypergammaglobulinaemia and a raised ESR. The majority of cases undergo spontaneous regression or show persistent localized disease. A small number of patients have died, usually with more widespread disease. They often show evidence of immune deficiency.

Lymph nodes show varying degrees of capsular fibrosis. The most prominent histological feature is expansion of the sinuses that are filled with histiocytes. These characteristically have abundant, often vacuolated cytoplasm, and rounded nuclei with coarse chromatin and often a single prominent nucleolus. A proportion of these histiocytes show varying numbers of lymphocytes, which do not appear degenerate, within their cytoplasm. Less commonly, erythrocytes, plasma cells and polymorphs appear to be engulfed. The intervening medullary cords show large numbers of plasma cells. Residual reactive germinal centres may be seen but these regress with time. In long-standing cases there is fibrous replacement of the involved nodes. This leads to persistence of the lymphadenopathy, and surgery may be needed for cosmetic reasons or to relieve obstruction.

A study using X-linked polymorphic loci has shown the histiocytes of Rosai–Dorfman disease to be polyclonal. The atypical histiocytes express CD68 and muramidase. In contrast to reactive sinus histiocytes, they are S100+, possibly indicating origin from antigen-presenting cells. They are negative for CD1a. Parvovirus B19 has been detected by immunohistochemistry in a small number of cases of nodal and extranodal Rosai–Dorfman disease (Mehraein *et al.* 2006). The virus was found in both B- and T-lymphocytes but not in histiocytes. Expression of HHV-6 antigens has also been reported in this disease (Luppi *et al.* 1998). It is unclear whether these viruses have an aetiological role. Rosai–Dorfman disease has been recorded in patients with ALPS (Maric *et al.* 2004).

Langerhans cell histiocytosis presenting as lymphadenopathy is uncommon and in this book has been included under 'Histiocytic and dendritic cell neoplasms' in Chapter 9 in keeping with the World Health Organization (WHO) classification. It is predominantly a sinus proliferation and thus enters into the differential diagnosis of Rosai–Dorfman disease. The histiocytes in the two diseases have different morphological characteristics. Both stain for S100 protein but only Langerhans cells express CD1a (Box 3.16 and Figures 3.64 to 3.67).

BOX 3.16: Rosai–Dorfman disease

- Most common in childhood and young adult life
- Cervical lymphadenopathy most common presentation
- May occur at other nodal and extranodal sites
- Fever, night sweats, weight loss, malaise and arthralgia may occur
- Anaemia, hypergammaglobulinaemia and raised erythrocyte sedimentation rate (ESR) frequent
- Capsular fibrosis
- Sinuses expanded by histiocytes with rounded nuclei, prominent nucleoli and abundant cytoplasm
- Histiocytes engulf lymphocytes and other haematopoietic cells
- Large numbers of plasma cells in medullary cords

Figure 3.64 Rosai–Dorfman disease. Lymph node showing marked sinus histiocytosis with compression of other lymph node compartments.

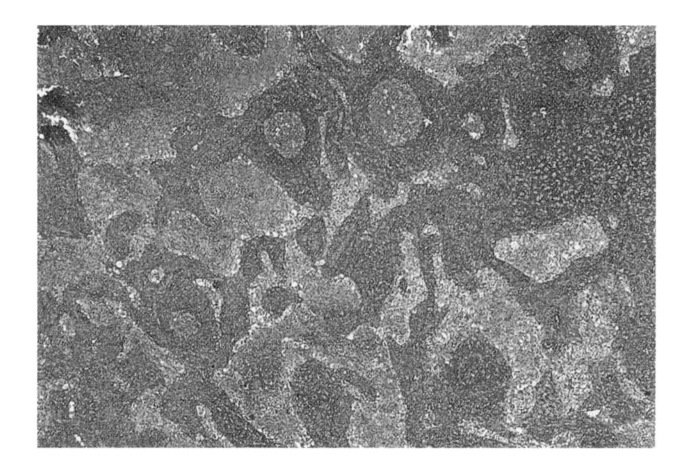

Figure 3.65 Rosai–Dorfman disease. Prominent sinus histiocytosis.

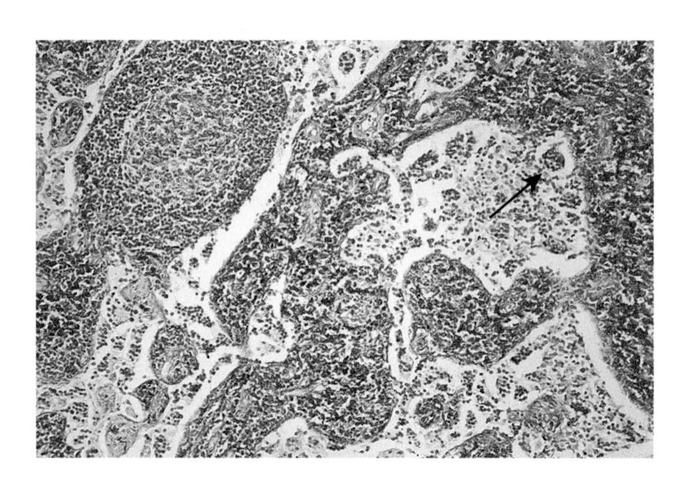

Figure 3.66 Rosai–Dorfman disease. Methyl green pyronin-stained section. The medullary cords contain many pyroninophilic plasma cells. Note the lymphophagocytic histiocytes in the sinuses (arrows).

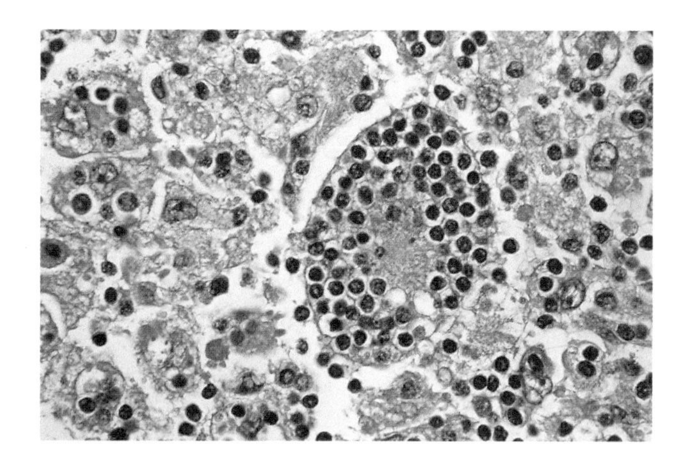

Figure 3.67 Rosai–Dorfman disease. High-power view showing lymphophagocytic histiocytes.

SINUS HISTIOCYTOSIS AND LIPOGRANULOMATOUS REACTION

Lipids of exogenous origin, such as parenteral nutrition fluids, contrast media used in lymphangiography, lipid-based substances and oils used as depot vehicles for the slow release of injected drugs, and endogenous lipids in patients who are obese, diabetic or hyperlipidaemic or who have fat necrosis, haematomas or cholesterol deposits, may stimulate a lipogranulomatous reaction in lymph nodes (Box 3.17 and Figures 3.68 to 3.72).

Lipid granulomas are formed as a result of the accumulation of histiocytes and foreign-body-type giant cells around lipid in the subcapsular and medullary sinuses. Phagocytosis of lipid by the macrophages and giant cells causes vacuolation of their cytoplasm. Epithelioid histiocytes may occur but they seldom aggregate to form discrete granulomas. Plasma cells, lymphocytes and sometimes eosinophils may accompany the histiocytic proliferation.

Lymph nodes at the porta hepatis

These nodes frequently show a lipogranulomatous reaction, often within the paracortex rather than in the sinuses.

Lymphangiogram effect

Haematopathologists became familiar with this reaction when staging laparotomies were performed as a part of the management of Hodgkin lymphoma. Lymphangiograms are now less frequently performed. The sinuses of the affected nodes are distended by a lipogranulomatous reaction to the oil-based contrast medium.

Silicone lymphadenopathy

Silicone in liquid form has been used for breast enhancement either by direct injection of silicone into the breast

BOX 3.17: Lipogranulomatous lymphadenitis

- Exogenous lipid (lymphangiogram, parenteral nutrition)
- Endogenous lipid (obesity, diabetes, hyperlipidaemia)
- Sinus histiocytosis
- Sinuses contain lipid vacuoles surrounded by foamy histiocytes, giant cells and epithelioid cells
- Much of the lipid in nodes from the porta hepatis appears in the paracortex

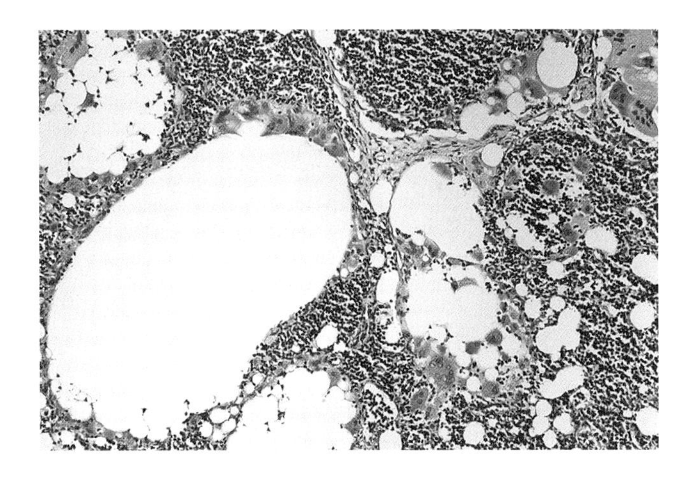

Figure 3.70 Lymphangiogram effect. High-power view showing histiocyte and giant cell reaction to the contrast medium.

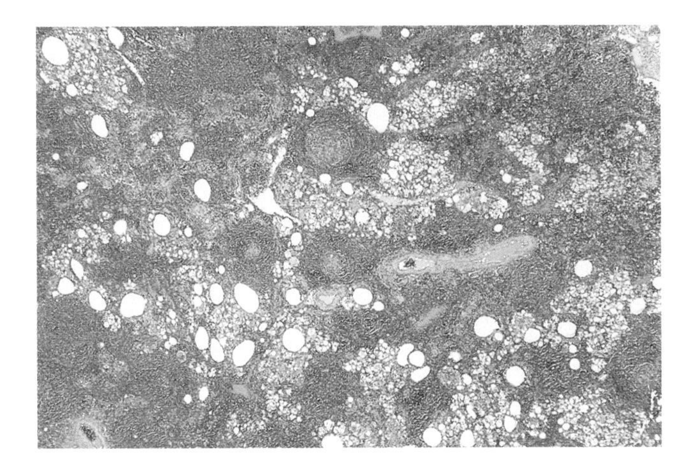

Figure 3.68 Lymph node from porta hepatis showing a lipogranulomatous reaction to lipid. Lipid vacuoles in the paracortex and sinuses are surrounded by histiocytes and giant cells.

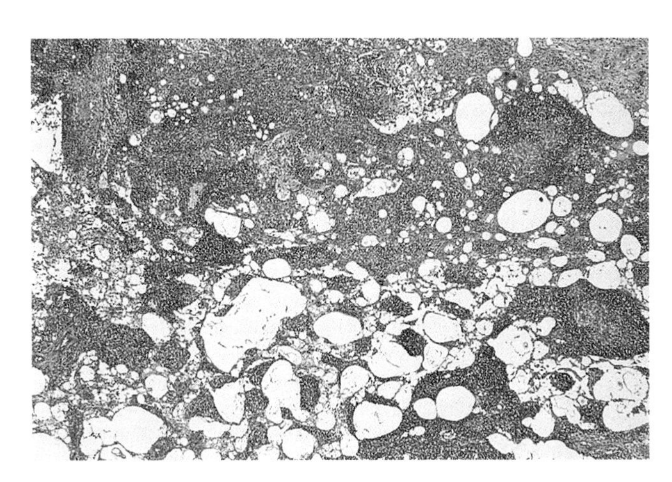

Figure 3.71 Whipple disease. The sinuses of the lymph node are filled with foamy histiocytes surrounding lipid vacuoles.

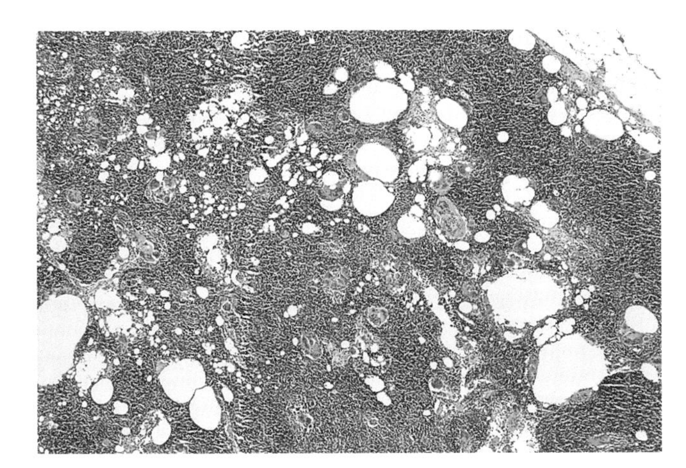

Figure 3.69 Lymphangiogram effect, para-aortic lymph node. The lipid-rich contrast medium has evoked a lipogranulomatous reaction in the lymph node sinuses.

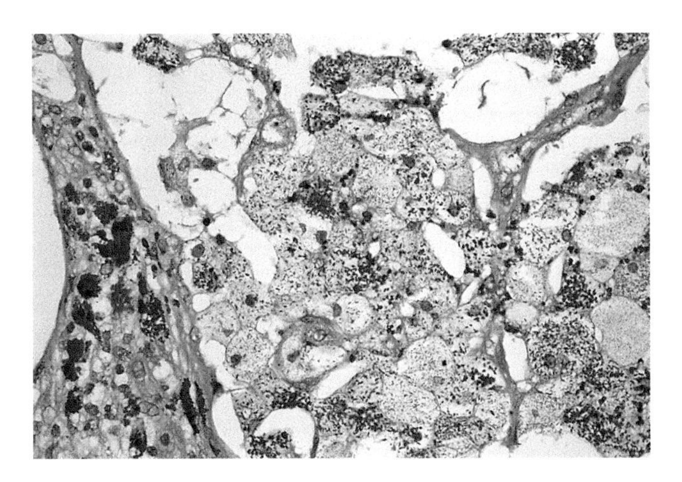

Figure 3.72 Whipple disease. Periodic acid–Schiff-stained section showing large numbers of bacilli in foamy histiocytes.

or, more commonly, by implanting a prosthesis containing silicone. The fluid causes a lipogranulomatous reaction followed by fibrosis in the breast and the draining lymph nodes.

Silicone in solid form causes a different type of lymphadenopathy. This is seen in patients who have received interphalangeal prostheses for severe rheumatoid arthritis. Small fragments of silicone carried to the lymph nodes cause a giant cell reaction in the sinuses or the paracortex. Refractile non-birefringent fragments of silicone are seen in these giant cells.

Infective causes

Whipple disease, a rare bacterial infection caused by *Tropheryma whipplei*, is associated with mesenteric lymphadenopathy but systemic lymphadenopathy is also seen in half the cases. Lipid vacuoles are characteristically present in the sinuses. The sinus histiocytes are distended by PAS-positive bacilli.

Numerous foamy histiocytes may be seen in lymph nodes from patients with lepromatous leprosy. These tend to be in the paracortex rather than in the sinuses. Modified Ziehl–Neelsen stain (Wade-Fite) shows numerous acid-fast leprosy bacilli.

Metabolic causes

Foamy macrophages are seen in Neimann–Pick and Fabry diseases. Genetic and biochemical tests are needed for their confirmation.

VASCULAR TRANSFORMATION OF SINUSES

This uncommon condition is most likely to be encountered in nodes removed at surgery for cancer. The cortical and medullary sinuses are filled with proliferating vascular channels of varying density. Vascular obstruction is thought to be the cause of this lesion, although angiogenic factors may play a role. The most important differential diagnosis is with Kaposi sarcoma. Kaposi sarcoma has a more solid structure and often involves the capsule, and the blood-filled clefts between the spindle cells are not lined by endothelial cells (Box 3.18 and Figure 3.73).

BOX 3.18: Vascular transformation of sinuses

- Most frequently encountered in lymph nodes draining carcinomas
- Probably caused by vascular obstruction
- Proliferation of endothelial-lined vascular spaces in subcapsular and medullary sinuses
- Sinuses often engorged with blood

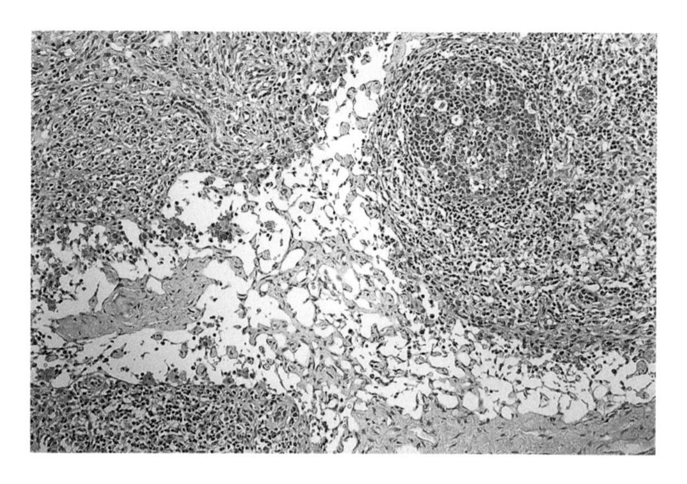

Figure 3.73 Vascular transformation of sinuses. The sinuses shown are filled with a meshwork of endothelial cell–lined vascular spaces.

GRANULOMATOUS LYMPHADENITIS

These are lymphadenopathies that are characterized by the presence of granulomas or localized aggregates of histiocytes as the most prominent feature. They may be conveniently divided into those with suppuration (i.e., necrosis and the presence of neutrophils) and those granulomas formed of epithelioid histiocytes. The latter group may be further divided into non-necrotizing or sarcoid-like granulomatous, and necrotizing granulomatous lymphadenitis. All groups of reaction, however, may be accompanied by varying degrees of follicular and paracortical hyperplasia, but the granulomatous reaction is the most prominent feature.

SUPPURATIVE GRANULOMATOUS LYMPHADENOPATHY

Suppurative granulomatous lymphadenitis generally represents a B-cell–associated granulomatous reaction that is different from the hypersensitivity type or epithelioid granulomatous lymphadenitis seen in tuberculosis, leprosy and sarcoidosis. Suppurative granulomas often appear to start as an accumulation of monocytoid B-cells followed by polymorph infiltration with necrosis developing in these foci, leading eventually to granuloma formation. This contrasts with the formation of epithelioid granulomas of the hypersensitivity type, which appears to be mediated by activated T-cells, dendritic cells and histiocytes without the participation of B-cells. Suppurative granulomas are distinguished from the hypersensitivity-type granulomas by the presence of neutrophils, prominence of monocytoid B-cells and a paucity of multinucleated giant cells. Immunostaining reveals a mixture of B- and T-cells mixed with histiocytes, polymorphs, immunoblasts and dendritic cells. In contrast, B-cells are few or absent within hypersensitivity-type granulomas.

BOX 3.19: Common aetiological agents in suppurative granulomatous lymphadenopathy

- *Bartonella henselae* (cat scratch disease)
- *Chlamydia trachomatis* (lymphogranuloma venereum)
- *Francisella tularensis* (tularaemia)
- *Yersinia enterocolitica, Y. pseudotuberculosis*
- *Listeria monocytogenes*
- *Pseudomonas mallei* (glanders), *P. pseudomallei* (melioidosis)
- *Corynebacterium ovis, C. pyogenes, C. ulcerans*

Included among the more common causes of bacterial suppurative granulomatous lymphadenitis are cat scratch disease, lymphogranuloma venereum, tularaemia and *Yersinia* infections (Box 3.19). There are no specific morphological features to distinguish among these various entities and they will be described together. The demonstration of the causative agent by histochemical and/or immunohistochemical stains and the use of serological tests, bacterial culture or molecular techniques such as PCR provides the definitive aetiological diagnosis.

The lymph nodes involved are usually regional nodes draining the portal of entry of the causative organism, but systemic dissemination may occur with progression of the infection. There is initially a florid follicular hyperplasia with abundant monocytoid B-cells in the sinuses and parafollicular areas while nodal architecture remains preserved. Subsequently, small suppurative granulomas develop. These comprise collections of histiocytes with and without central aggregates of neutrophils, and occur in close proximity to the monocytoid B-cell clusters from which they may be difficult to separate. In later stages there is coalescence of the enlarging granulomas, which display central stellate areas of necrosis that contain neutrophil debris (nuclear dust) and fibrinoid material and are surrounded by a rim of palisading histiocytes, often with proliferating fibroblasts. The monocytoid B-cells may be less prominent at this stage. Multinucleated cells are rare (Boxes 3.20 and 3.21).

Cat scratch disease

The causative organism of cat scratch disease is *Bartonella henselae*, a Gram-positive pleomorphic coccobacillus that can be found in clumps in the foci of necrosis and around blood vessels and within sinusoidal macrophages, especially in the late phase of the disease. The organisms can be demonstrated in over 60 per cent of cases with the Warthin–Starry stain or Dieterle stain, or immunohistochemically. Cat scratch disease is the most common cause of suppurative granulomatous lymphadenitis and may show systemic involvement including suppurative hepatosplenic granulomas and osteomyelitis (Figures 3.74 and 3.75).

BOX 3.20: Common causes of epithelioid granulomatous lymphadenitis

NECROTIZING GRANULOMAS

- *Mycobacterium tuberculosis, M. avium intracellulare, M. lepra*
- Systemic fungal infections, especially *Cryptococcus neoformans, Histoplasma capsulatum, Coccidioides immitis, Blastomycosis dermatitidis*

NON-NECROTIZING GRANULOMAS

- Sarcoidosis
- Berylliosis
- Crohn's disease
- Lymph nodes draining carcinoma

BOX 3.21: Suppurative granulomatous lymphadenitis

- Usually involves lymph nodes draining the portal of entry of the causative organism
- Follicular hyperplasia
- Monocytoid B-cell hyperplasia with a variable infiltrate of polymorphs
- Areas of necrosis develop, surrounded by palisaded histiocytes but few giant cells
- Areas of necrosis assume a serpiginous or stellate shape
- Aetiological agent may be demonstrated by special stains, culture or the polymerase chain reaction

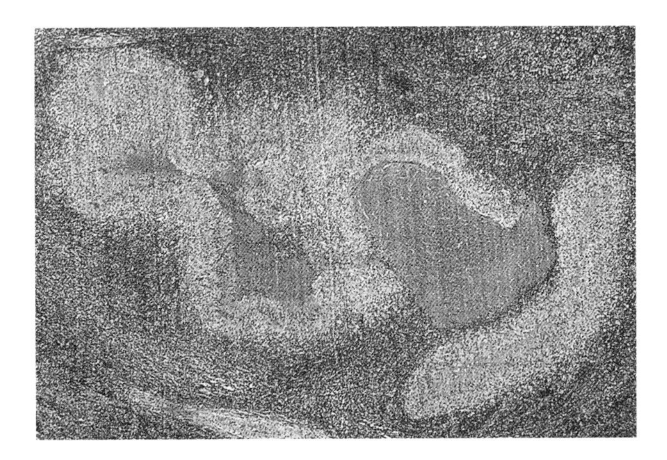

Figure 3.74 Cat scratch disease. Lymph node showing serpiginous areas of necrosis surrounded by palisaded epithelioid histiocytes.

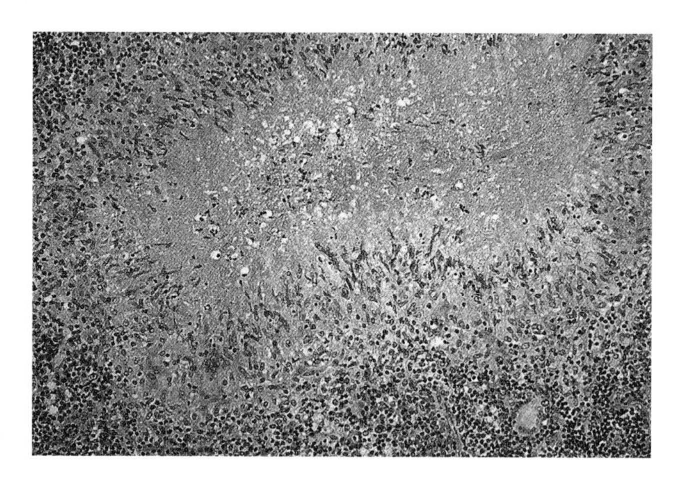

Figure 3.75 Cat scratch disease. High-power view showing palisaded epithelioid histiocytes around an area of necrosis. Some of the nuclear debris at the centre of the area of necrosis is from polymorph leukocytes.

Lymphogranuloma venereum

Lymphogranuloma venereum may show vacuolated macrophages containing *Chlamydia trachomatis,* as fine, sand-like infective organisms, in and around the suppurative areas, as demonstrated with the Warthin–Starry stain. There is often accompanying oedema with inflammation around the lymph node (perilymphadenitis).

Yersinia lymphadenitis

Yersinia lymphadenitis shows no distinguishing features from the other forms of suppurative granulomatous lymphadenitis other than its mesenteric location, and may be associated with changes in the terminal ileum. Confirmation requires culture or serological testing.

Other causes of suppurative granulomatous lymphadenitis

Tularaemia, caused by *Francisella tularensis*, is transmitted by ticks and other biting arthropods and produces a necrotizing granulomatous lymphadenitis that is indistinguishable from the preceding diseases. It may be associated with Langhans-type giant cells. Other rare causes of suppurative granulomatous lymphadenitis include infection by *Listeria monocytogenes, Pseudomonas mallei* (glanders) and *P. pseudomallei* (melioidosis), which cannot be distinguished except by identification of the causative organism in the tissue section, culture or PCR. Corynebacteria, especially *Corynebacterium ovis, C. ulcerans, C. diphtheriae* and *C. pyogenes*, can produce suppurative granulomas in draining lymph nodes.

EPITHELIOID GRANULOMATOUS LYMPHADENOPATHY (HYPERSENSITIVITY-TYPE GRANULOMATOSIS)

The common causes of epithelioid or hypersensitivity-type granulomatous lymphadenopathy are listed in Box 3.20. They may be divided into necrotizing and non-necrotizing groups, with tuberculosis and sarcoidosis, respectively, representing the prototypes of these two reactions (Box 3.22 and Figures 3.76 to 3.95).

Necrotizing epithelioid granulomatous lymphadenopathy

TUBERCULOUS LYMPHADENITIS

Tuberculous lymphadenitis is the prototype of necrotizing epithelioid granulomatous lymphadenitis, with central areas of caseous necrosis surrounded by palisaded

BOX 3.22: Epithelioid granulomatous lymphadenopathy

- Histological appearances vary from loose collections of epithelioid cells or solitary giant cells to well-formed granulomas
- Histological appearance varies with the causative agent and the host's immune status
- Special stains (periodic acid–Schiff, Grocott, Ziehl–Neelsen and Giemsa) aid the identification of many organisms

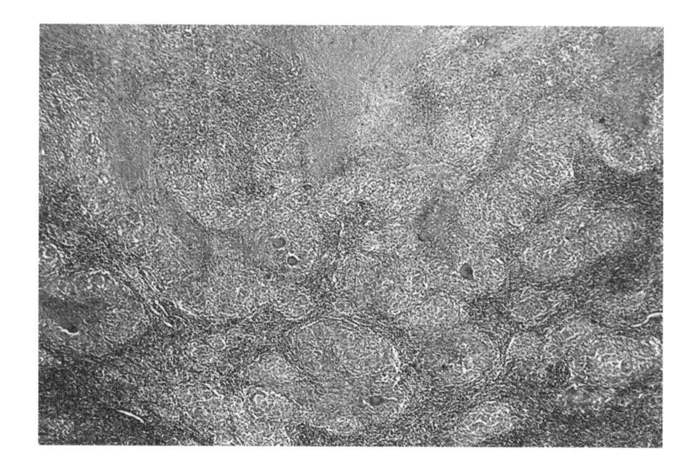

Figure 3.76 Tuberculous lymph node showing necrotizing epithelioid cell/giant cell granulomas. Note the tendency of the granulomas to coalesce.

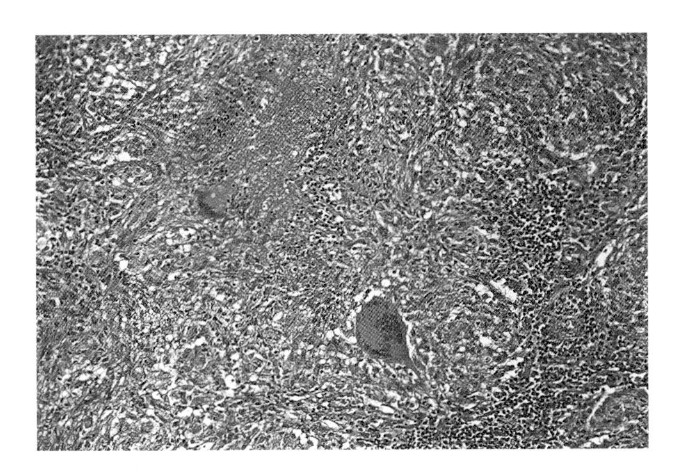

Figure 3.77 Tuberculous lymph node showing caseating granulomas containing Langhans-type giant cells.

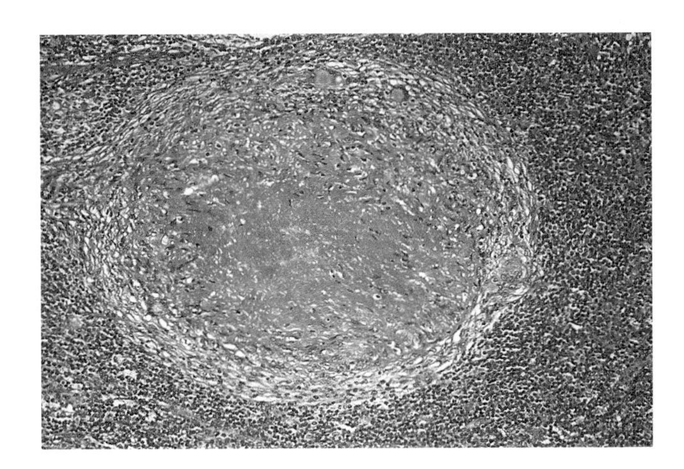

Figure 3.78 Bacille Calmette–Guérin (BCG)–infected lymph node from an immunocompetent patient showing a well-defined necrotizing granuloma.

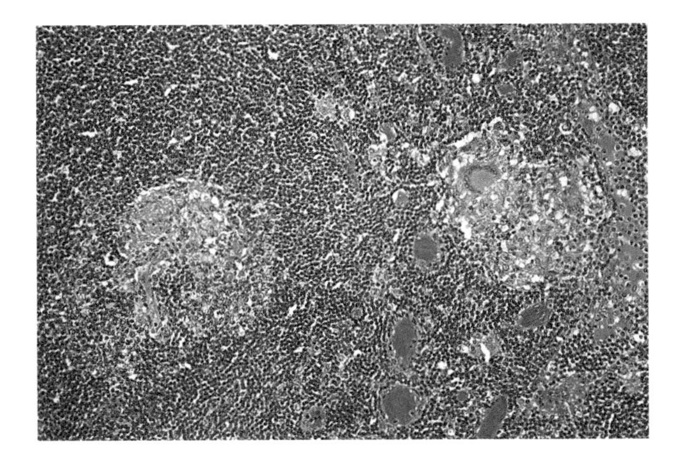

Figure 3.79 Abdominal lymph node from a patient with Crohn's disease showing two granulomas.

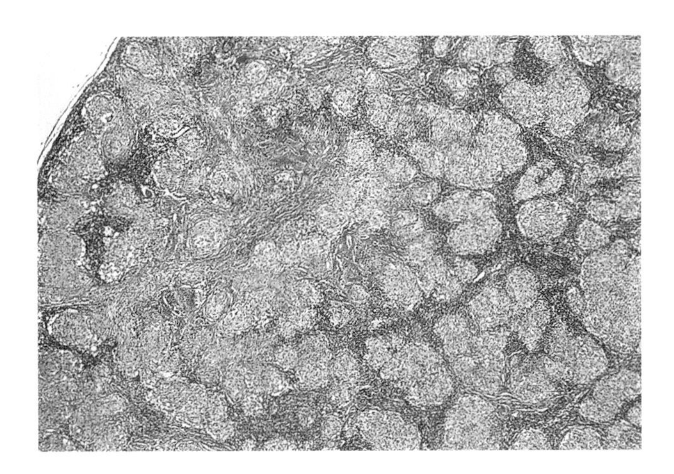

Figure 3.80 Sarcoidosis of lymph node. The granulomas tend to remain discrete and to undergo progressive fibrosis.

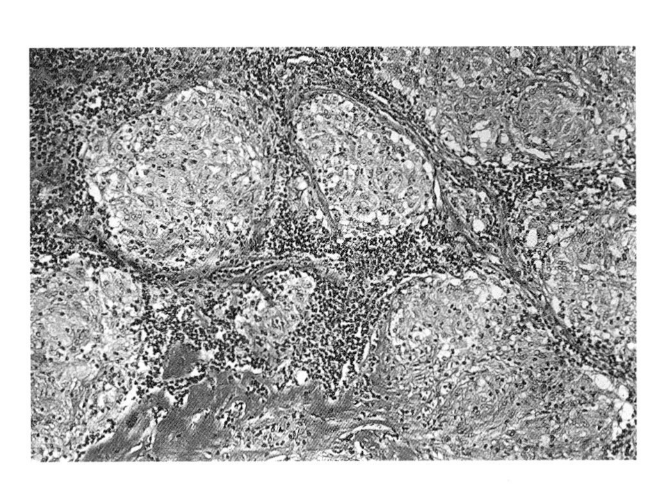

Figure 3.81 Sarcoidosis of lymph node showing well-defined non-necrotizing epithelioid cell granulomas undergoing early sclerosis.

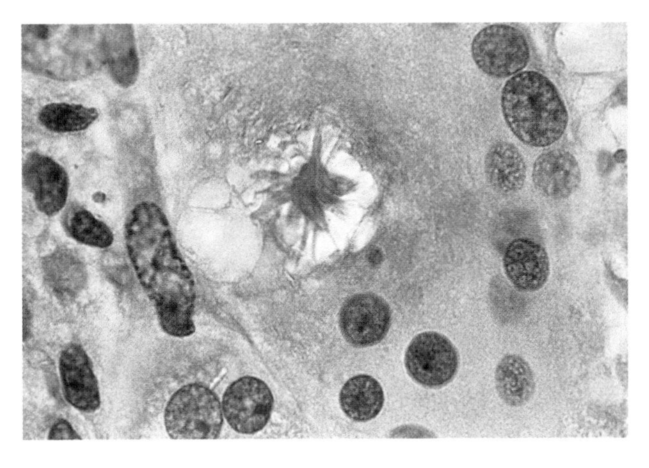

Figure 3.82 Sarcoidosis of lymph node showing an asteroid body in a giant cell.

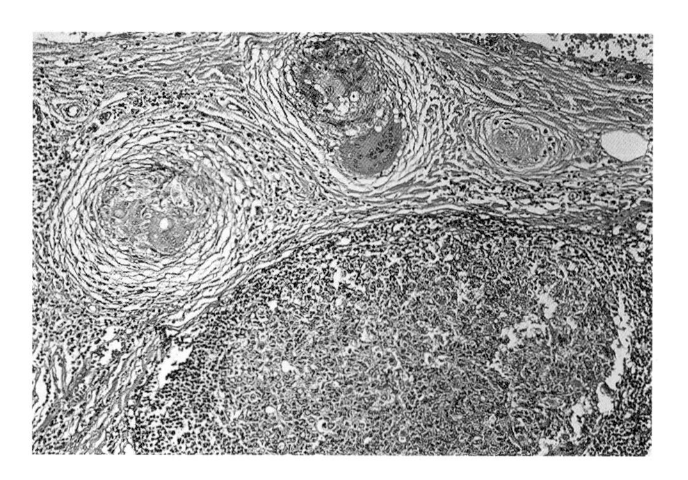

Figure 3.83 Lymph node from a patient with schistosomiasis. There is marked capsular fibrosis with several granulomas surrounding schistosome ova.

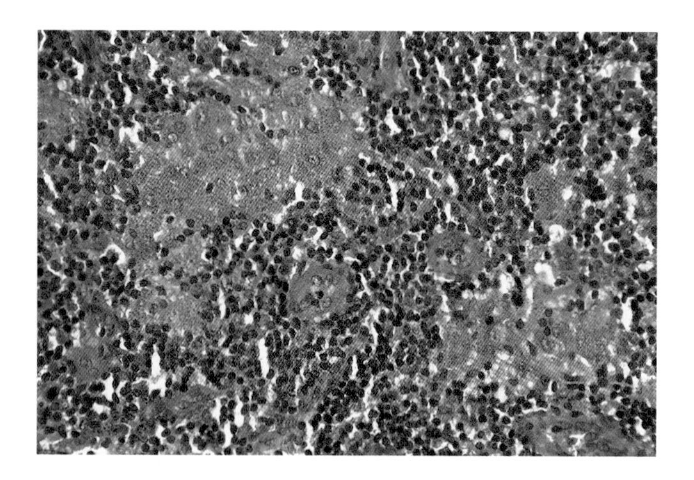

Figure 3.84 Leishmaniasis of lymph node. Organisms are visible in the cytoplasm of the histiocytes.

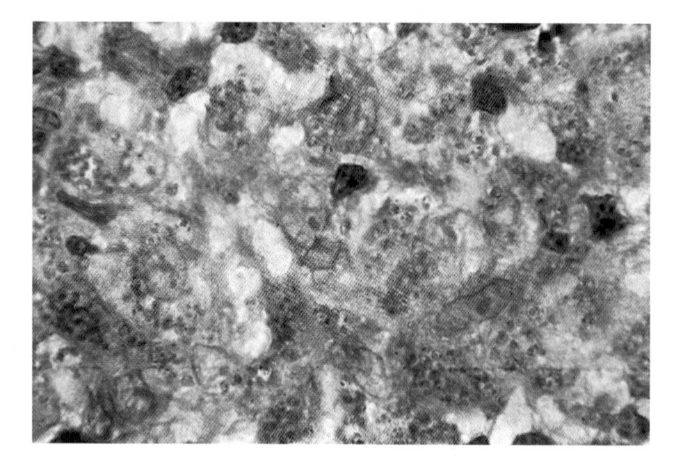

Figure 3.85 Leishmaniasis of lymph node, Giemsa stain. The organisms are seen more easily with this stain than with haematoxylin and eosin.

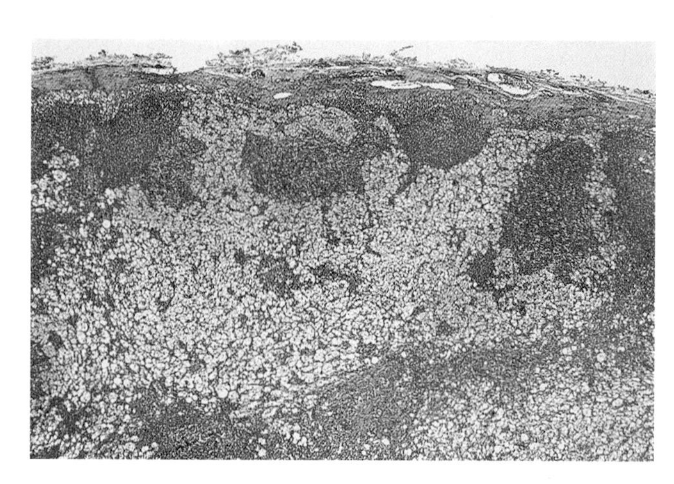

Figure 3.86 Lepromatous leprosy. Foamy histiocytes (lepra cells) fill the paracortex of the node.

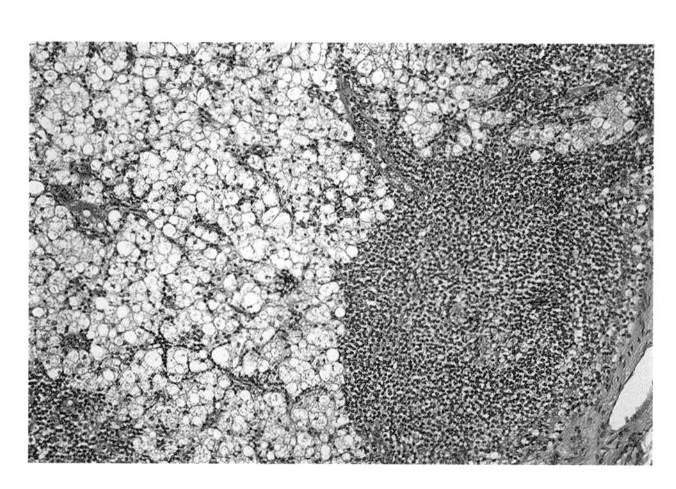

Figure 3.87 Lepromatous leprosy. High-power view of lepra cells.

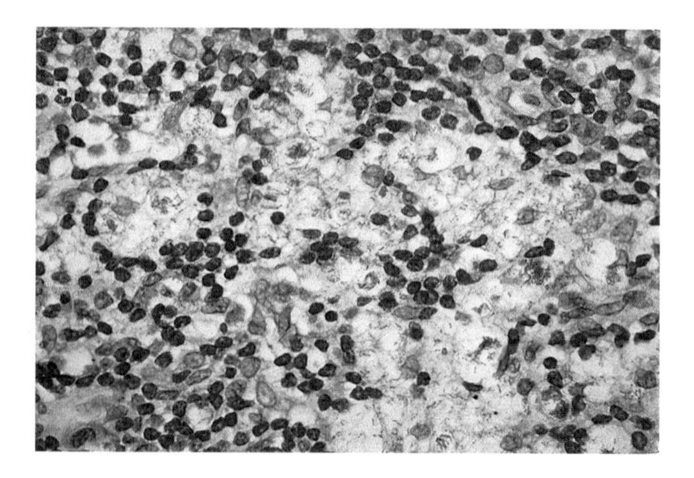

Figure 3.88 Lepromatous leprosy. Stain for acid-fast bacilli shows large numbers of leprosy bacilli in lepra cells.

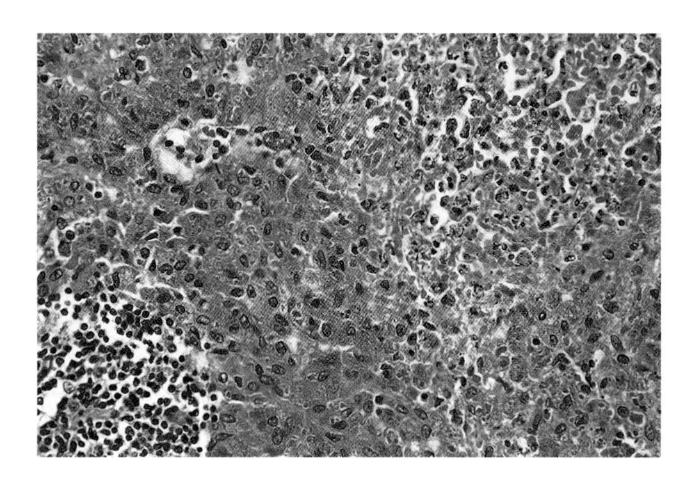

Figure 3.89 Bacille Calmette-Guérin (BCG)–infected lymph node from an immunosuppressed patient showing an area of necrosis surrounded by plump histiocytes.

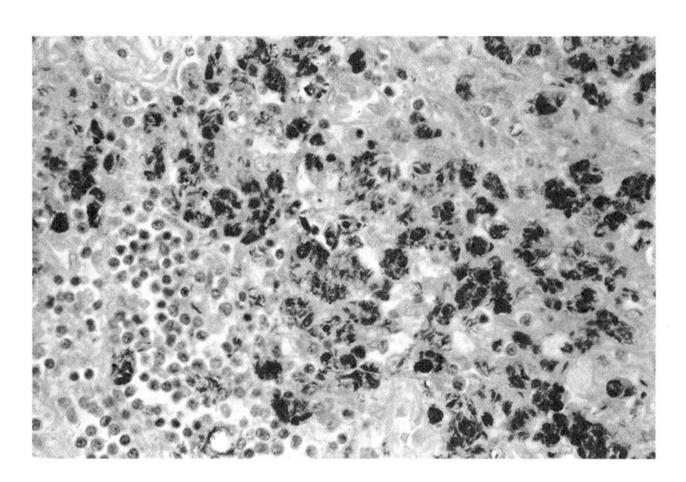

Figure 3.90 Bacille Calmette–Guérin (BCG)–infected lymph node from an immunosuppressed patient stained by the Ziehl-Neelsen method. Large numbers of acid-fast bacilli fill the plump histiocytes.

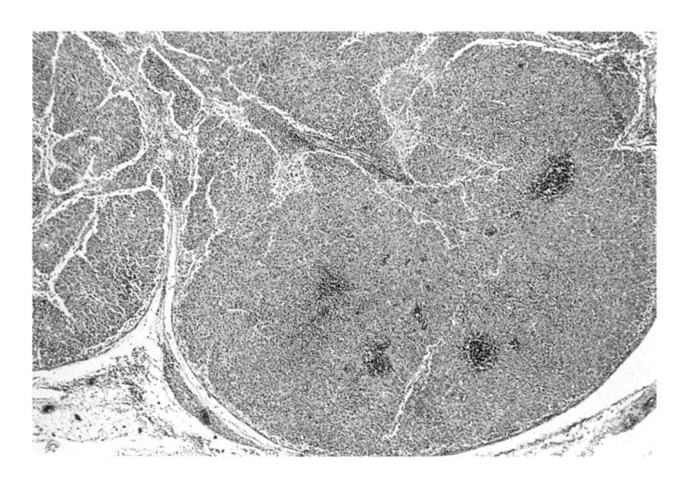

Figure 3.91 *Mycobacterium avium-intracellulare* infection in an immunosuppressed patient. Sheets of pale-staining histiocytes replace the lymph node.

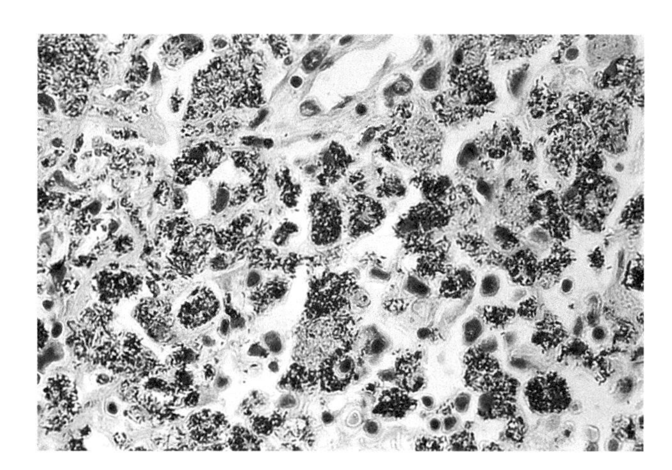

Figure 3.92 *Mycobacterium avium-intracellulare* infection. Ziehl–Neelsen–stained section, showing histiocytes filled with acid-fast bacilli.

epithelioid histiocytes and Langhans giant cells. Variable numbers of T-cells are found around the palisading epithelioid cells. Proliferating fibroblasts may be present around the entire granuloma. There is a tendency for tuberculous granulomas to coalesce, forming confluent areas of necrosis and granulomatous inflammation. Tubercle bacilli demonstrated with the Ziehl–Neelsen stain may be difficult to find; they are usually most easily detected within the necrotic areas or in the cytoplasm of giant cells. Other mycobacteria including *Mycobacterium scrofulaceum, M. bovis, M. kansasii, M. marinum, M. ulcerans* and *M. fortuitum* can produce a similar appearing lymphadenitis, as can bacillus Calmette–Guérin (BCG).

M. avium-intracellulare may produce an indistinguishable necrotizing granulomatous response but may also show a wide spectrum of histological changes that include a striking histiocytosis with a spindle-cell proliferation of short fascicles and a storiform pattern that may be mistaken for fibrous histiocytoma, inflammatory pseudotumour or smooth muscle tumour. In all these reactions, the atypical mycobacteria are readily demonstrated with a Ziehl–Neelsen, methenamine silver, PAS or Gram stain. The stains may often reveal myriad bacilli similar to the globi of lepromatous leprosy that can be seen as basophilic cytoplasmic streaks in H&E sections.

In tuberculoid leprosy, the nodes show discrete epithelioid granulomas with a variable number of Langhans giant cells but without necrosis. The granulomas may undergo fibrosis and hyalinization and may mimic sarcoidosis. The lepromatous form of the disease gives rise to a striking paracortical histiocytosis with large clusters of the bacilli or globi within the foam cells, also called lepra cells or Virchow cells. In borderline leprosy, the nodal morphology may be that of either tuberculoid or lepromatous disease (Box 3.23).

BOX 3.23: Infective lymphadenopathy in patients with impaired or absent immune response

- Sheets of histiocytes without well-defined granuloma formation
- Histiocytes have abundant eosinophilic or foamy cytoplasm
- Large numbers of organisms demonstrable within the histiocytes
- Areas of necrosis may occur within the histiocyte aggregates

FUNGAL LYMPHADENOPATHY

Fungal infections of lymph nodes can be divided into primary and opportunistic infections, the latter infecting immune-compromised patients. This division is somewhat artificial, as even the primary or deep mycoses require some degree of depression of immunity to produce systemic infection. Fungal infections may result in a spectrum of changes ranging from histiocytosis to epithelioid granulomas with or without suppurative necrosis, and the specific diagnosis is based on identification of the fungus by special stains or culture. The following systemic mycotic infections can involve lymph nodes: cryptococcosis (*Cryptococcus neoformans*), histoplasmosis (*Histoplasma capsulatum*), coccidioidomycosis (*Coccidioides immitis*) and blastomycosis (*Blastomyces dermatitidis*).

Lymph node involvement in cryptococcosis occurs most commonly in mediastinal nodes secondary to pulmonary infections. The extent of granulomatous reaction is very variable, ranging from minimal cellular reaction to full-blown granulomas, but the yeast-like organisms are often numerous. They are refractile and of variable size and have a clear halo around the cell wall. This halo represents the mucopolysaccharide capsule, which is demonstrated by PAS or Grocott stains.

Nodal involvement by histoplasmosis generally follows pulmonary infection and shows pathological features that resemble tuberculosis. The enlarged nodes contain tuberculoid granulomas and often undergo caseous necrosis. The necrotic areas are frequently surrounded by a fibrous reaction and may calcify. The histological reaction in patients with HIV/AIDS and other forms of immunodeficiency lacks granuloma formation. Sheets of histiocytes containing numerous organisms replace the lymph node. The encapsulated budding yeasts are best visualized with the Grocott or PAS stain within the areas of necrosis, and within histiocytes and giant cells.

Other pathogenic fungi produce a range of cellular responses. In some situations, the lymph node reaction may be non-specific, especially in nodes draining infected sites where non-specific germinal centre and paracortical hyperplasia are seen without granuloma formation. Although many fungi are visible in H&E-stained sections, the PAS and Grocott stains allow better visualization of the organism and, whenever fungal lymphadenitis is suspected, fungal culture should be performed.

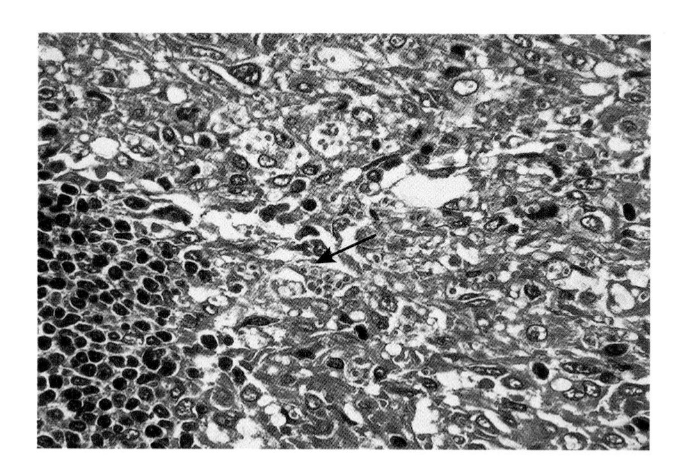

Figure 3.93 Histoplasmosis. Clusters of yeasts are seen within the cytoplasm of histiocytes (arrow).

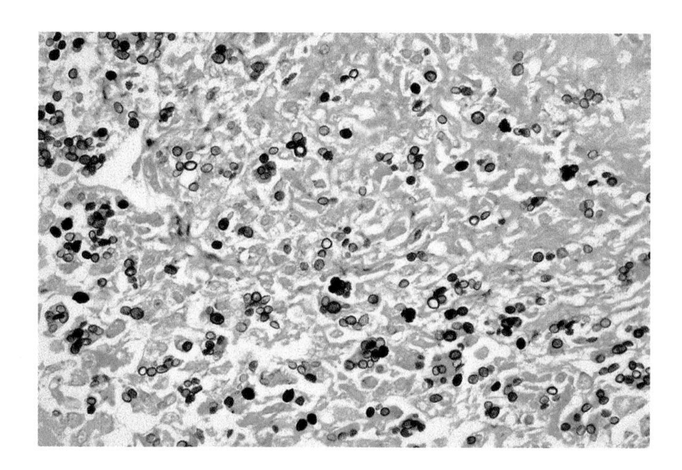

Figure 3.94 Histoplasmosis. Grocott stain highlights the intracellular organisms.

Non-necrotizing granulomatous lymphadenopathy

SARCOIDOSIS

The granulomas of sarcoidosis are composed of clusters of epithelioid histiocytes with occasional giant cells surrounded by variable numbers of T-cells. They show a lesser tendency than tuberculous granulomas to coalesce and do not show caseous necrosis, although small areas of central necrosis may be seen. The multinucleated giant cells may contain birefringent rounded concretions known as Schaumann bodies. Calcium and iron can be demonstrated in these layered inclusions. Less frequently, asteroid bodies are present. These are inclusions with varying numbers of curved processes radiating from the centre.

Although striking in appearance, neither of these inclusions is specific and they may be found in other reactive conditions. With disease progression the granulomas become surrounded by, and are eventually replaced by, fibrous tissue (Figures 3.80–3.82).

BERYLLIOSIS

The granulomatous lymphadenitis caused by beryllium is histologically indistinguishable from sarcoid.

Protozoan and nematode infections

Lymphadenitis caused by protozoa, nematodes, larval nematodes and ova are generally uncommon and represent more a curiosity; however, in endemic areas, this form of lymph node involvement may be routinely encountered. These infections are discussed under granulomatous reactions because they do not conform to any specific reaction pattern and a granulomatous response is the most common reaction to the parasitic antigens. Follicular hyperplasia may also be present and eosinophilia is a frequent accompaniment of such infections (Figures 3.83, 3.84, and 3.93 to 3.95).

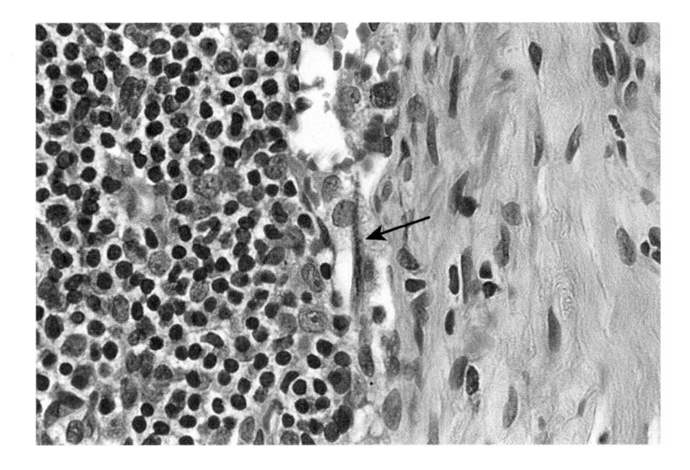

Figure 3.95 Lymph node from a patient with onchocerciasis. There is fibrosis of the capsule and a microfilaria is seen in the subcapsular sinus (arrow). Since the organism was alive at the time of biopsy, it has not excited a granulomatous reaction.

LYMPHADENITIS CAUSED BY ENTAMOEBA HISTOLYTICA

Amoebiasis caused by *E. histolytica* may involve lymph nodes with the protozoa being carried to draining lymph nodes. As the amoebae closely mimic the appearance of foamy macrophages and even carcinoma cells, and as they tend to be located in the sinuses in the first instance, they should not be mistaken for these cells. Both macrophages and amoebae may contain ingested red cells and may be positive with PAS stain but the amoebae also stain with colloidal iron. Carcinoma cells are readily excluded on the basis of immunostaining for cytokeratin.

LEISHMANIA LYMPHADENITIS

The protozoal genus *Leishmania* is endemic in many areas of the world, where it is transmitted by sandflies. Ulcers may develop at the site of sandfly bites. Local lymph nodes are frequently involved and lymphadenopathy may be the presenting feature. Disseminated disease gives rise to visceral leishmaniasis (kala-azar) and is also seen in HIV/AIDS patients.

Lymph nodes show reactive changes, with loose collections of histiocytes in the paracortex. As the disease progresses, these become more organized as epithelioid giant cell granulomas. In the early stages of the disease and in immunodeficient individuals, the parasite amastigotes are easily found within histiocytes. When well-formed granulomas are present, it is often not possible to find organisms, although they are detectable by PCR.

LYMPHADENITIS CAUSED BY METAZOAN PARASITES

Metazoan parasites may occasionally be found in lymph nodes. Among these are *Ascaris lumbricoides* (the common roundworm), *Toxocara cati* and *T. canis* (cat and dog ascarids), *Strongyloides stercoralis, Ancylostoma duodenale, Necator americanus* (hookworms), *Trichinella spiralis* (pork worm), *Oxyuris vermicularis* (threadworm), *Trichuris trichura* (whipworm) and *Wuchereria bancrofti* (filariasis), *Brugia malayi* (filariasis), *Onchocerca volvulus* (river blindness) and *Dracunculus medinensis* (Guinea worm). The adult and larval forms of these nematodes may enter blood vessels and lymphatics during their migration. When they lodge in the lymph node and die, they produce a granulomatous reaction, often with the nematode in the centre. Invariably there is accompanying necrosis, and multinucleated giant cells and eosinophils and the Splendore–Hoeppli reaction may occur around the dead nematode. This granulomatous reaction is eventually replaced by fibrosis, which may cause lymphatic obstruction and lymphoedema.

Lymph nodes draining malignancy

Lymph nodes draining malignancy, most frequently carcinoma, may show a sarcoid-like granulomatous response in the absence of tumour deposits in the node. Such epithelioid granulomas are seen in a variety of carcinomas and are the result of stimulation by antigens from the tumour carried to the regional nodes.

Lymphadenopathy caused by foreign materials

Detritic lymphadenitis is a granulomatous reaction to the cementing material and metallic debris that migrate from joint prostheses to draining lymph nodes. The flakes of foreign material are pigmented and the cementing substances are both birefringent and refractile, and are present in macrophages within the sinuses and paracortex.

Tattoo pigment is usually seen in the paracortex of nodes draining an area of tattooing. Black granules are seen in macrophages. Careful examination of this pigment will usually reveal other (red, green and blue) pigments. These are often birefringent and are highlighted by visualization in polarized light. Tattoo pigment is sometimes associated with areas of necrosis and/or granuloma formation, presumably resulting from hypersensitivity to one of the components (Box 3.24 and Figures 3.96 to 3.101).

BOX 3.24: Lymphadenopathy caused by foreign materials

- Reaction to inhaled, injected or implanted foreign materials
- Commonly encountered causes include tattoo pigment and debris from joint prostheses
- Material may be pigmented
- Material may be birefringent
- There may be no reaction to the foreign material
- Reaction may take the form of epithelioid cell giant cell granulomas or of foreign-body-type giant cells
- Necrosis may be present

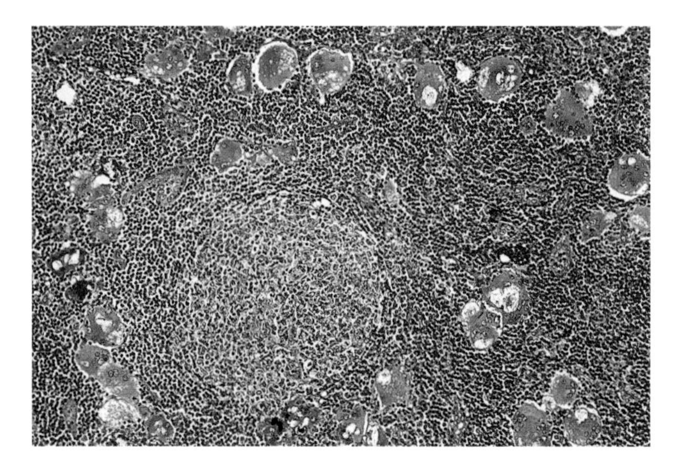

Figure 3.96 Silicone lymphadenopathy. This paracortical giant cell reaction is to particulate silicone from a joint prosthesis.

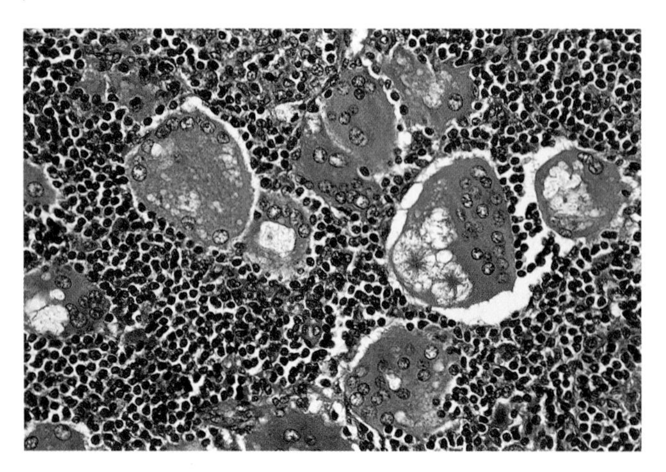

Figure 3.97 Silicone lymphadenopathy. High-power view showing asteroid bodies and refractile silicone particles in the giant cells.

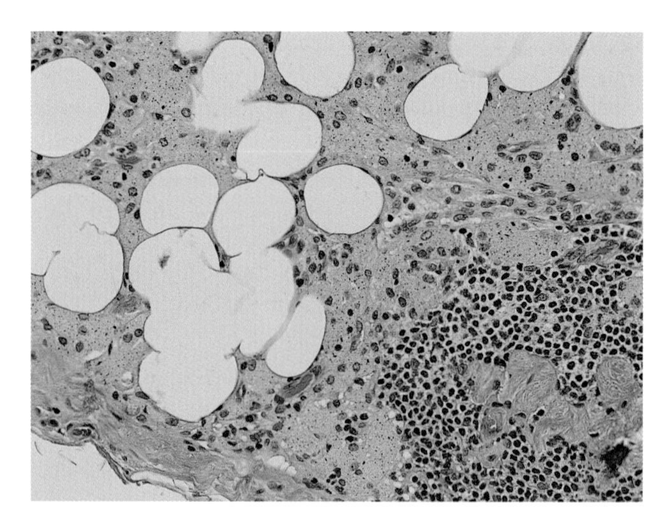

Figure 3.98 Enlarged groin lymph node from a patient with both hip and knee prostheses demonstrating numerous histiocytes in the paracortex.

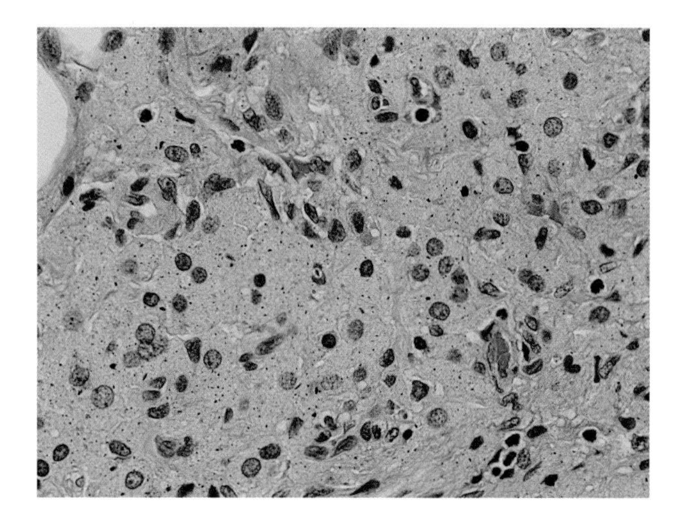

Figure 3.99 High-power view of the lymph node in Figure 3.98 from a patient with both hip and knee prostheses demonstrating fragments of metal material in histiocytes.

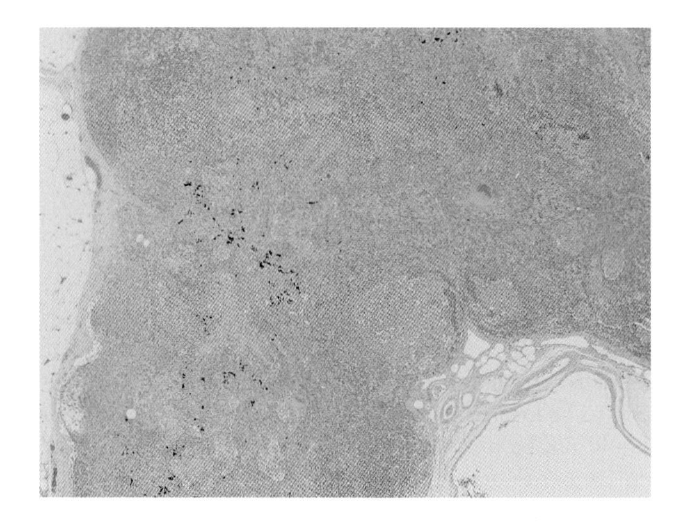

Figure 3.100 Tattoo pigment in a hyperplastic lymph node. The pigment is mainly distributed in the paracortex.

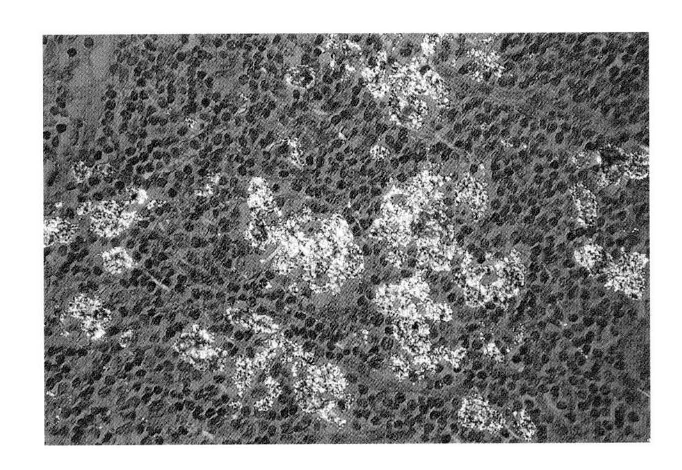

Figure 3.101 Tattoo pigment viewed in polarized light showing multiple colours.

REFERENCES

Arber DA, Kamel OW, van de Rijn M, *et al.* 1995 Frequent presence of the Epstein-Barr virus in inflammatory pseudotumor. *Human Pathology* **26:** 1093–1098.

Bateman AC, Ashton-Key MR, Jogai S. 2015 Lymph node granulomas in immunoglobulin G4-related disease. *Histopathology* **67:** 557–561.

Cheuk W, Chan JKC, Shek TWH, *et al.* 2001 Inflammatory pseudotumor–like follicular dendritic cell tumor. *American Journal of Surgical Pathology* **25:** 721–731.

Deshpande V, Zen Y, Chan J, *et al.* 2012 Consensus statement on the pathology of IgG4-related disease. *Modern Pathology* **25**: 1181–1192.

Du MQ, Liu H, Diss TC, *et al.* 2001 Kaposi sarcoma-associated herpes virus infects monotypic (IgMλ) but polyclonal naïve B cells in Castleman disease and associated lymphoproliferative disorders. *Blood* **97:** 2130–2136.

Facchetti F, Incardona P, Lonardi S, *et al.* 2009 Nodal inflammatory pseudotumor caused by luetic infection. *American Journal of Surgical Pathology* **33**: 447–453.

Ferry JA. 2013 IgG4-related lymphadenopathy and IgG4-related lymphoma: moving targets. *Diagnostic Histopathology* **19**: 128–139.

Grimm KE, Barry TS, Chizhevsky V, *et al.* 2012. Histopathological findings in 29 lymph node biopsies with increased IgG4 plasma cells. *Modern Pathology* **25**: 480–491.

Isaacson PG, Schmid C, Pan L, *et al.* 1992 Epstein-Barr virus latent membrane protein expression by Hodgkin and Reed–Sternberg-like cells in acute infectious mononucleosis. *Journal of Pathology* **167:** 267–271.

Lim MS, Straus SE, Dale JK, *et al.* 1998 Pathological findings in human autoimmune lymphoproliferative syndrome. *American Journal of Pathology* **153:** 1541–1550.

Logani S, Lucas DR, Cheng JD, *et al.* 1999 Spindle cell tumors associated with mycobacteria in lymph nodes of HIV-positive patients: 'Kaposi sarcoma with mycobacteria' and 'mycobacterial pseudotumor'. *American Journal of Surgical Pathology* **23**: 656–661.

Luppi M, Barozzi P, Garber R, *et al.* 1998 Expression of human herpes virus-6 antigens in benign and malignant lymphoproliferative diseases. *American Journal of Pathology* **153:** 815–823.

Maric I, Pittaluga S, Dale J, *et al.* 2004 Sinus histiocytosis with massive lymphadenopathy in patients with autoimmune lymphoproliferative syndrome. *Modern Pathology* **17:** 258A.

Mehraein Y, Wagner M, Remberger K, *et al.* 2006 Parvovirus B19 detected in Rosai–Dorfman disease in nodal and extranodal manifestations. *Journal of Clinical Pathology* **59:** 1320–1326.

Moran CA, Suster S, Abbondanzo SL 1997 Inflammatory pseudotumor of lymph nodes: A study of 25 cases with emphasis on morphological heterogeneity. *Human Pathology* **28:** 332–338.

Perrone T, De Wolf-Peeters C, Frizzera G 1988 Inflammatory pseudotumor of lymph nodes. A distinctive pattern of nodal reaction. *American Journal of Surgical Pathology* **12:** 351–361.

Pileri SA, Facchetti F, Ascani S, *et al.* 2001 Myeloperoxidase expression by histiocytes in Kikuchi's and Kikuchi-like lymphadenopathy. *American Journal of Pathology* **159:** 915–924.

Reynolds DJ, Banks PM, Gulley ML 1995 New characterization of infectious mononucleosis and a phenotypic comparison with Hodgkin's disease. *American Journal of Pathology* **146:** 379–388.

Rosai J, Dorfman RT 1969 Sinus histiocytosis with massive lymphadenopathy; a newly recognised benign clinicopathological entity. *Archives of Pathology* **87:** 63–70.

Rushin JM, Riordan GP, Heaton RB, *et al.* 1990 Cytomegalovirus-infected cells express Leu-M1 antigen. A potential source of diagnostic error. *American Journal of Pathology* **136:** 989–995.

Sneller MC, Wang J, Dale JK, *et al.* 1997 Clinical, immunologic, and genetic features of an autoimmune lymphoproliferative syndrome associated with abnormal lymphocyte apoptosis. *Blood* **89:** 1341–1348.

Straus SE, Jaffe ES, Puck JM, *et al.* 2001 The development of lymphomas in families with autoimmune lymphoproliferative syndrome with germline Fas mutations and defective lymphocyte apoptosis. *Blood* **98:** 194–200.

Teachey DT, Seif AE, Grupp SA 2009 Advances in the management and understanding of autoimmune lymphoproliferative syndrome (ALPS). *British Journal of Haematology* **148:** 205–216.

Van der Werff ten Bosch J, Delabie J, Bohler T, *et al.* 1999 Revision of the diagnosis of T-zone lymphoma in the father of a patient with autoimmune lymphoproliferative syndrome type II. *British Journal of Haematology* **106:** 1045–1048.

Younes M, Podesta A, Helie M, Buckley P 1991 Infection of T but not B lymphocytes by cytomegalovirus in lymph node. An immunophenotypic study. *American Journal of Surgical Pathology* **15:** 75–80.

4

Precursor B- and T-cell lymphomas

PRECURSOR B-LYMPHOBLASTIC LEUKAEMIA/LYMPHOBLASTIC LYMPHOMA

B-lymphoblastic leukaemia and B-lymphoblastic lymphoma are essentially one disease process. Those patients with predominantly blood and bone marrow involvement are designated as having leukaemia; those with nodal or extranodal disease and less than 25 per cent lymphoblasts in the bone marrow are designated as having lymphoma. In practice, the majority of cases are leukaemic. Although B-lymphoblastic leukaemia/lymphoma is approximately four times as common as T-lymphoblastic leukaemia/lymphoma, the T-cell neoplasms present more frequently as a solid tumour (lymphoma).

B-lymphoblastic lymphoma presents most commonly as skin nodules, often multiple, bone or soft tissue tumours and lymphadenopathy. The tumour occurs most commonly in childhood (75 per cent below 6 years of age) with fewer young adult cases.

Morphologically, B-lymphoblastic and T-lymphoblastic lymphoma are indistinguishable. The tumour is composed of medium-sized blast cells that may be very uniform or show varying degrees of anisocytosis. Nuclei may be rounded or convoluted. The latter feature is best seen in imprints, very thin sections or plastic-embedded tissue. Convoluted nuclei do not have any phenotypic or clinical significance. In formalin-fixed tissues, the nuclear chromatin appears smooth or like fine dust. Nucleoli are usually small and inconspicuous. Mitotic figures are usually easily seen. The cytoplasm shows only weak basophilia in Giemsa-stained preparations and is indistinct in haematoxylin and eosin (H&E)–stained sections.

Lymphoblastic lymphomas infiltrate lymph nodes in the manner of a leukaemia, leaving the reticulin structure relatively intact and often leaving isolated germinal centres. Where the tumour cells infiltrate the capsule, hilum and around blood vessels, they often adopt an Indian file formation between connective tissue fibres. Scattered histiocytes may give the tumour a 'starry-sky' appearance, but this is never as prominent as in Burkitt lymphoma and the histiocytes contain less apoptotic debris (Boxes 4.1 and 4.2 and Figures 4.1 to 4.5).

BOX 4.1: B-lymphoblastic lymphoma: Clinical features

- Predominantly a childhood disease with occasional adult patients
- Common sites of involvement are skin, bone and lymph nodes
- Central nervous system and testes are common sites of relapse
- Designated as lymphoblastic leukaemia if the bone marrow contains more than 25 per cent lymphoblasts

BOX 4.2: Lymphoblastic lymphoma: Morphology

- Infiltrates lymph node leaving reticulin architecture partially intact
- Residual reactive germinal centres may be present
- Infiltration of capsule and around blood vessels often shows Indian file pattern
- Smooth or dust-like nuclear chromatin
- Inconspicuous nucleoli
- Cytoplasm inconspicuous
- Nuclei may be convoluted
- Mitotic figures frequent

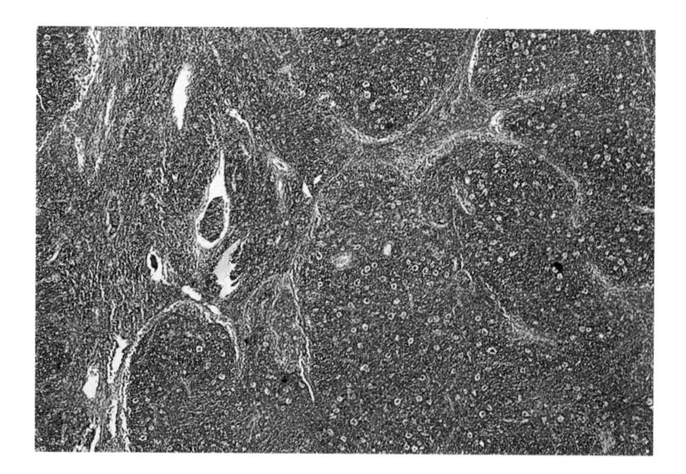

Figure 4.1 Low-power photograph of lymphoblastic lymphoma. Note preservation of much of the underlying nodal structure. Numerous 'starry-sky' macrophages are seen in this tumour.

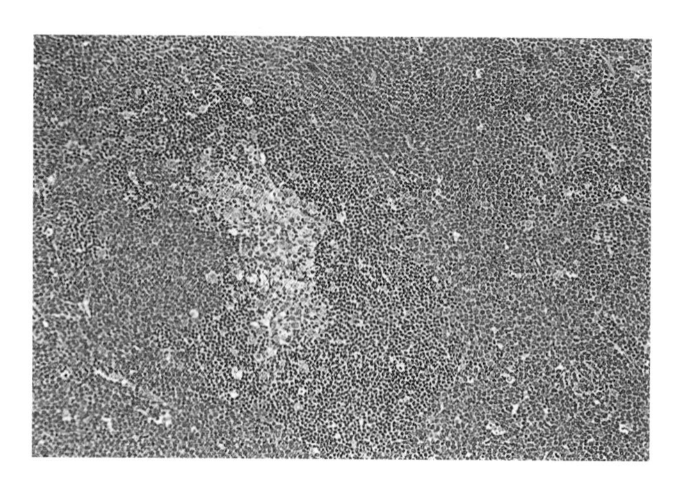

Figure 4.2 Lymphoblastic lymphoma. The uniform tumour cells surround and partially invade a residual reactive follicle.

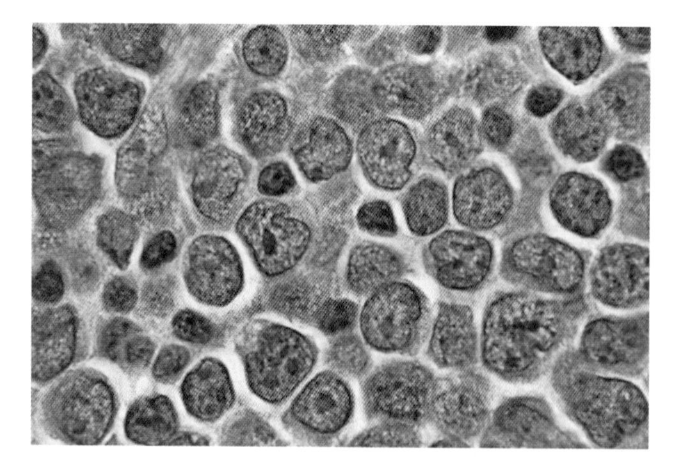

Figure 4.3 High-power view of lymphoblastic lymphoma. The tumour cells are relatively uniform in size, have fine nuclear chromatin and show one or more small nucleoli.

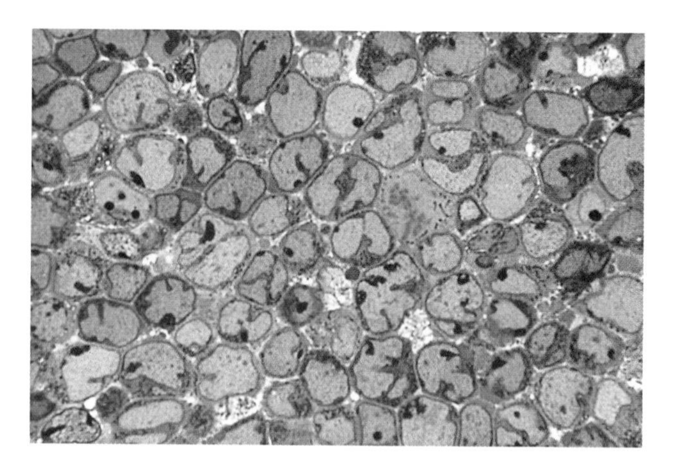

Figure 4.4 Plastic-embedded section of lymphoblastic lymphoma showing deep fissuring of the tumour cell nuclei, a feature seen in some but not all cases.

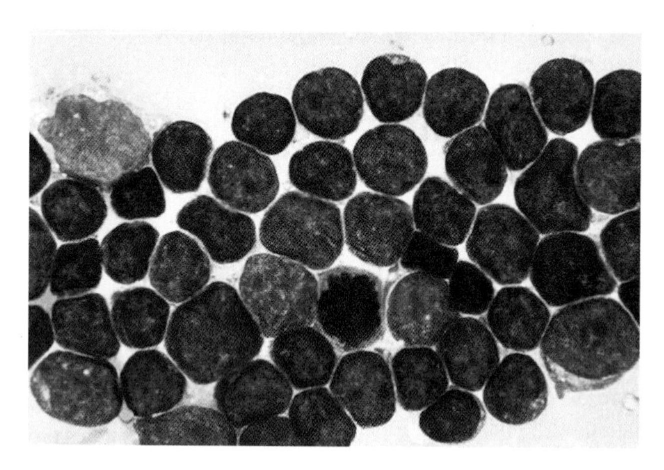

Figure 4.5 Imprint preparation of lymphoblastic lymphoma showing uniform size of tumour cells, fine nuclear chromatin and multiple nucleoli. The cytoplasmic rim is narrow and inconspicuous on most cells.

IMMUNOHISTOCHEMISTRY

Many B-lymphoblastic lymphomas do not express CD45 detectable in paraffin sections, a potential pitfall for the unwary in the differential diagnosis of small blue cell tumours. Similarly, many cases do not express CD20. The most reliable B-cell markers expressed on B-lymphoblastic lymphoma are CD19, cytoplasmic CD22 and cytoplasmic CD79a. It should be noted that this is also expressed by some T-lymphoblastic lymphomas (Pilozzi *et al.* 1998). Conversely, B-lymphoblastic lymphoma does not express T-cell antigens, although CD43 (sometimes erroneously considered to be a T-cell specific antigen) labels B-lymphoblastic lymphoma. Most cases of B-lymphoblastic lymphoma express CD10; CD34 labels about half of the cases. PAX5 is a sensitive marker of B-lineage but may be positive in acute myeloid leukaemia (AML) (Valbuena *et al.* 2006). The precursor status of B-lymphoblastic lymphoma is established by positive nuclear staining for terminal deoxynucleotidyl transferase (TdT).

Acute B-lymphoblastic leukaemia/lymphoblastic lymphoma (B-ALL/LBL) can be separated into three categories according to degree of differentiation as follows:

- Precursor B-ALL/LBL: CD19+, cytoplasmic CD22+, cytoplasmic CD79a+, TdT+.
- Common ALL/LBL: CD10+ (in addition to above).
- Pre-B-ALL/LBL: cytoplasmic μ-chains (in addition to above) (Box 4.3 and Figures 4.6 to 4.9).

BOX 4.3: B-lymphoblastic lymphoma: Immunohistochemistry

- Often CD45−
- Often CD20−
- CD79a+, CD3−
- CD10+ (not in the precursor B-cell stage)
- CD43+
- CD34+ in 50 per cent of cases
- TdT+

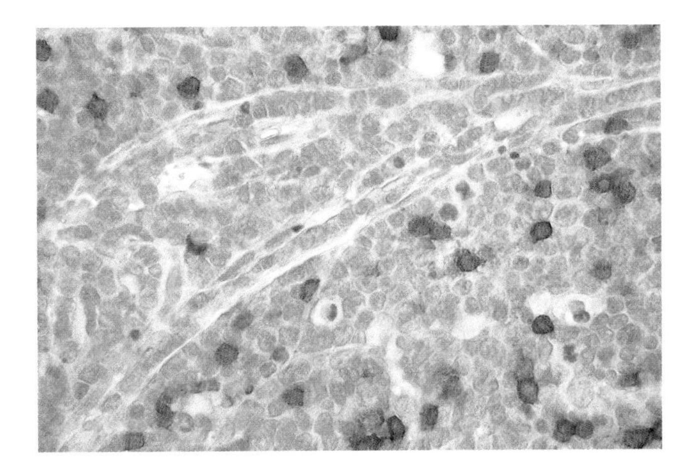

Figure 4.6 B-lymphoblastic lymphoma stained for CD20. Reactive B-lymphocytes show positivity; the tumour cells are negative.

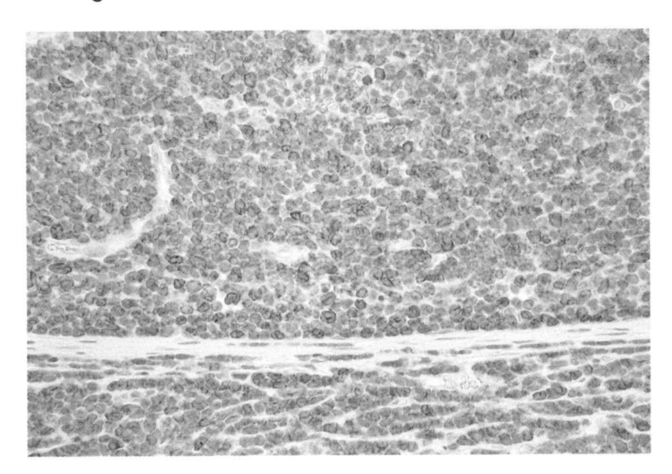

Figure 4.7 B-lymphoblastic lymphoma stained for CD79a. The tumour cells are positive. Note 'Indian file' arrangement of the cells outside the capsule.

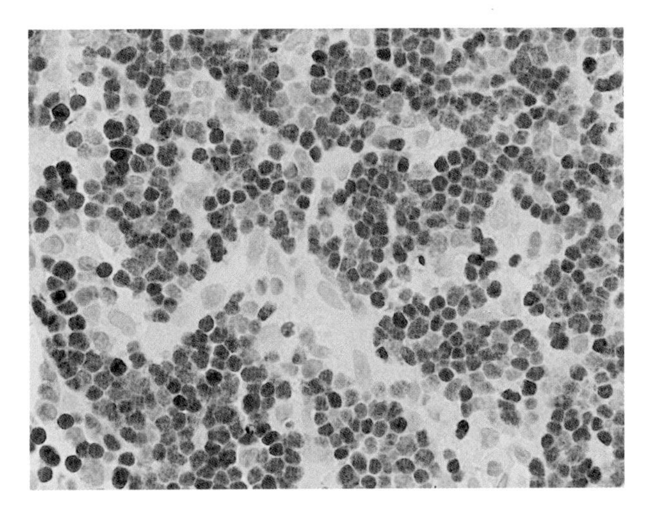

Figure 4.8 Lymphoblastic lymphoma stained for terminal deoxynucleotidyl transferase (TdT). The lymphoma cells show nuclear positivity.

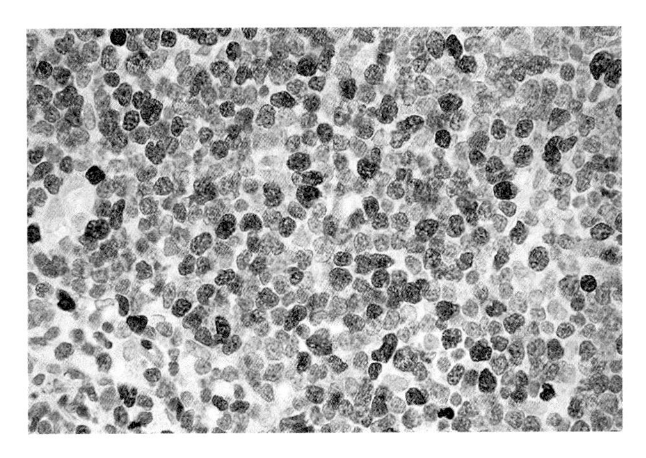

Figure 4.9 Lymphoblastic lymphoma stained for Ki67. The proliferation fraction in this tumour is in the region of 70 per cent.

GENETICS

The majority of cases of B-lymphoblastic lymphoma show clonal, but unmutated, rearrangements of the immunoglobulin genes. Of these, two thirds show clonal immunoglobulin heavy chain (IgH) rearrangements, and two thirds show clonally rearranged IgH and light chain genes. A total of 60–80 per cent of cases also show T-cell receptor (TCR) gene rearrangements (Van der Velden *et al.* 2004).

The cytogenetic abnormalities detected in B-lymphoblastic leukaemia/lymphoma (mainly established on leukaemic cases) are varied and have prognostic significance. The current World Health Organization (WHO) classification identifies seven cytogenetic categories of B-ALL/LBL. The two commonest abnormalities, together accounting for more than half of all cases, are t(12;21) and hyperdiploidy, usually with trisomy of chromosomes 4 and 10, and have a favourable prognosis. The Philadelphia chromosomal abnormality t(9;22) occurs in a small number of cases of B-lymphoblastic leukaemia; these cases are typically CD10+, CD19+, TdT+

and often express myeloid-associated antigens such as CD13, CD33 or CD25 (Borowitz and Chan 2008, Cortelazzo *et al.* 2011). Cases with an MLL gene translocation, typically t(4;11), are usually CD19+ and CD10− and express CD15 (Cortelazzo *et al.* 2011).

DIFFERENTIAL DIAGNOSIS

B-lymphoblastic lymphoma cannot be distinguished from T-lymphoblastic lymphoma on morphology alone. Clinical features may help in that T-lymphoblastic lymphoma patients frequently have mediastinal (thymic) tumours, a rare feature in B-lymphoblastic lymphoma. Immunohistochemical profiles will usually separate the two entities. B-lymphoblastic lymphoma expresses CD79a but not CD3, whereas T-lymphoblastic lymphoma may express CD79a but also expresses CD3. Gene rearrangement studies show clonal immunoglobulin heavy and/or light chain rearrangements in B-lymphoblastic lymphoma but may also show rearrangement of one or more of the TCR genes. T-lymphoblastic lymphoma shows clonal rearrangement of the TCR genes but may show IgH rearrangements in 10–25 per cent of cases.

Myeloid sarcoma may mimic lymphoblastic lymphoma. It may be identified by the presence of granulated cells. The expression of myelomonocytic markers on these tumours can usually be demonstrated by immunohistochemistry. See Chapter 9 for more detail regarding myeloid sarcoma.

The blastoid variant of mantle cell lymphoma may closely resemble lymphoblastic lymphoma and probably accounts for many of the cases of adult lymphoblastic lymphoma reported in the past. Immunohistochemistry will clearly separate the two entities, with lymphoblastic lymphoma being TdT+ and mantle cell lymphoma cyclin D1+.

Burkitt lymphoma has in the past been confused with lymphoblastic lymphoma, and leukaemic cases were included in the French–American–British (FAB) classification of lymphoblastic lymphomas. Morphologically, Burkitt lymphoma cells have granular nuclear chromatin and show intense cytoplasmic basophilia. They exhibit a mature B-cell phenotype and are TdT−.

Rarely, diffuse large B-cell lymphoma will come into the differential diagnosis of lymphoblastic lymphoma if the tumour cells appear small and have smooth chromatin with inconspicuous nucleoli. As with Burkitt lymphoma, these will exhibit a mature B-cell phenotype and will not express TdT.

PRECURSOR T-LYMPHOBLASTIC LEUKAEMIA/LYMPHOBLASTIC LYMPHOMA

Leukaemia and lymphoma are regarded as two related facets of precursor T-lymphoblastic neoplasia. T-lymphoblastic neoplasia is less common than B-lymphoblastic neoplasia,

BOX 4.4: T-lymphoblastic lymphoma: Clinical features

- Childhood and young adult disease with a peak incidence in adolescence
- Male predominance
- Involves anterior mediastinum most frequently may cause mediastinal obstruction and be associated with pleural effusions
- Also involves lymph nodes, liver, spleen and skin
- Central nervous system and testes are common sites of relapse
- Designated as lymphoblastic leukaemia if the bone marrow lymphoblasts exceed 25 per cent

but a larger proportion of patients present with solid tumours and without evidence of leukaemia. The arbitrary cut-off point for leukaemia, as with B-lymphoblastic neoplasia, is less than 25 per cent bone marrow lymphoblasts. T-lymphoblastic lymphoma occurs at an older age than B-lymphoblastic lymphoma and is more frequent in males. The peak incidence is in adolescent males.

Anterior mediastinal (thymic) tumours are common in T-lymphoblastic lymphoma/leukaemia; they often cause mediastinal obstruction and may be associated with pleural effusions. Lymph nodes and extranodal sites may be involved. The central nervous system and gonads may be involved at presentation and are common sites of relapse.

Morphologically, T-lymphoblastic lymphoma is indistinguishable from B-lymphoblastic lymphoma. It is composed of medium-sized blast cells that infiltrate lymph nodes, leaving much of the underlying reticulin structure intact. Infiltration into connective tissue structures often shows an Indian file pattern. Tumour cell nuclei may be rounded or convoluted with fine nuclear chromatin and small inconspicuous nucleoli. The mitotic index is usually high. The tumour cell cytoplasm is weakly basophilic in Giemsa-stained preparations and not clearly defined in H&E-stained sections. Scattered histiocytes may be present, giving a 'starry-sky' pattern (Boxes 4.2 and 4.4).

IMMUNOHISTOCHEMISTRY

T-lymphoblastic lymphoma shows variable expression of CD1a, CD2, CD3, CD4, CD5, CD7 and CD8. Of these, CD3 is the most frequently expressed and is the most reliable lineage marker. CD43 is positive and myeloid-associated antigens may be expressed. T-lymphoblastic lymphoma may express CD10, and CD79a is positive in 10 per cent of the biopsies. Most, but not all, cases are TdT+. Other markers that indicate the precursor nature of the tumour cells are CD1a, CD34 and CD99. Of these, CD99 is the most reliable (Box 4.5 and Figures 4.10 and 4.11).

BOX 4.5: T-lymphoblastic lymphoma: Immunohistochemistry

- Variable expression of CD1a, CD2, CD3, CD4, CD5, CD7 and CD8
- CD3 is most frequently expressed and is the most reliable lineage marker
- CD43+
- May express some myeloid antigens
- May express CD10
- May express CD79a
- TdT usually positive

Other markers of precursor T-lymphoblasts are CD1a, CD34 and CD99.

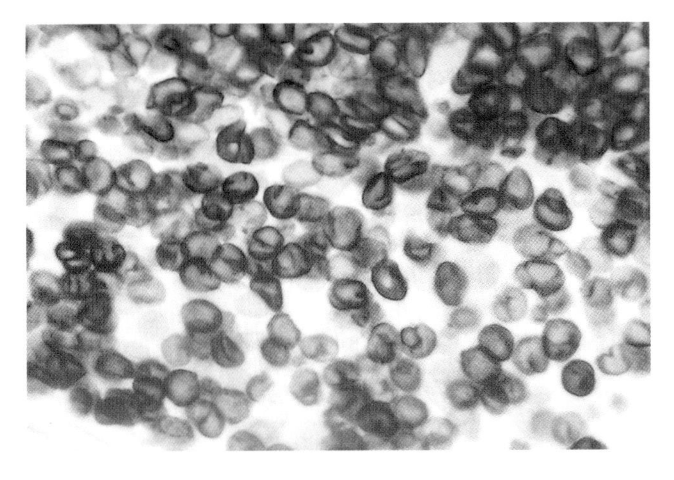

Figure 4.10 T-lymphoblastic lymphoma stained for CD3. The strong cytoplasmic positivity highlights nuclear clefts in this case.

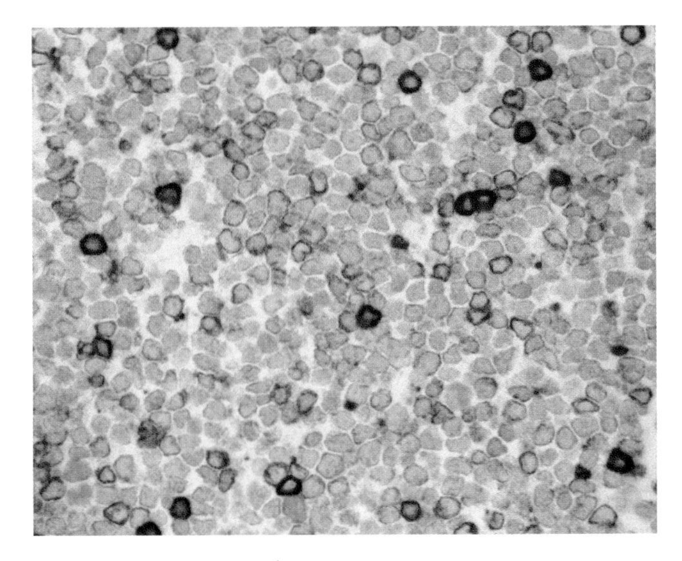

Figure 4.11 T-lymphoblastic lymphoma stained for CD79a. Scattered non-neoplastic B-cells are strongly stained. The majority of the lymphoblastic cells show variable cytoplasmic staining.

GENETICS

T-lymphoblastic leukaemia/lymphoma usually demonstrates clonal rearrangement of one or more of the TCR genes but, since these occur with high frequency in B-lymphoblastic leukaemia/lymphoma, they are not lineage specific. About one third of cases show translocations between one of the TCR genes and a number of oncogenes. Loss of 9p occurs in about one third of the cases, resulting in loss of the tumour suppressor gene CDKN2A. The TAL-1 gene involved in haematopoietic growth control is dysregulated by deletions in its regulatory region in 25 per cent of cases.

DIFFERENTIAL DIAGNOSIS

T-lymphoblastic lymphoma and B-lymphoblastic lymphoma are indistinguishable morphologically, although they show clinical differences. They also show overlap in their immunohistochemical profile, with many T-lymphoblastic lymphomas weakly expressing CD79a. The most reliable distinguishing marker is CD3, expressed only in T-lymphoblastic lymphoma.

The *WHO Classification of Tumours of Haematopoietic and Lymphoid Tissues* (Borowitz and Chan 2008) refers to two single-case reports of indolent T-lymphoblastic proliferations (Velankar *et al.* 1999, Strauchen 2001). In both of these reports the proliferations caused tumours on the oropharynx and nasopharynx with minor spread to the cervical lymph nodes. The proliferations had the morphology and immunophenotype of T-lymphoblastic lymphoma but lacked clonality of the TCR genes. Both patients had multiple recurrences without dissemination and with prolonged survival. One patient developed myasthenia gravis that appeared to be temporally related to tumour recurrence. It is possible that these lymphoid proliferations are cortical thymocytes attracted to the upper airways by nasopharyngeal epithelium and that they have no relationship to lymphoblastic lymphoma. It should be noted that TdT positivity does not equate to neoplasia and is expressed by precursor T-cells found in the thymus and precursor B-cells (haematogones) in the bone marrow (Chantepie *et al.* 2013). Occasional TdT+ cells may be found adjacent to sinuses in reactive lymph nodes (Behm *et al.* 2002).

MYELOID AND LYMPHOID NEOPLASMS WITH FGFR1 ABNORMALITIES

This pluripotential haematopoietic stem cell neoplasm (also known as 8p11 myeloproliferative syndrome) is described here because it often manifests initially in lymph nodes as a lymphoblastic lymphoma (usually T-cell, less commonly B-cell) (Abruzzo *et al.* 1992, Macdonald *et al.* 2002, Bain *et al.* 2008). This is usually accompanied by eosinophilia and bone marrow myeloproliferation. The tumours have the morphology and immunophenotype of lymphoblastic lymphomas but in addition show an infiltrate of eosinophils. The tumour has a poor prognosis, often progressing to AML (Goradia *et al.* 2008, Jackson *et al.* 2010).

Currently 65 cases of this neoplasm have been described (Jackson *et al.* 2010). The age range is from 3 to 84 years. It is characterized by translocations of, or insertions into, the fibroblast growth factor receptor 1 (FGFR1) gene at 8p11. Ten translocation partners have been reported, the most common being the ZNF gene at 13q11–12. The finding of a lymphoblastic lymphoma associated with eosinophilia and/or myeloproliferative disease should raise suspicions for this syndrome.

REFERENCES

Abruzzo LV, Jaffe ES, Cotelingam JD, *et al.* 1992 T-cell lymphoblastic lymphoma with eosinophilia associated with subsequent myeloid malignancy. *American Journal of Surgical Pathology* **16:** 236–245.

Bain BJ, Gilliland DG, Horny H-P, Vardiman JW 2008 Myeloid and lymphoid neoplasms with eosinophilia and abnormalities of PDGFRA, PDGFRB or FGFR1. In Swerdlow SH, Campo E, Harris NL, *et al.* (Eds) *WHO Classification of Tumours of Haematopoietic and Lymphoid Tissues.* Lyon: IACR Press.

Behm FG, Henry EC, Lorsbach RB, Onciu M 2002 Terminal deoxynucleotidyl transferase–positive lymphoid cells in reactive lymph nodes from children with malignant tumors: incidence, distribution pattern, and immunophenotype in 26 patients. *American Journal of Clinical Pathology* **118:** 248–254.

Borowitz MJ, Chan JKC 2008 B lymphoblastic leukaemia/lymphoma with recurrent genetic abnormalities. In Swerdlow SH, Campo E, Harris NL, *et al.* (Eds) *WHO Classification of Tumours of Haematopoietic and Lymphoid Tissues.* Lyon: IACR Press.

Chantepie SP, Cornet E, Salaün V, Reman O 2013 Hematogones: An overview. *Leukemia Research* **37**:1404–1411.

Cortelazzo S, Ponzoni M, Ferreri, Hoelzer D 2011 Lymphoblastic lymphoma. *Critical Reviews in Oncology/Haematology* **79**: 330–343.

Goradia A, Bayerl M, Cornfield D 2008 The 8p11 myeloproliferative syndrome: Review of literature and an illustrative case report. *International Journal of Experimental Pathology* **1:** 448–456.

Jackson CC, Medeiros LJ, Miranda RN 2010 8p11 myeloproliferative syndrome: a review. *Human Pathology* **41:** 461–476.

Macdonald D, Reiter A, Cross NCP 2002 The 8p11 myeloproliferative syndrome: A distinct clinical entity caused by constitutive activation of FGFR1. *Acta Haematologica* **107:** 101–107.

Pilozzi E, Pulford K, Jones M 1998 Co-expression of CD79a (JCB117) and CD3 by lymphoblastic lymphoma. *Journal of Pathology* **186:** 140–143.

Strauchen JA 2001 Indolent T-lymphoblastic proliferation. Report of a case with an 11 year history and association with myasthenia gravis. *American Journal of Surgical Pathology* **25:** 411–415.

Valbuena JR, Medeiros LJ, Rassidakis GZ, *et al.* 2006 Expression of B cell–specific activator protein/PAX5 in acute myeloid leukemia with t(8;21)(q22;q22). *American Journal of Clinical Pathology* **126:** 235–240.

Van der Velden VH, Bruggmann M, Hoogeveen PG, *et al.* 2004 TCRB gene rearrangements in childhood and adult precursor-B-ALL: Frequency applicability as MRD-PCR target, and stability between diagnosis and relapse. *Leukemia* **18:** 1971–1980.

Velankar MM, Nathwani BN, Schlutz MJ, *et al.* 1999 Indolent T-lymphoblastic proliferation: Report of a case with a 16-year course without cytotoxic therapy. *American Journal of Surgical Pathology* **23:** 977–984.

5

Mature B-cell lymphomas

OVERVIEW

Mature B-cells are characterized by the synthesis, expression and sometimes secretion of immunoglobulin molecules. The almost infinite diversity of these molecules is achieved by rearrangement (shuffling) of the constant, joining, diversity and variable regions of the immunoglobulin genes. Further diversity occurs in follicle centres by the process of somatic mutation of the variable region genes. Thus, at the genetic level, mature B-cells are characterized by rearranged immunoglobulin genes. Follicle centre and post–follicle centre cells have mutated variable region genes.

Mature B-cell lymphomas will show clonal rearrangement of their immunoglobulin genes. Those that have arisen at the post-follicular stage of B-cell development will also show mutated variable region (v-region) genes. Lymphomas arising at the follicle centre stage of B-cell maturation show ongoing mutations of the v-region genes (i.e., multiple mutations within a single clone). Analysis of the immunoglobulin genes provides a valuable means for the identification and subdivision of mature B-cell lymphomas.

B-cell lymphomas may be divided into those with a low growth fraction (B-cell chronic lymphocytic leukaemia/small lymphocytic lymphoma [B-CLL/SLL], lymphoplasmacytic lymphoma, mantle cell lymphoma, marginal zone lymphoma and follicular lymphoma) and those with high growth fraction (diffuse large B-cell lymphoma [DLBCL] and Burkitt lymphoma). A proportion of low-growth fraction lymphomas transform into high-growth fraction lymphomas. Low-growth fraction lymphomas exhibit alterations in genes controlling apoptosis, whereas those with a high growth fraction have abnormalities of genes involving proliferation control (Sanchez-Beato *et al.* 2003).

There are reliable lineage-specific immunohistochemical markers for the identification of B-cells. The two most commonly used for paraffin-embedded tissues are CD20 and CD79a. Of these, CD79a is the most reliable, since it is expressed on cells from the pre-B-cell stage of maturation through to plasma cells. CD20 has a more limited range of expression and may not be expressed on lymphomas representing the early and late stages of B-cell differentiation. However, because of the use of anti-CD20 antibodies in the treatment of many B-cell lymphomas, it is important for patient management to determine the CD20 expression status. In addition to these lineage-specific markers, there are a number of antigens that are useful in identifying subtypes of B-cell lymphoma. These include CD5 (expressed on B-CLL/SLL and mantle cell lymphomas), CD10 (expressed on follicle centre cell lymphomas, lymphoblastic lymphoma and Burkitt lymphoma) and CD23 (expressed on B-CLL/SLL and primary mediastinal [thymic] large B-cell lymphoma). The transcription factor BCL-6 is expressed in the nuclei of follicle centre cell–derived lymphomas. HGAL and LMO2 are germinal centre markers that are helpful in identifying follicular lymphoma and may be expressed when other germinal centre markers, such as CD10 and BCL-6, are absent. HGAL is expressed on the cell membrane and LMO2 is expressed in the nuclei (Zhang and Aguilera 2014).

The majority of mature B-cell lymphomas show characteristic chromosomal abnormalities, often translocations involving the immunoglobulin genes. These may provide markers for the various subtypes of B-cell lymphoma (Table 5.1). They may be identified by conventional cytogenetics, fluorescent *in-situ* hybridization (FISH) or the polymerase chain reaction (PCR). A gene product overexpressed as a result of the translocation may be detectable by immunohistochemistry and provide a marker for the translocation—for example, cyclin D1 overexpressed in mantle cell lymphomas as a result of the 11;14 translocation.

Gene expression profiling of lymphomas provides useful information that can identify different subgroups of

Table 5.1 Common cytogenetic abnormalities found in mature B-cell lymphomas

Lymphoma	Translocation/mutation	Genes involved
Lymphoplasmacytic lymphoma		
Over 90% of cases	MYD88 L265P	NF-κB pathway
Follicular lymphoma	t(14;18)(q32;q21)	IgH, BCL-2
Mantle cell lymphoma	t(11;14)(q13;q32)	BCL-1, IgH
Hairy cell leukaemia	BRAF V600E	
Diffuse large B-cell lymphoma		
Approximately 30%	t(14;18)(q32;q21)	IgH, BCL-2
Approximately 30%	t(3;14)(q27;q32)	BCL-6, IgH
Very rare	t(2;17)(p23;q23)	ALK, clathrin
Burkitt lymphoma		
Approximately 80%	t(8;14)(q24;q32)	c-*myc*, IgH
Approximately 10%	t(2;8)(q11;q24)	k light chain, c-*myc*
Approximately 10%	t(8;22)(q24;q11)	c-*myc*, λ light chain

ALK, anaplastic lymphoma kinase; IgH, immunoglobulin heavy chain.

lymphomas within a single diagnostic category. Thus, a study of SLLs was able to categorize borderline cases among B-CLL/SLL, mantle cell lymphoma and splenic marginal zone lymphoma (Thieblemont *et al.* 2004). Gene expression profiles of DLBCLs have identified good prognostic subtypes with expression profiles of germinal centre B-cells and poor prognosis subtypes with expression profiles of activated B-cells (Alizadeh *et al.* 2000). Gene expression profiling is an expensive and technically sophisticated process that requires the availability of unfixed tissue. Fortunately, the technique identifies genes of discriminatory value that may then be demonstrated in fixed tissues using immunohistochemical techniques (Chang *et al.* 2004, Thieblemont *et al.* 2004).

These techniques, together with clinical features and morphology, have provided reliable methods for the identification and subclassification of mature B-cell lymphomas. This group comprises approximately 90 per cent of malignant lymphomas seen in most parts of the world. DLBCL is the commonest subtype, accounting for 30–40 per cent of all cases. In the developed world, follicular lymphomas account for over 20 per cent of cases.

B-CELL CHRONIC LYMPHOCYTIC LEUKAEMIA/SMALL LYMPHOCYTIC LYMPHOMA

SLL is usually regarded as the tissue manifestation of B-CLL. Patients diagnosed as having B-CLL on blood and bone marrow examination often have some degree of lymphadenopathy. If lymph nodes from such patients are biopsied, they show the features of SLL, being diffusely infiltrated by small lymphocytes with scattered prolymphocytes and paraimmunoblasts. Some patients present with prominent lymphadenopathy, and a lymph node biopsy may be the diagnostic procedure. In this 'tumour-forming' variant of B-CLL, the paraimmunoblasts are often aggregated into clusters, forming so-called proliferation centres. Most patients with tumour-forming B-CLL will have, or will subsequently develop, the blood and bone marrow manifestations of this disease. A recent study of cases of B-CLL/SLL with and without prominent proliferation centres found no difference between these groups with respect to clinical features and immunophenotype (Asplund *et al.* 2002). In many cases of B-CLL/SLL, there is some preservation of the underlying nodal structure. Reticulin stains will often show preservation of sinus structures, and residual reactive follicles may be seen.

Occasional cells in B-CLL/SLL may show immunoglobulin inclusions, often appearing to be intranuclear (Dutcher bodies). These eosinophilic inclusions are usually of the immunoglobulin M (IgM) isotype and stain strongly with periodic acid–Schiff (PAS). Occasionally, a substantial number of the cells contain inclusions. Such cases were designated as lymphoplasmacytoid lymphomas in the Kiel classification. In all other respects, however, these cases have the morphology and immunophenotype of B-lymphocytic lymphoma and they should be regarded as a variant of this disease.

In rare cases of SLL, paraimmunoblasts may form the predominant cell type. This has been referred to as the paraimmunoblastic variant of SLL. It should not be confused with B-prolymphocytic leukaemia, which is clinically and phenotypically a different disease. Few cases of the paraimmunoblastic variant of SLL have been studied and it is not known whether the prognosis of such cases is different from that of the more usual form of SLL.

B-CLL/SLL may transform to a large B-cell lymphoma, often referred to as Richter syndrome. This transformation is associated with a poor prognosis. Transformation to Hodgkin lymphoma may also be observed. More commonly isolated Reed–Sternberg (RS) cells, which may be Epstein–Barr virus (EBV) positive, are found scattered among the B-CLL/SLL cells. These should be recorded but not interpreted as transformation to Hodgkin lymphoma. Such transformation is identified by the finding of Hodgkin/Reed–Sternberg (H/RS) cells in the appropriate cellular setting for Hodgkin lymphoma (Boxes 5.1 and 5.2 and Figures 5.1 to 5.8).

BOX 5.1: B-cell chronic lymphocytic leukaemia/small lymphocytic lymphoma: Clinical features

- Median age 65 years
- Male predominance 2:1
- Most patients have bone marrow involvement (stage IV)
- Presenting features:
  - Asymptomatic (incidental finding)
  - Fatigue
  - Autoimmune haemolytic anaemia
  - Infections
  - Splenomegaly
  - Lymphadenopathy

BOX 5.2: B-cell chronic lymphocytic leukaemia/small lymphocytic lymphoma: Morphology

- Architecture of node partially preserved
- Diffuse infiltrate of lymphocytes with clumped chromatin
- Prolymphocytes or paraimmunoblasts, either singly or forming 'proliferation centres'
- Immunoglobulin inclusions (uncommon)
- Transformation (rare)
  - Diffuse large B-cell lymphoma (Richter syndrome)
  - Hodgkin lymphoma

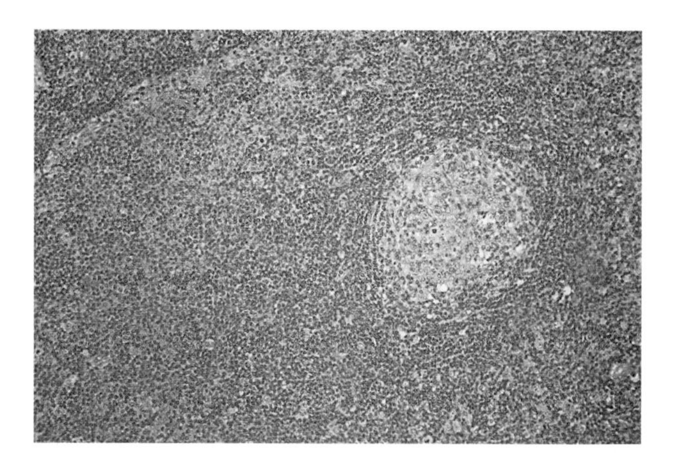

Figure 5.1 B-cell chronic lymphocytic leukaemia/small lymphocytic lymphoma. Low-power view showing residual germinal centre surrounded by tumour cells.

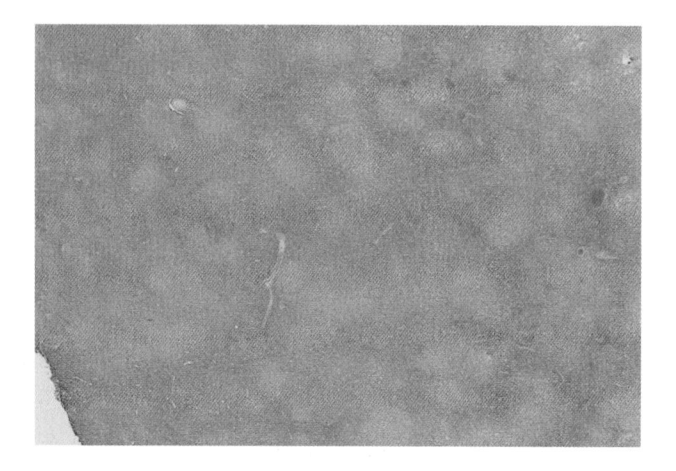

Figure 5.2 B-cell chronic lymphocytic leukaemia/small lymphocytic lymphoma. Low-power view showing pale proliferation centres.

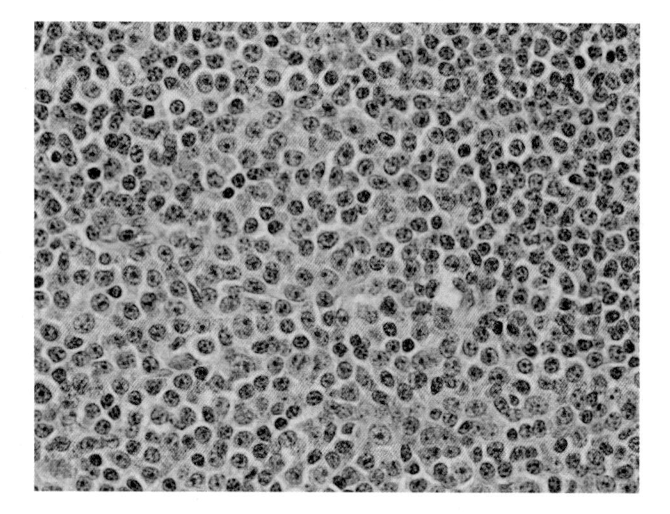

Figure 5.3 B-cell chronic lymphocytic leukaemia/small lymphocytic lymphoma showing a proliferation centre containing many prolymphocytes and paraimmunoblasts.

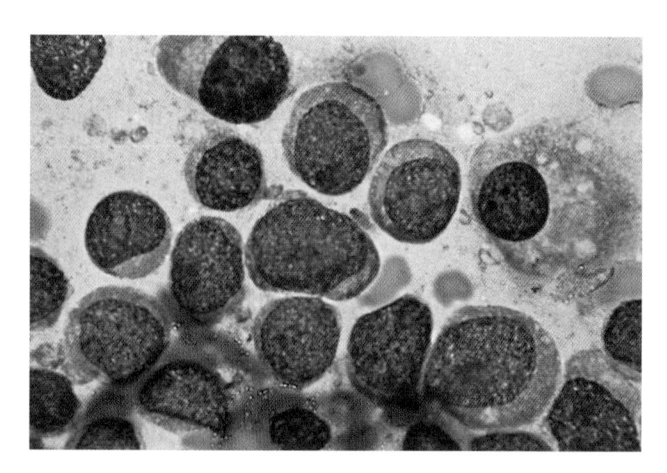

Figure 5.4 B-cell chronic lymphocytic leukaemia/small lymphocytic lymphoma. Paraimmunoblastic variant imprint preparation showing fine chromatin, visible nucleoli and cytoplasmic basophilia.

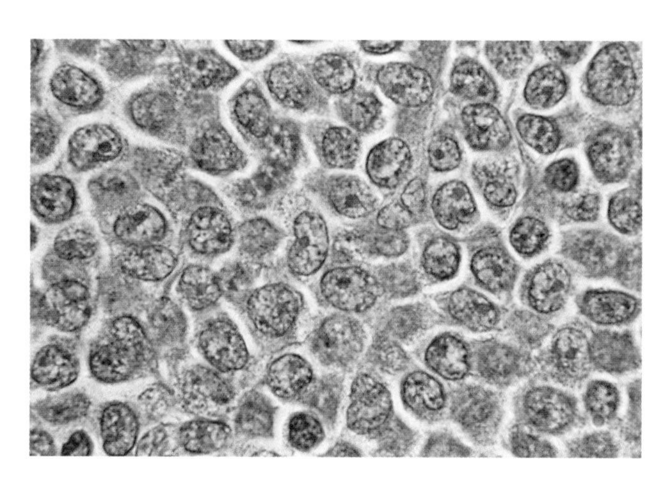

Figure 5.5 B-cell chronic lymphocytic leukaemia/small lymphocytic lymphoma paraimmunoblastic variant in which most of the cells resemble paraimmunoblasts.

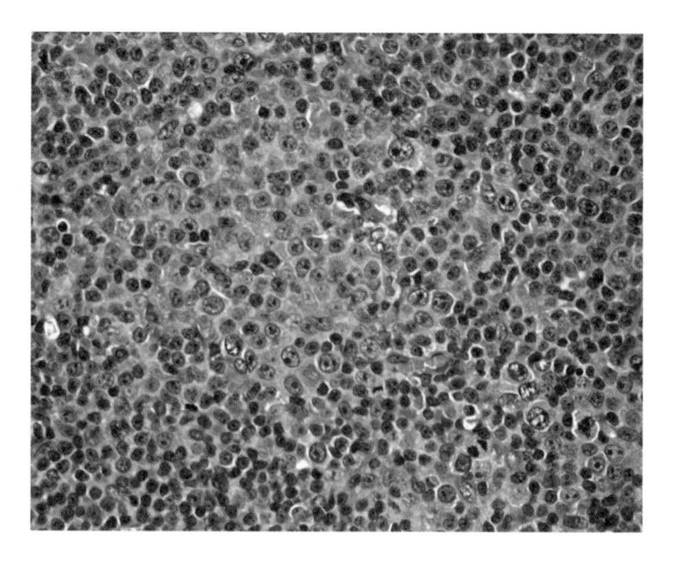

Figure 5.6 Proliferation centre showing paraimmuno-blasts with small lymphocytes at the periphery.

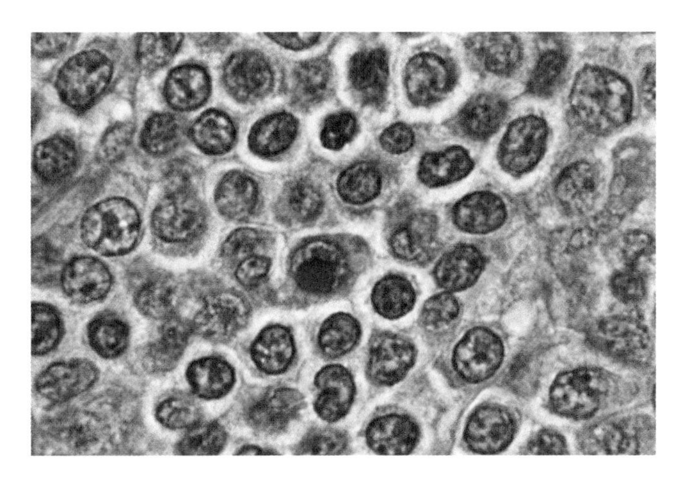

Figure 5.7 B-cell chronic lymphocytic leukaemia/small lymphocytic lymphoma. Periodic acid–Schiff stain showing intranuclear inclusion of immunoglobulin M (IgM; Dutcher body).

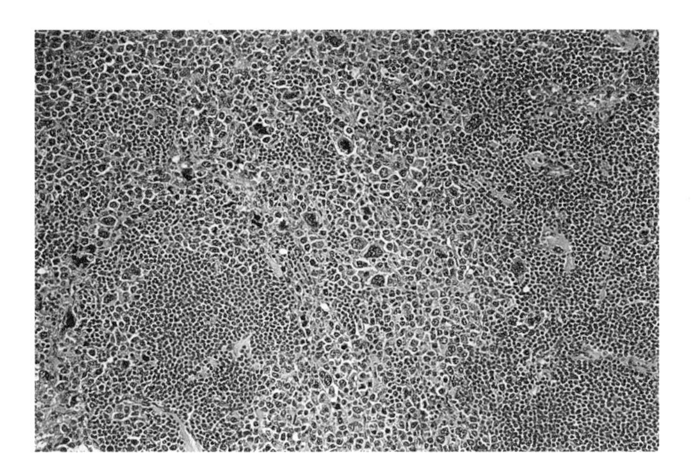

Figure 5.8 B-cell chronic lymphocytic leukaemia/small lymphocytic lymphoma showing transformation to high-grade lymphoma (Richter syndrome).

IMMUNOHISTOCHEMISTRY

B-CLL/SLL cells usually show weak expression of surface immunoglobulins by flow cytometry. However, clear expression of perinuclear IgM and sometimes IgD together with light chains can usually be identified using immunohistochemistry. The B-cell–associated antigens CD20 and CD79a are expressed. The tumour cells are negative for CD10 but express CD5, CD23, LEF1 and CD43. CD23 is fixation sensitive and often most strongly expressed on paraimmunoblasts. CD23 and LEF1 are useful in the separation of B-lymphocytic lymphoma from other small B-cell lymphomas including mantle cell lymphoma. ZAP-70 (a tyrosine kinase) can be demonstrated by flow cytometry and immunohistochemistry in cases of B-CLL/SLL with unmutated immunoglobulin genes (see later) (Box 5.3 and Figures 5.9 to 5.14).

BOX 5.3: B-cell chronic lymphocytic leukaemia/small lymphocytic lymphoma: Immunohistochemistry

- IgM together with light chain detectable in many cases
- IgD may be detected
- CD20+
- CD79a+
- CD5+
- CD23+
- LEF1+
- ZAP-70+ in most cases with unmutated immunoglobulin genes

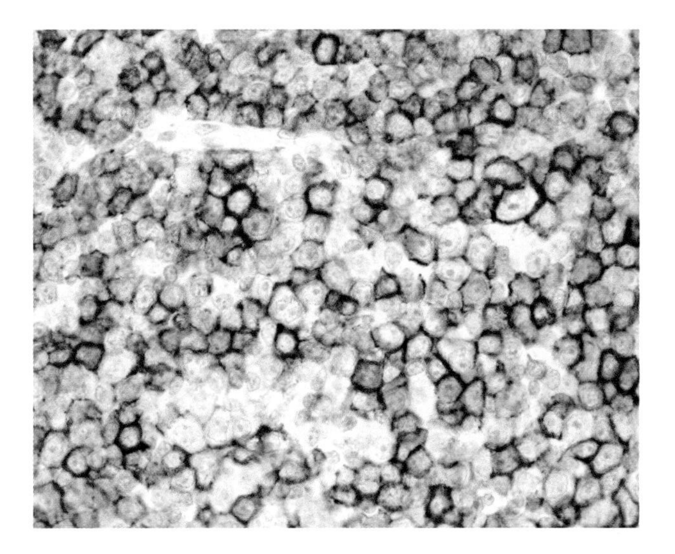

Figure 5.9 Membrane staining of B-cells for CD23. Strongest staining seen on paraimmunoblasts.

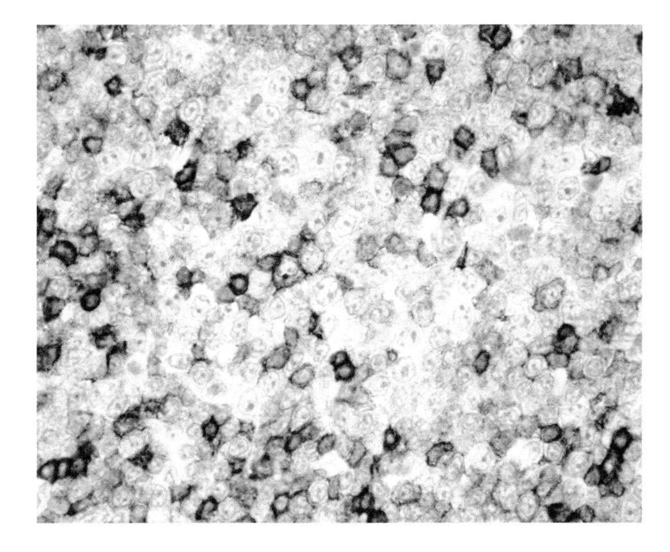

Figure 5.10 Proliferation centre showing strong CD5 positivity of small T-lymphocytes with variable membrane staining of paraimmunoblasts.

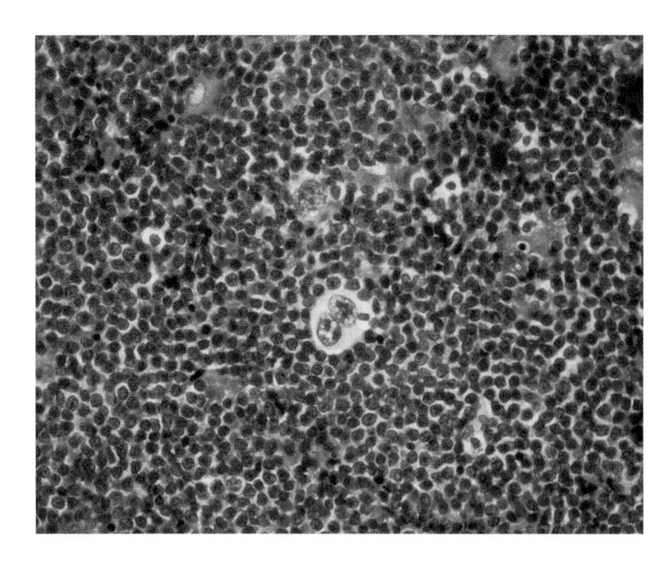

Figure 5.11 Section of B-cell chronic lymphocytic leukaemia/small lymphocytic lymphoma (B-CLL/SLL) showing Hodgkin/Reed–Sternberg (H/RS) cells. There are scattered histiocytes but the stromal component is not sufficient to categorize this as Hodgkin lymphoma. It should be categorized as B-CLL/SLL containing RS cells.

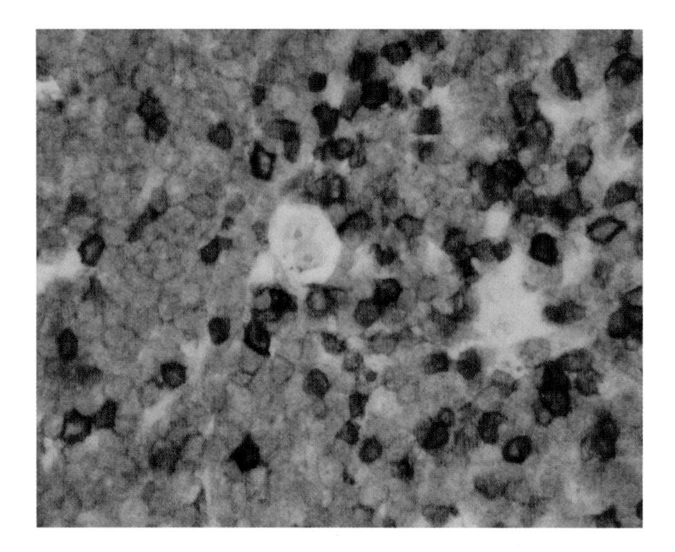

Figure 5.12 Same case as that seen in Figure 5.11 stained for CD5. The B-cell chronic lymphocytic leukaemia/small lymphocytic lymphoma cells are positively stained. The more strongly staining cells are reactive T-cells. These were increased in number around the Hodgkin/Reed–Sternberg (H/RS) cells.

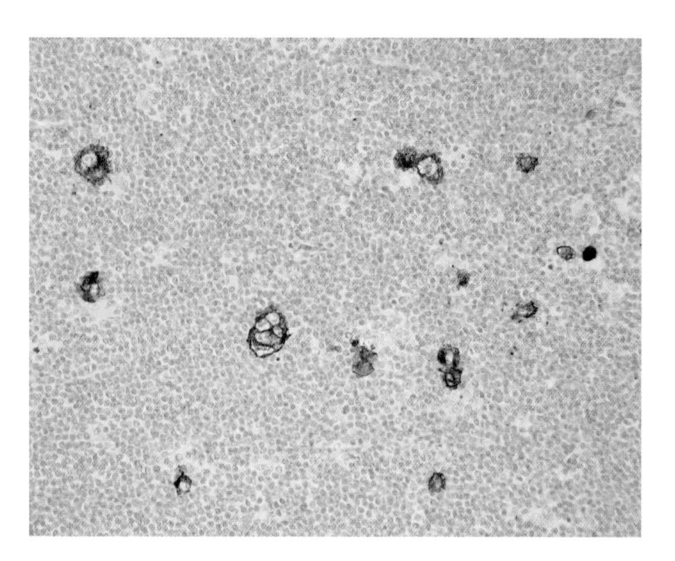

Figure 5.13 Section stained for CD30 showing Hodgkin/Reed–Sternberg (H/RS) cells in a monomorphic population of B-cell chronic lymphocytic leukaemia/small lymphocytic lymphoma cells.

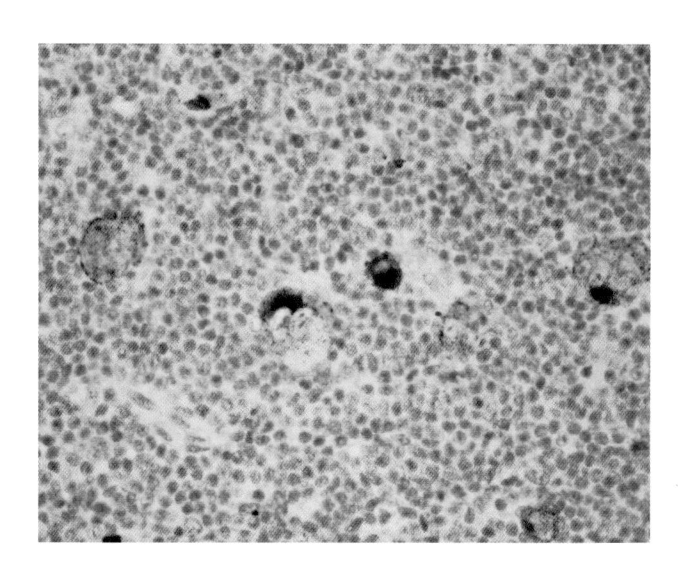

Figure 5.14 Section stained for CD15 showing positive Hodgkin/Reed–Sternberg (H/RS) cells.

GENETICS

Most genetic studies of B-CLL have been performed on blood and marrow with few studies of tissue infiltrates. The disease appears to be heterogeneous. Trisomy 12 is found in 10–15 per cent of cases by cytogenetics and in 20 per cent of cases using interphase FISH. Only a proportion of the clone identified by the phenotype exhibits trisomy 12, suggesting that this is probably a secondary genetic event. Twenty per cent of B-CLLs show deletions of chromosome 13q14 by routine cytogenetics and 60 per cent show this deletion by interphase FISH. Deletions of chromosome 11q13 are found in less than 5 per cent of patients by conventional cytogenetics but in 20 per cent using interphase FISH. Deletions of 11q and 17p carry a poor prognosis. Mutations of the gene for p53 occur in 10–15 per cent of cases.

Rearranged immunoglobulin heavy and light chain genes are found in B-CLL. Mutational patterns in the variable region genes of over half the cases suggest that these cases have been exposed to the mutational influence of the germinal centre. Cases with mutated immunoglobulin genes are more likely to be Binet stage A with stable disease and typical lymphocyte morphology and to have deletions of 13q14. Cases lacking mutations (naive cells) are associated with progressive disease, atypical morphology and trisomy 12. Gene expression profiling has shown that ZAP-70 (expressed in unmutated cases) is the gene that best distinguishes these two prognostic subtypes (Weistner *et al.* 2003) although there is significant discordance between mutational status and ZAP-70 expression. It appears that ZAP-70 expression, irrespective of mutational status, carries a poor prognosis (Kienle *et al.* 2010).

DIFFERENTIAL DIAGNOSIS

The main differential diagnosis of B-CLL/SLL is with mantle cell lymphoma. In the latter, the cells are usually more angulated than small lymphocytes but may appear rounded and can be morphologically indistinguishable from B-CLL/SLL. The presence of paraimmunoblasts, either singly or in clusters, identifies SLL. The expression of CD23 or LEF1 differentiates B-CLL/SLL from mantle cell lymphoma (Zhang and Aguilera 2014). Nuclear expression of cyclin D1 identifies mantle cell lymphomas. Cases of SLL with prominent proliferation centres may be mistaken for follicle centre cell lymphoma on low-power inspection. SLL shows some preservation of the underlying reticulin structure of the node and does not show the nodular reticulin pattern characteristic of follicle centre cell lymphoma. The paraimmunoblasts in proliferation centres have more delicate nuclear chromatin and a more uniform nuclear size than the centroblasts of follicle centre cell lymphoma. Immunohistochemistry will differentiate B-CLL/SLL (CD5+, CD23+, LEF1+) from follicle centre cell lymphoma (CD10+, BCL-6+, HGAL+, LMO2+).

B-CELL PROLYMPHOCYTIC LEUKAEMIA

B-cell prolymphocytic leukaemia (B-PLL) is a rare B-cell neoplasm in patients with a mean age of 70 years and has a male predominance of 1.6:1. Patients usually present with marked splenomegaly and a high white cell count; lymphadenopathy is uncommon. In tissue sections, the tumour cells are approximately twice the size of lymphocytes. The nuclei are generally rounded with finely granular chromatin and a single prominent central nucleolus. When lymph nodes are involved they show a diffuse or vaguely nodular infiltration.

B-PLL cells express surface IgM with or without IgD. The B-cell antigens CD20 and CD79a are positive. CD5 is expressed in approximately one third of the cases and CD23 is usually negative.

Approximately 20 per cent of cases of B-PLL have been reported to show t(11;14)(q13;q32), but these are now regarded as examples of the blastoid variant of mantle cell lymphoma.

DIFFERENTIAL DIAGNOSIS

B-PLL is not the same as B-CLL with large numbers of prolymphocytes and paraimmunoblasts (paraimmunoblastic variant of B-CLL). These two diseases can usually be distinguished by their clinical and immunophenotypic features.

LYMPHOPLASMACYTIC LYMPHOMA/ IMMUNOCYTOMA

A number of B-cell lymphomas may show plasma cell differentiation. This is a common and characteristic feature of both nodal and extranodal marginal zone lymphomas. It is rarely seen in follicle centre cell lymphomas and is not seen in mantle cell lymphomas. Such cases showing plasma cell differentiation may in the past have been categorized as lymphoplasmacytic lymphomas. The term 'immunocytomas (lymphoplasmacytoid type)' was used in the Kiel classification for lymphocytic tumours containing large numbers of immunoglobulin inclusions. These tumours, which are CD5+ and CD23+, are now regarded as a variant of B-CLL/SLL. Thus the diagnosis of lymphoplasmacytic lymphoma should be made only in the absence of features of other B-cell lymphomas, such as neoplastic follicles or marginal zone B-cells. In those cases in which a lymphoplasmacytic tumour cannot be clearly categorized, it should be reported as an indeterminate small B-cell lymphoma with plasmacytic differentiation and a differential diagnosis provided.

Lymphoplasmacytic lymphoma is a clonal proliferation of small lymphocytes and plasma cells with intermediate plasmacytoid forms. Patients most commonly present with bone marrow and splenic disease; presentation with lymphadenopathy is less common. Lymph nodes involved by lymphoplasmacytic lymphoma show a diffuse infiltrate of small lymphocytes, plasma cells and intermediate forms. These cells are best seen in Giemsa-stained preparations. The plasma cells are not uniformly distributed, but often aggregate adjacent to sinuses or fibrous trabeculae. Cytoplasmic and intranuclear inclusions of immunoglobulin (Dutcher bodies) are often present in a proportion of the plasma cells and are highlighted by PAS staining.

Many lymphoplasmacytic lymphomas show preservation of the underlying nodal architecture and frequently exhibit marked dilatation of the sinuses often associated with haemosiderin-laden macrophages. The dilated sinuses are filled with eosinophilic proteinaceous fluid as well as tumour cells. Scattered mast cells are a characteristic feature of lymphoplasmacytic lymphoma infiltrates at all sites. Variable numbers of epithelioid histiocytes may be present. Andriko *et al.* (2001) described three histological patterns among 20 cases of lymphoplasmacytic lymphoma.

Seven cases had the aforementioned histological pattern, four showed hyperplastic follicles and nine showed diffuse effacement of the nodal structure with variable fibrosis. Amyloid was present in two cases.

Some cases of lymphoplasmacytic lymphoma are associated with Waldenstrom macroglobulinaemia. Waldenstrom macroglobulinaemia should not, however, be regarded as a specific histological diagnosis, since it is also rarely associated with splenic marginal zone lymphoma, B-CLL and extranodal marginal zone lymphoma.

Lymphoplasmacytic lymphomas often contain variable numbers of immunoblasts and may show progression to a DLBCL of immunoblastic type (immunoblastic lymphoma). Lin *et al.* (2003) reported progression to large B-cell lymphoma in 12 of 92 patients with lymphoplasmacytic lymphoma. This transformation was associated with a very poor prognosis (Boxes 5.4 and 5.5 and Figures 5.15 to 5.18).

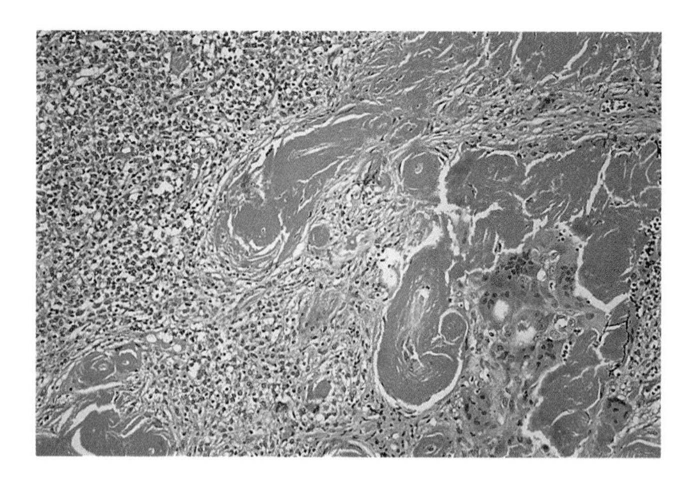

Figure 5.15 Lymphoplasmacytic lymphoma showing amyloid deposition. Note the giant cell reaction to the amyloid.

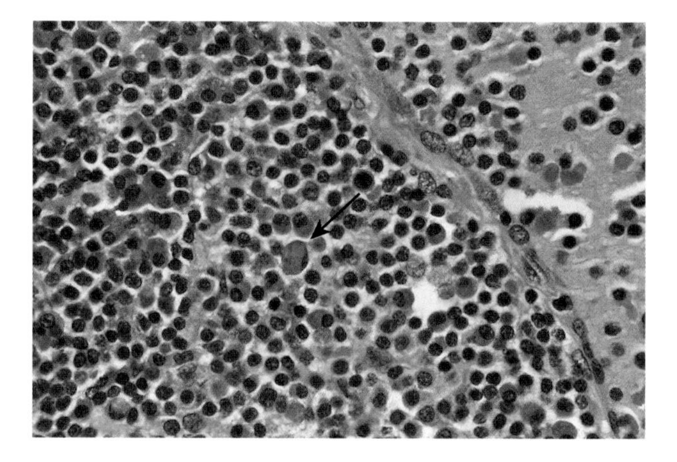

Figure 5.16 Lymphoplasmacytic lymphoma showing a dilated sinus containing pink proteinaceous material. The adjacent cells are lymphocytes and plasma cells. The arrow indicates a Dutcher body.

BOX 5.4: Lymphoplasmacytic lymphoma/ immunocytoma: Clinical features

- Median age 63 years
- Slight male predominance
- Sites of involvement:
  - Bone marrow
  - Blood
  - Spleen
  - Lymph nodes
- Most patients have monoclonal immunoglobulin M (IgM) paraprotein (Waldenstrom macroglobulinaemia); may have hyperviscosity syndrome, neuropathy, IgM deposits in skin and gastrointestinal tract
- Cases expressing IgG or more rarely IgA do not have hyperviscosity syndrome
- Reported association with hepatitis C infection

BOX 5.5: Lymphoplasmacytic lymphoma/ immunocytoma: Morphology

- Nodal architecture often partially preserved
- Infiltrate composed of small lymphocytes, plasma cells and intermediate lymphoplasmacytoid forms
- Intranuclear IgM inclusions (Dutcher bodies)
- Sinuses dilated and filled with eosinophilic proteinaceous material
- Haemosiderosis
- Variable fibrosis
- Amyloid deposition (uncommon)
- Increased numbers of mast cells

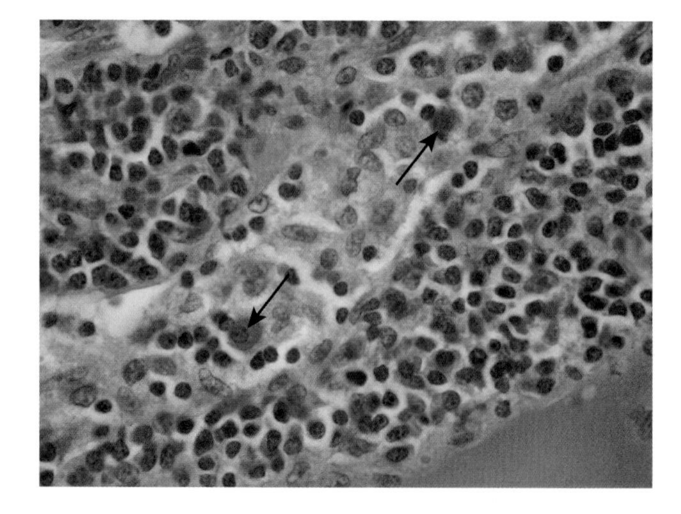

Figure 5.17 High-power view of lymphoplasmacytic lymphoma showing parasinusoidal distribution of lymphocytes and plasma cells. Mast cells are present in the sinus (arrows).

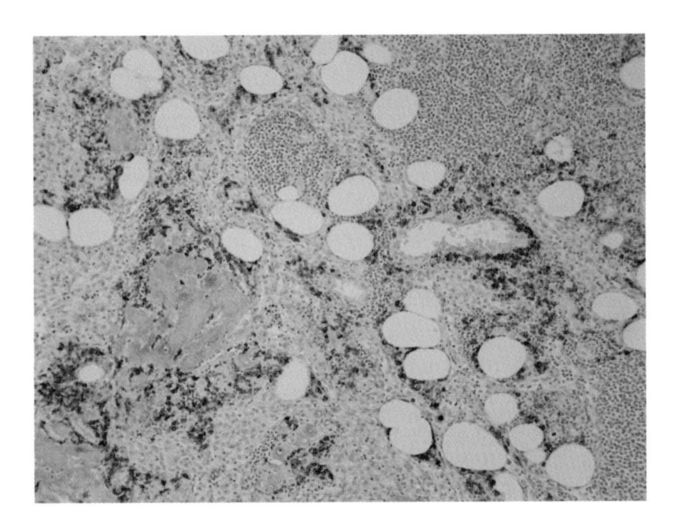

Figure 5.18 Lymphoplasmacytic lymphoma showing strong kappa positivity of plasma cells that accumulate around vessels and fibrous trabeculae.

BOX 5.6: Lymphoplasmacytic lymphoma/ immunocytoma: Immunohistochemistry

- Cytoplasmic immunoglobulin M (IgM)+, expression of IgG and IgA less common
- Cytoplasmic/intranuclear IgM inclusions
- CD20+, but often negative on plasma cells
- CD79a+
- CD5−, CD10−, CD23−
- CD38+ and CD138+ on plasma cells

IMMUNOHISTOCHEMISTRY

The tumour cells express surface and cytoplasmic immunoglobulin of the IgM isotype, and less commonly of the IgG or IgA isotype, showing light chain restriction. They express the B-cell markers CD20 and CD79a but are negative for CD5 and CD10. It should be noted that CD20 is often not expressed on the plasma cell component of the lymphoma. The plasma cell markers CD38 and CD138 are usually positive (Box 5.6).

GENETICS

The immunoglobulin genes are rearranged and mutated, as in post-germinal centre cells. The somatic mutation MYD88 L265P is present in over 90 per cent of cases of lymphoplasmacytic lymphoma, and molecular analysis for this mutation can help distinguish lymphoplasmacytic lymphoma from other small B-cell lymphomas (Landgren and Tageja 2014). The t(9;14)(p13;q32) chromosomal translocation which juxtaposes the PAX5 gene at 9q13 with the regulatory elements of the immunoglobulin heavy chain gene at 14q32 was previously thought to be characteristic of lymphoplasmacytic lymphoma but, in fact, occurs rarely in this lymphoma.

DIFFERENTIAL DIAGNOSIS

Differentiation from reactive lymphadenopathy with many plasma cells (e.g., as in rheumatoid disease) can usually be made on the basis of the more obvious reactive features in the latter. The identification of light chain restriction in lymphoplasmacytic lymphoma supports the neoplastic nature of the infiltrate. Other B-cell lymphomas showing plasma cell differentiation, such as marginal zone lymphomas, can usually be differentiated from lymphoplasmacytic lymphomas by the presence of other characteristic cell types, such as marginal zone B-cells and colonized reactive follicles. Immunohistochemical staining for immunoglobulin usually shows the plasma cells to be more zonal in their distribution in marginal zone B-cell lymphomas than is seen in lymphoplasmacytic lymphoma (Hisada *et al.* 2007). In some cases, amyloidosis may be so extensive that the appearances suggest primary amyloidosis. In such cases, a careful search should be made for residual islands of lymphoplasmacytic lymphoma.

HAIRY CELL LEUKAEMIA

Hairy cell leukaemia is an uncommon B-cell neoplasm that primarily involves the bone marrow, blood and spleen. The name is derived from the appearance of the lymphoid cells in peripheral blood smears, where they show hairy cytoplasmic projections. The disease has a median age of 55 years and a male to female ratio of 5:1. Patients most commonly present with pancytopenia and splenomegaly. Hairy cell leukaemia is usually an indolent disease, responding well to α-interferon and purine analogues, although survivors have an increased incidence of Hodgkin and non-Hodgkin lymphoma and other neoplasms. Lymph node involvement in hairy cell leukaemia is rare, but lymphadenopathy caused by intercurrent opportunistic infections, as a consequence of the immunodeficiency associated with this disease, may occur. When lymph nodes are involved by hairy cell leukaemia, the infiltrate is diffuse, involving the paracortex and eventually the whole node.

The cells of hairy cell leukaemia are about twice the size of small lymphocytes. Their nuclear chromatin is less heterochromatic than that of B-CLL cells. In tissue sections, their nuclei are usually oval or bean shaped and are surrounded by an ill-defined zone of clear cytoplasm ('fried-egg' appearance). Obstruction of the microcirculation may lead to blood lakes and pseudo- angiomatous formations (Figures 5.19 and 5.20).

IMMUNOHISTOCHEMISTRY

The tumour cells express surface immunoglobulin and B-cell–associated antigens. They are usually negative for CD5, CD10 and CD23. They strongly express CD11c and CD25, but these are not specific. CD103, which also labels epitheliotropic mucosal T-cells, is more specific but not applicable to fixed tissue sections. The most specific markers for hairy cell leukaemia are annexin A1 and VE1, which

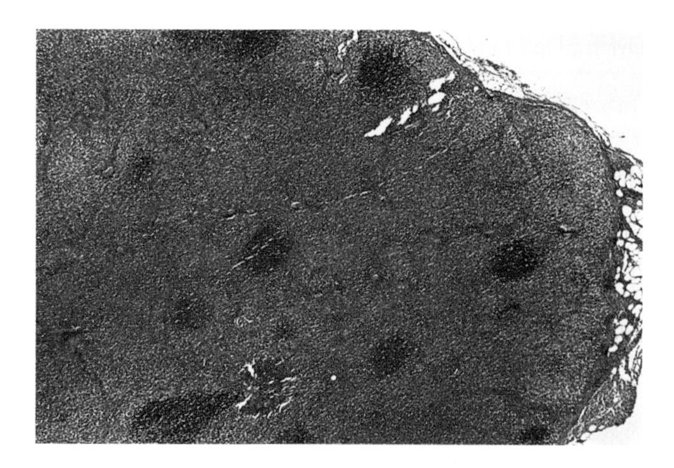

Figure 5.19 Hairy cell leukaemia. Splenic hilar lymph node showing residual follicles pushed apart by pale-staining hairy cells.

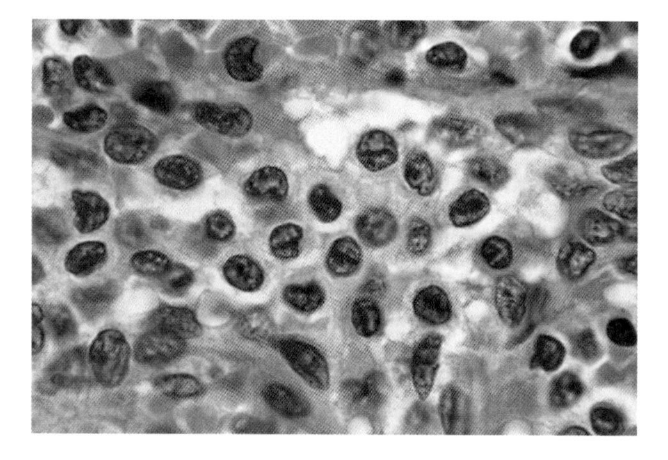

Figure 5.20 Hairy cell leukaemia. High-power view showing the grooved bean-shaped nuclei of the hairy cells and their abundant pale cytoplasm.

is a BRAF V600E mutation–specific antibody (Sherman *et al.* 2011, Andrulis *et al.* 2012). The monoclonal antibody DBA44 labels these cells in fixed tissue sections but is not entirely specific.

GENETICS

The immunoglobulin heavy and light chain genes are rearranged and mutated. Cyclin D1 is weakly expressed in over half the cases of hairy cell leukaemia, but this is not associated with translocation of the BCL-1 gene.

EXTRAMEDULLARY PLASMACYTOMA

Most pathologists will be familiar with bone marrow-derived plasma cell tumours presenting as single or multiple osteolytic tumours (multiple myeloma). Soft tissue plasmacytomas do not appear to be related to multiple myelomatosis. They occur most frequently in the upper aerodigestive tract and may spread to cervical lymph nodes. Primary plasmacytoma of the lymph nodes is uncommon and nodal involvement from a soft tissue plasmacytoma should be excluded before this diagnosis is made.

Lymph nodes may be completely or partially replaced by the neoplastic infiltrate. The tumour cells may have the typical cytological appearances of mature plasma cells. The plasmablastic form has a more blastic morphology with an open chromatin structure and visible nucleoli. Anaplastic plasmacytomas may show considerable morphological variability and need to be differentiated from large cell lymphomas and other anaplastic neoplasms. Tumour cells may contain immunoglobulin inclusions in the form of globules or, more rarely, crystalloids. Extracellular eosinophilic material is commonly seen and it may form the most prominent histological feature. In some cases, but not all, this material is amyloid, giving positive staining with Congo red stain and showing characteristic anomalous colours in polarized light. Amyloid frequently elicits a foreign body giant cell reaction (Figures 5.21 to 5.23).

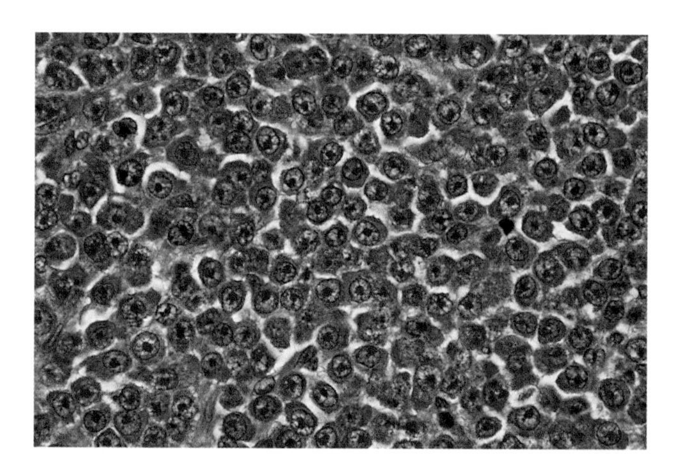

Figure 5.21 Plasmacytoma, plasmablastic type showing prominent nucleoli and dark amphophilic cytoplasm.

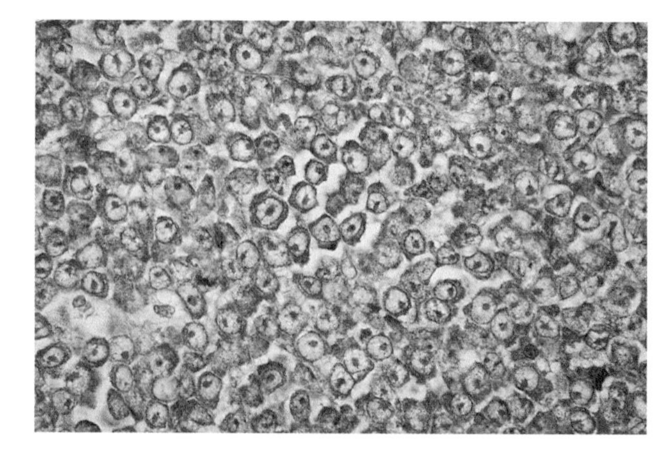

Figure 5.22 Plasmacytoma, plasmablastic stained by methyl green pyronin to demonstrate cytoplasmic RNA.

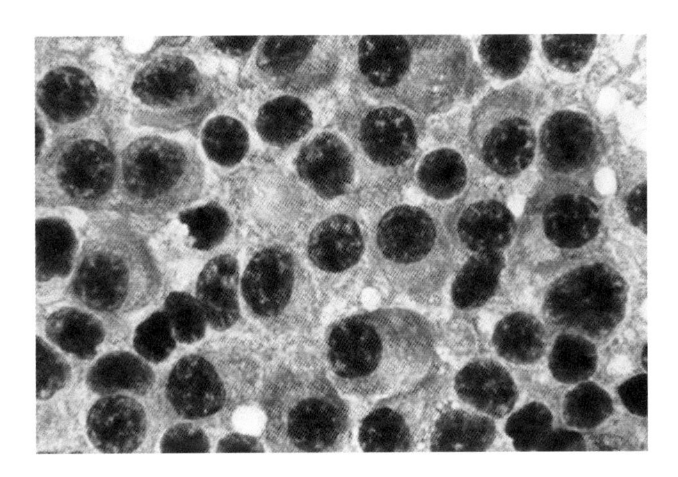

Figure 5.23 Imprint preparation of plasmacytoma showing the appearance of mature plasma cells.

IMMUNOHISTOCHEMISTRY

Plasma cells lose many of the surface markers that characterize mature B-cells. Thus they are frequently negative for CD45 and CD20, but usually do express CD79a. The tumour cells synthesize immunoglobulin, most commonly of the IgA or IgG isotype, and show light chain restriction. A number of antibodies relatively specific for plasma cells, such as CD38, MUM1 and CD138, may be of diagnostic value. Normal and neoplastic plasma cells commonly express epithelial membrane antigen (EMA), and rare cases may express cytoplasmic cytokeratin. Yu *et al.* (2011) reported that the majority of extramedullary plasmacytomas express CD19 whereas this was not seen in myeloma. Unlike myeloma, in which it is often expressed, CD56 expression is infrequent in extramedullary plasmacytoma (Kremer *et al.* 2005).

GENETICS

There are numerous cytogenetic studies of myeloma. It is uncertain how these relate to soft tissue or, more specifically, lymph node plasmacytomas. All cases of myeloma exhibit translocations involving the immunoglobulin genes. Some of these are recurrent, whereas others appear random and rare. Many involve the switch region of the immunoglobulin heavy chain gene, consistent with the expression of IgA or IgG by many of these tumours. The translocation t(11;14)(q13;q32) occurs in 20–25 per cent of myelomas, leading to overexpression of cyclin D1, but this is not seen in extramedullary plasmacytomas. However, this should be borne in mind when using antibodies to cyclin D1 to confirm the diagnosis of mantle cell lymphoma.

DIFFERENTIAL DIAGNOSIS

Intense plasma cell proliferation may occur in reactive and inflammatory lymphadenopathies. The plasma cells are typically well differentiated, with condensed nuclear chromatin and basophilic cytoplasm. They are most characteristically distributed in the medullary cords of the lymph node. The association with other inflammatory cells and reactive features will usually distinguish these proliferations from plasmacytoma. The demonstration of light chain restriction will usually make the distinction in difficult cases. Plasmacytoma and lymphoplasmacytic lymphoma should always be excluded in lymph nodes showing amyloidosis.

Castleman disease of the plasmacytic subtype can usually be differentiated from plasmacytoma by the presence of characteristic hyalinized follicles. Some patients with this disease show λ light chain restriction. The presence of small lymphocytes, plasma cells and intermediate forms differentiates lymphoplasmacytic lymphoma from the more homogeneous plasma cell infiltrates of plasmacytoma.

Marginal zone lymphomas may show extreme plasma cell differentiation, leading to confusion with plasmacytoma. They can usually be differentiated by the presence of characteristic marginal zone cells and sometimes by the presence of large numbers of epithelioid histiocytes.

The differentiation of blastic and anaplastic plasmacytomas from large cell lymphomas may be difficult, particularly those showing the characteristic morphology of immunoblastic lymphoma, with large single nucleoli and basophilic cytoplasm. The expression of CD45 and CD20 would favour the diagnosis of large B-cell lymphoma, whereas the absence of these markers and the presence of EMA and plasma cell antigens would favour plasmacytoma. The absence of leukocyte common antigen (CD45) and the presence of EMA and, very rarely, cytokeratin in an anaplastic plasmacytoma might lead the unwary into a diagnosis of anaplastic carcinoma. Anaplastic lymphoma kinase (ALK)–positive large B-cell lymphoma should be considered in the differential diagnosis of plasmablastic tumours that lack lineage markers and express EMA.

NODAL MARGINAL ZONE LYMPHOMA

Following the description of monocytoid B-cell lymphoma at nodal sites (Sheibani *et al.* 1986), it became apparent that these tumours were morphologically and immunophenotypically similar to extranodal marginal zone (mucosa-associated lymphoid tissue [MALT]) lymphoma. Analysis of several series of nodal monocytoid lymphomas showed an association with synchronous or metachronous extranodal marginal zone lymphoma in many, but not all, cases. It thus appears that the presence of a nodal marginal zone B-cell lymphoma is often a manifestation of an underlying overt or occult MALT lymphoma. It was subsequently shown that a small proportion of nodal marginal zone lymphomas have the immunophenotype of splenic marginal zone lymphoma rather than extranodal marginal zone lymphoma (splenic marginal zone lymphomas often express IgD and are CD43–, whereas extranodal marginal zone lymphomas are IgD– and often CD43+). These cases, however, did not have evidence of splenic disease at the time of lymph node biopsy (Campo *et al.* 1999).

The cells of nodal marginal zone B-cell lymphoma have small to medium-sized nuclei that are rounded, indented or irregular. They have abundant pale or clear cytoplasm. It is this cytoplasmic pallor that makes clusters of monocytoid B-cells stand out in histological sections. The neoplastic marginal zone cells surround reactive follicles and show increasing follicular colonization with disease progression.

These lymphomas frequently show plasma cell differentiation, the plasma cells exhibiting the same monoclonal immunoglobulin within their cytoplasm as is found on the surface of the marginal zone B-cells. Epithelioid cell clusters are also commonly found in marginal zone B-cell lymphomas, particularly those associated with autoimmune sialadenitis. Epithelioid cells frequently surround small groups of marginal zone B-cells (Boxes 5.7 and 5.8 and Figures 5.24 to 5.32).

BOX 5.7: Nodal marginal zone lymphoma: Clinical features

- Less than 2 per cent of all non-Hodgkin lymphomas
- 30–50 per cent have associated extranodal marginal zone lymphomas
- Rare cases associated with splenic marginal zone lymphoma
- Sites of involvement:
  - Peripheral lymph nodes, localized or generalized
  - Bone marrow
  - Peripheral blood

BOX 5.8: Nodal marginal zone lymphoma: Morphology

- Partial preservation of nodal architecture
- Marginal zone B-cells infiltrate interfollicular areas, within or adjacent to sinuses
- Follicular colonization
- Nuclei rounded, oval or cleaved
- Clear cytoplasm
- Plasma cell differentiation in some cases
- Infiltrate of epithelioid histiocytes in some cases
- Scattered blast cells
- May transform to large B-cell lymphoma

Figure 5.24 Nodal marginal zone lymphoma. The pale marginal zone cells follow the distribution of the lymph node sinuses.

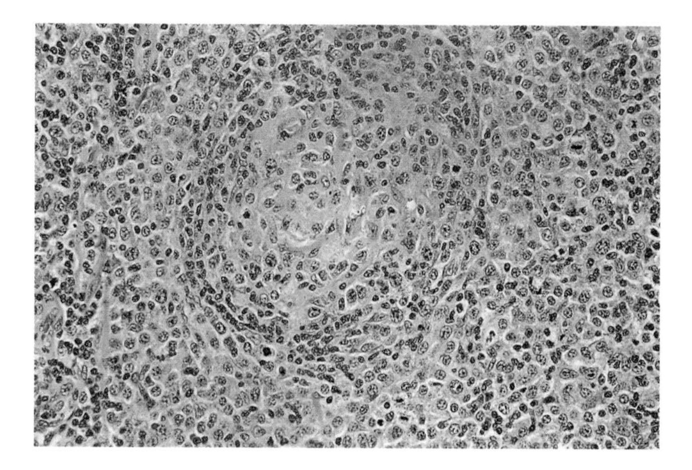

Figure 5.25 Higher power view of nodal marginal zone lymphoma showing a reactive germinal centre surrounded by an attenuated mantle zone of small lymphocytes with an outer zone of marginal zone cells.

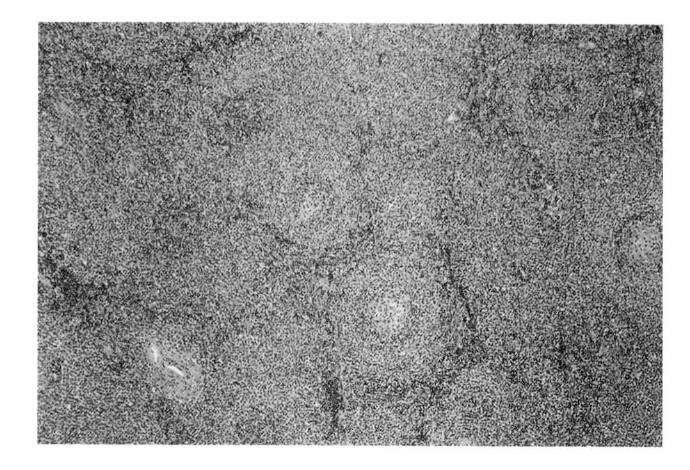

Figure 5.26 Nodal marginal zone lymphoma showing nodular growth pattern with reactive follicles at the centre of the nodules. Note the paler staining marginal zone cells around the follicles.

Figure 5.27 Nodal marginal zone lymphoma, low power. Showing residual reactive follicles and islands of marginal zone cells surrounded by epithelioid histiocytes.

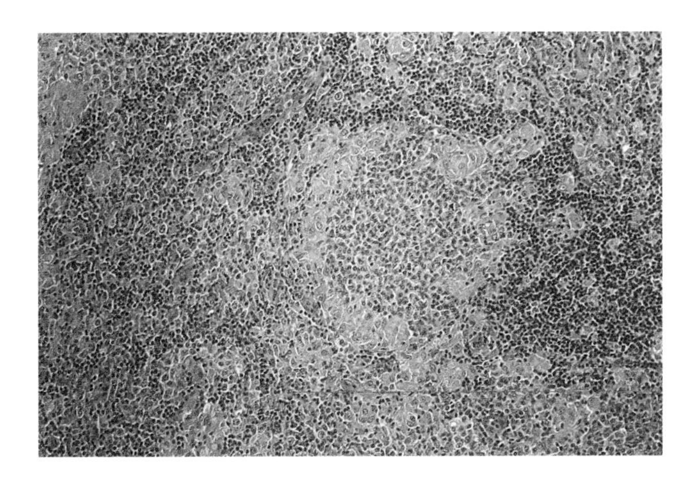

Figure 5.28 Nodal marginal zone lymphoma. High-power view of marginal zone cells surrounded by epithelioid histiocytes.

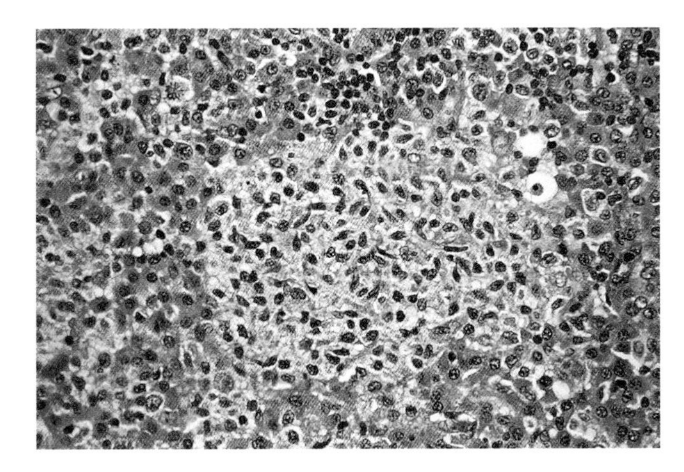

Figure 5.29 Nodal marginal zone lymphoma showing plasmacytic differentiation.

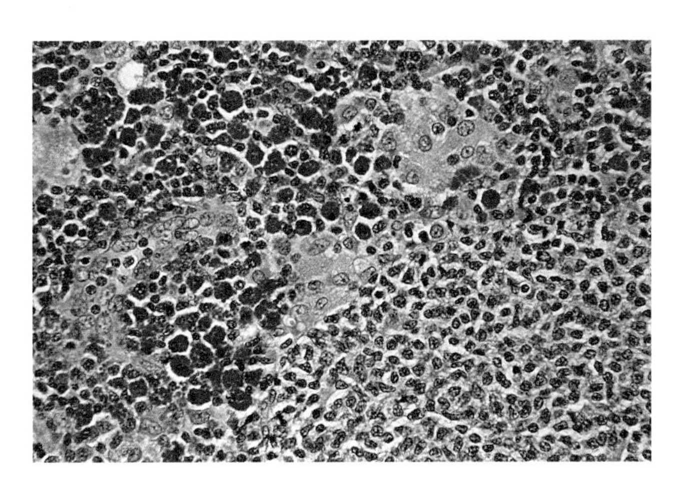

Figure 5.30 Nodal marginal zone lymphoma with plasmacytic differentiation. Periodic acid–Schiff stain showing inclusions of immunoglobulin within the plasma cells.

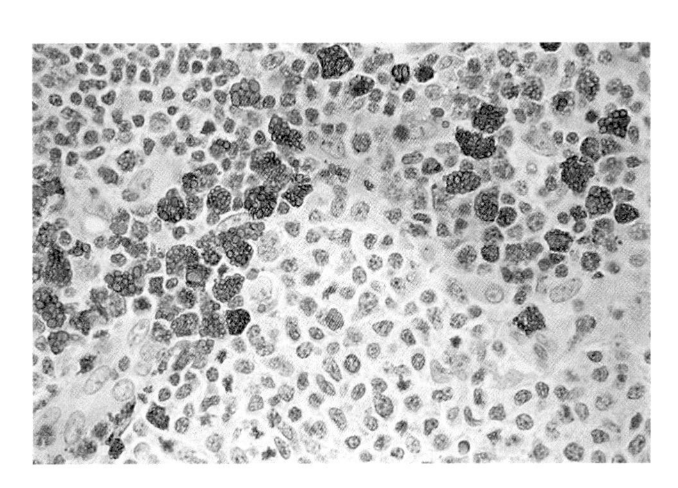

Figure 5.31 Same case as Figure 5.30 stained for immunoglobulin M (IgM), demonstrating that the inclusions are IgM.

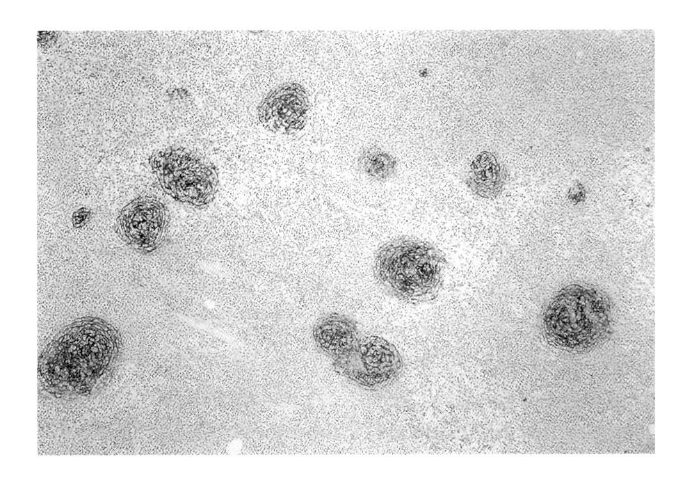

Figure 5.32 Low-power view of nodal marginal zone lymphoma stained for CD21, showing residual follicular dendritic cells.

BOX 5.9: Nodal marginal zone lymphoma: Immunohistochemistry

- Surface/cytoplasmic immunoglobulin M (IgM) or, less commonly, IgA or IgG
- IgD negative except in splenic marginal zone type
- CD20+ and CD79a+
- CD5 usually negative
- CD10− and CD23−
- CD43 usually positive except in splenic marginal zone type
- Tumour cells BCL-2+; residual reactive follicles BCL-2−

IMMUNOHISTOCHEMISTRY

Tumour cells express surface immunoglobulin, usually of the IgM class, sometimes in association with IgD; IgG and IgA are less commonly expressed. Cells showing plasma cell differentiation show cytoplasmic immunoglobulin of the same class and express plasma cell–associated antigens. The marginal zone cells express the B-cell–associated antigens CD20 and CD79a. They are usually negative for CD5, although rare CD5+ cases, usually with blood and bone marrow involvement, have been reported (Ballesteros *et al.* 1998). They rarely express CD23 or CD10 (van den Brand and van Krieken 2013). Abnormal expression of apoptosis regulator proteins has been reported (Camacho *et al.* 2003). CD21 and CD23 used to identify residual dendritic cells in colonized follicles can be useful markers (Box 5.9).

GENETICS

The translocations that characterize extranodal marginal zone lymphoma are not found in nodal marginal zone lymphoma. Trisomies 3, 7 and 18 have been reported.

DIFFERENTIAL DIAGNOSIS

Nodal marginal zone B-cell lymphoma must be separated from reactive lymphadenopathies associated with monocytoid B-cell hyperplasia. These include toxoplasma lymphadenitis and human immunodeficiency virus (HIV)–associated lymphadenitis. In the original descriptions of nodal marginal zone B-cell lymphoma, it was stated that demonstration of immunoglobulin light chain restriction might be necessary to make this distinction (Sheibani *et al.* 1986, 1988). The presence of characteristic marginal zone cells distinguishes this lymphoma from lymphoplasmacytic lymphoma. The presence of numerous epithelioid cells is characteristic of many marginal zone B-cell lymphomas but less common in lymphoplasmacytic lymphoma. Rare follicular lymphomas showing marginal zone differentiation may mimic nodal marginal zone B-cell lymphoma. In these tumours, the follicles have the characteristics of neoplastic follicles, including BCL-2 positivity, in contrast to the reactive follicles present in nodal marginal zone B-cell lymphoma.

PAEDIATRIC NODAL MARGINAL ZONE LYMPHOMA

This condition affects males predominantly (M20:F1) as stage 1 disease, usually of the head and neck region. The morphology and immunophenotype are similar to those of the adult disease except that occasional follicles showing progressive transformation may be present (Taddesse-Heath *et al.* 2003). The detection of monotypic immunoglobulin synthesis by immunohistochemistry is not sufficient for the diagnosis of neoplasia because light chain restriction may be seen in cases of paediatric marginal zone hyperplasia in the absence of clonal immunoglobulin gene rearrangements (Attygalle *et al.* 2004). The prognosis of this condition is excellent.

FOLLICULAR LYMPHOMA

Follicular lymphomas are the commonest subtype of low-grade non-Hodgkin lymphoma in the Western world, being most prevalent in the United States. They are less common in Africa and Asia. The lymphoma has its maximum incidence in the sixth and seventh decades of life; it is rare in childhood, although a paediatric variant is now recognized (Liu *et al.* 2013). Follicular lymphoma usually presents with lymphadenopathy. Involvement of the bone marrow and the spleen is common and most patients present with stage III or stage IV disease. Patients with bone marrow involvement may show overspill of lymphoma cells into the peripheral blood. Involvement of extranodal sites, such as the tonsil, intestine and skin, is relatively uncommon and may be secondary or primary. The disease usually runs a progressive course, though even without therapy it may undergo spontaneous remission for varying lengths of time. Chemotherapy may be used to control the disease but it is rarely curative (Boxes 5.10 and 5.11 and Figures 5.33 to 5.49).

HISTOLOGY

The follicles of follicular lymphoma are usually closely packed, extend throughout the node and have relatively uniform rounded outlines. They may extend into and replace the fat in the hilum of the node, frequently seen in axillary lymph nodes, and they also extend into the perinodal fat. The pre-existing nodal architecture is distorted and destroyed with loss of the sinus structure.

Follicular lymphomas are derived from follicle centre B-cells and are composed of variable mixtures of centroblasts and centrocytes. In contrast to reactive follicles, in which the centoblasts and centrocytes show zoning, in neoplastic follicles these cells are randomly distributed. The neoplastic follicles, like their reactive counterparts, contain dendritic reticulum cells and reactive follicle centre T-cells

BOX 5.10: Follicular lymphoma: Clinica features

- 35 per cent of all non-Hodgkin lymphomas in United States; smaller percentage in the rest of the world
- Median age 59 years
- Male to female ratio 1:1.7
- Usually presents with stage III or IV disease
- Sites of involvement:
  - Lymph nodes
  - Bone marrow
  - Peripheral blood
  - Spleen
  - Waldeyer's ring
- Primary follicular lymphomas of the gastrointestinal tract and skin are clinically different from nodal follicular lymphomas

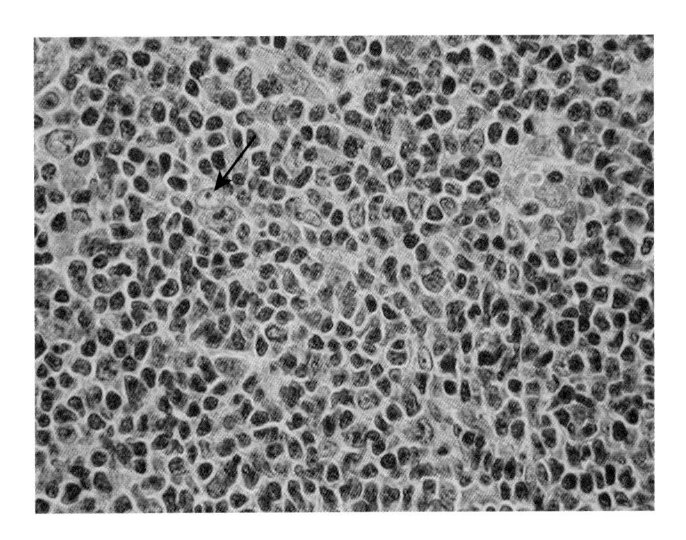

Figure 5.33 Grade 1 follicular lymphoma showing a single centroblast. The nucleus adjacent to the centroblast (arrow) is of a dendritic reticulum cell.

BOX 5.11: Follicular lymphoma: Morphology

- Loss of lymph node architecture
- Follicles composed of a mixture of centroblasts and centrocytes
- Regular-shaped follicles throughout node
- Attenuated mantle zones
- Follicles poorly defined
- Sclerosis common
- Grade 1, 0–5 blasts per high-power field (HPF); grade 2, 6–15 blasts per HPF; grade 3, >15 blasts per HPF
- Grade 3A centroblasts and centrocytes present, grade 3B centroblasts only.
- Growth patterns:
  - Follicular >75 per cent follicular
  - Follicular and diffuse 25–75 per cent follicular
  - Minimally follicular <25 per cent follicular
- Variants (see text)

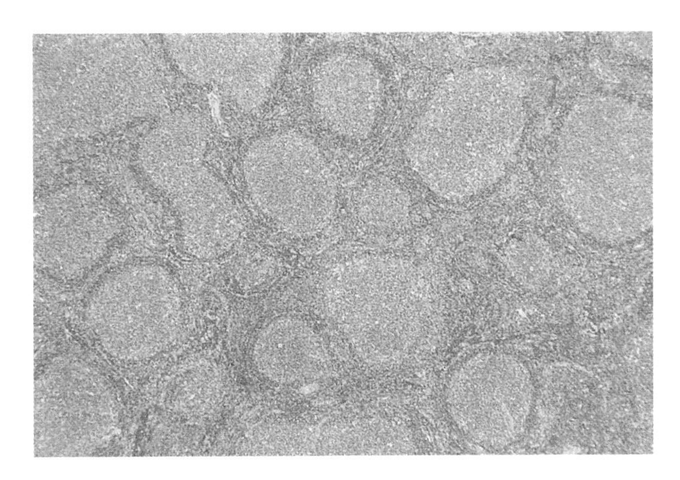

Figure 5.34 Follicular lymphoma. Relatively uniform follicles with attenuated mantles. No sinus structures seen.

(CD3+, CD4+, CD57+, PD-1+). Neoplastic follicles often acquire a mantle of non-neoplastic small B-cells, though this is usually more attenuated and less well defined than that of reactive follicles. There is a tendency for follicular lymphomas to become more blastic (acquire a larger proportion of centroblasts) with time, a trend that is often associated with the acquisition of further cytogenetic abnormalities. The trend towards a more blastic morphology is also associated with the development of areas of diffuse growth, so that the lymphoma may become follicular and diffuse or completely diffuse.

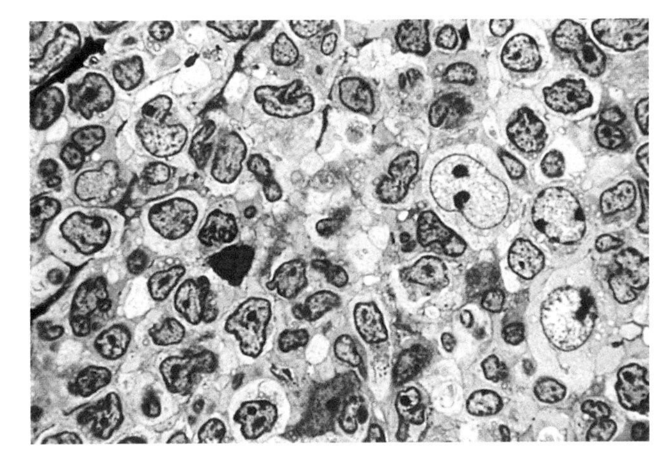

Figure 5.35 Plastic-embedded section of follicular lymphoma showing centrocytes and three centroblasts.

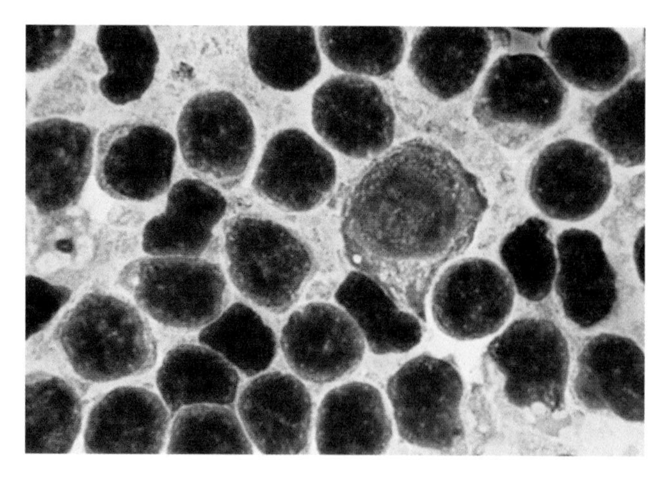

Figure 5.36 Imprint cytology of a grade 1 follicular lymphoma showing centrocytes and a centroblast.

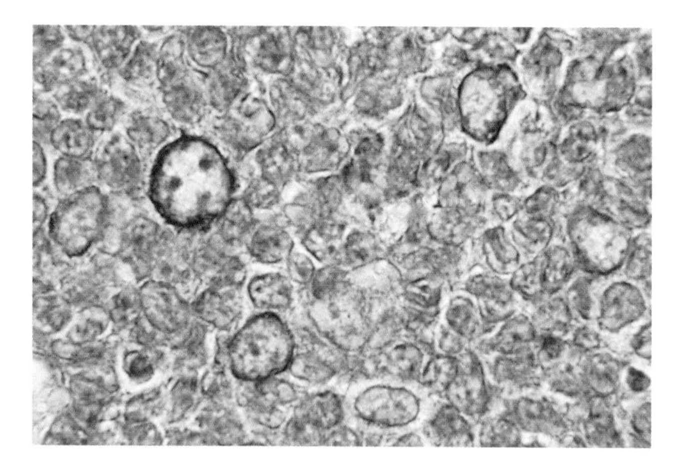

Figure 5.37 Follicular lymphoma. Giemsa stain showing basophilia of centroblast cytoplasm.

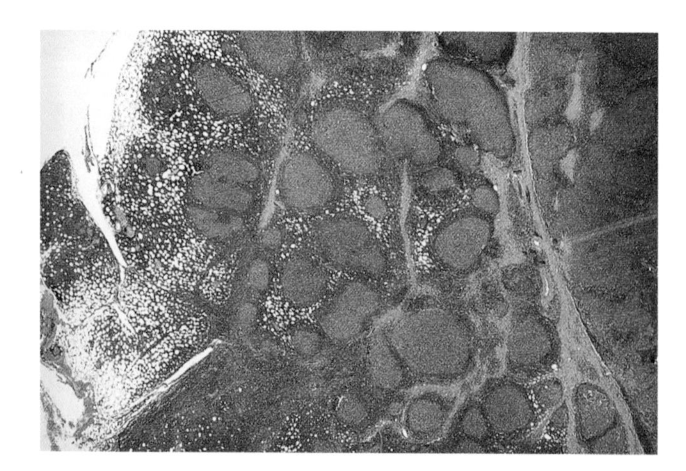

Figure 5.38 Follicular lymphoma showing extension into perinodal fat.

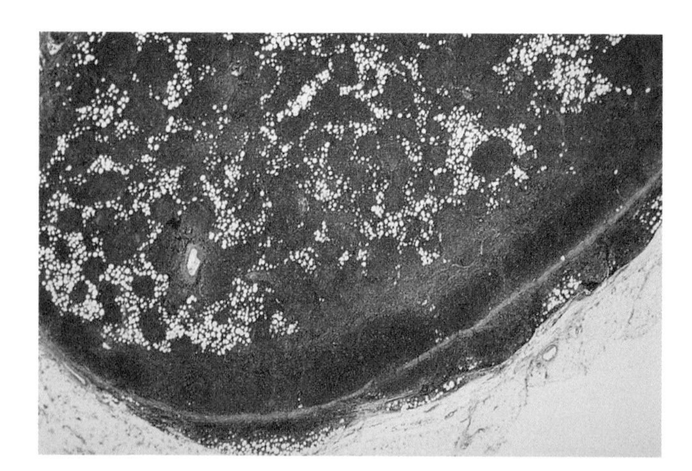

Figure 5.39 Follicular lymphoma involving axillary lymph node. Neoplastic follicles have invaded the fatty hilum and also extend outside the capsule.

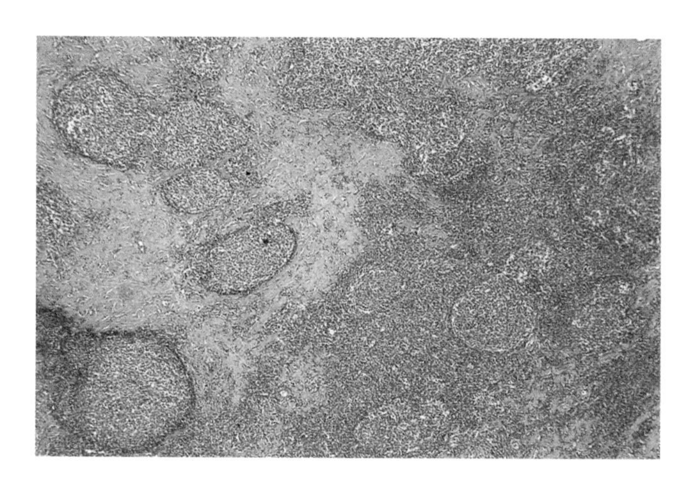

Figure 5.40 Follicular lymphoma showing sclerosis around follicles.

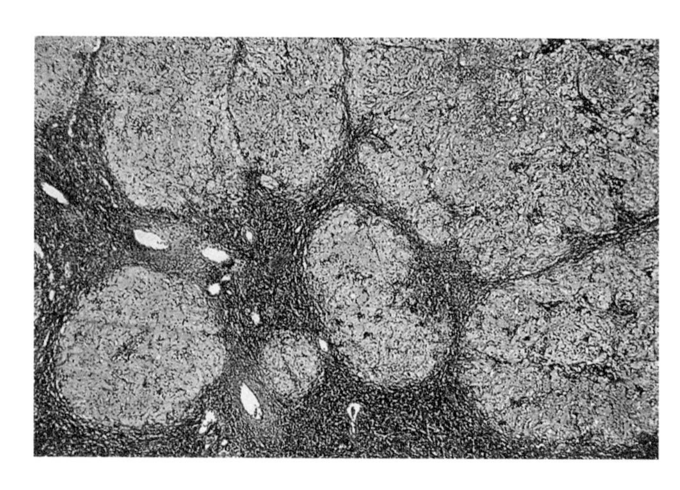

Figure 5.41 Follicular lymphoma showing sclerosis of follicle centres (reticulin stain).

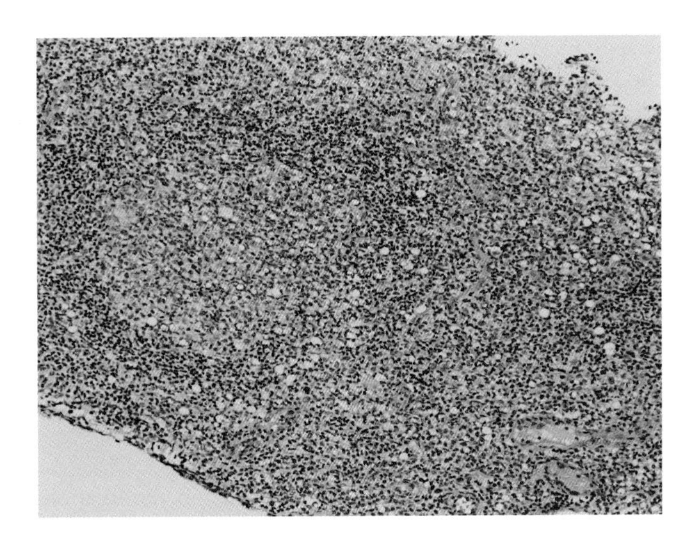

Figure 5.42 Follicular lymphoma showing signet ring change. Many of the cells appear vacuolated in this variant of follicular lymphoma.

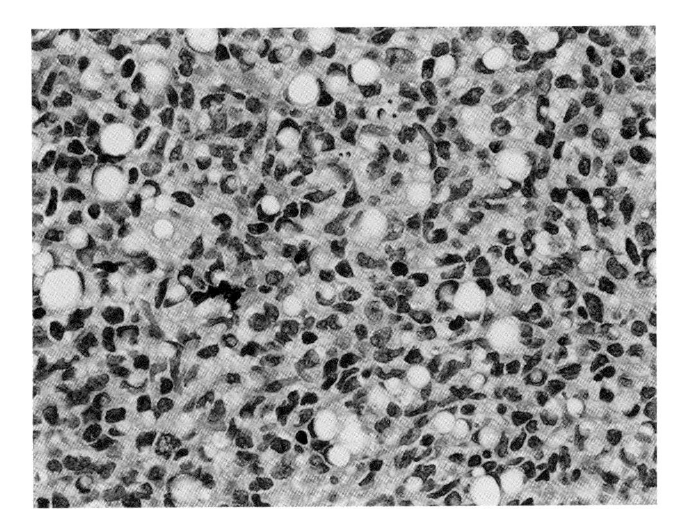

Figure 5.43 Follicular lymphoma showing signet ring change. The nucleus of many of the tumour cells is compressed to form a small crescent.

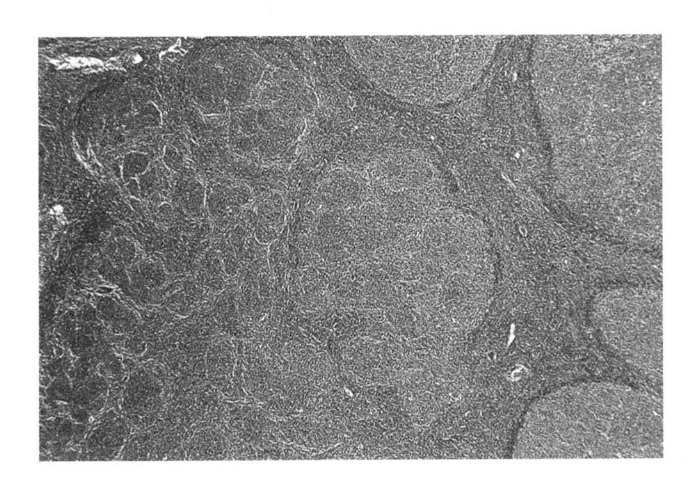

Figure 5.44 Follicular lymphoma in which one neoplastic follicle displays a floret pattern.

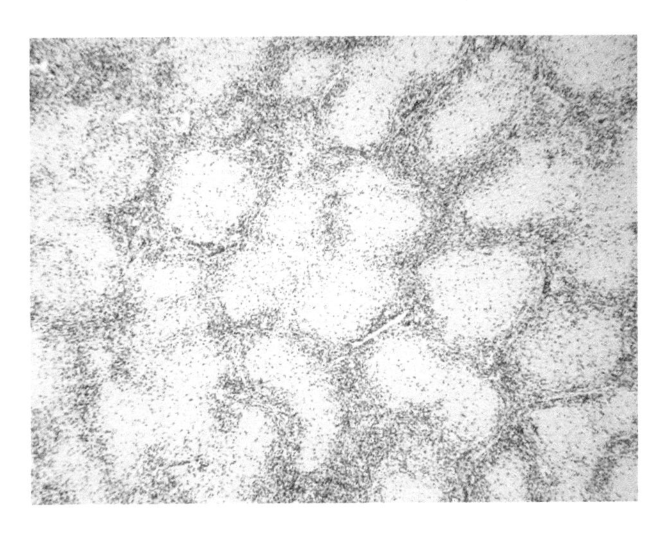

Figure 5.45 Follicular lymphoma showing follicles outlined by CD3+ T-cells.

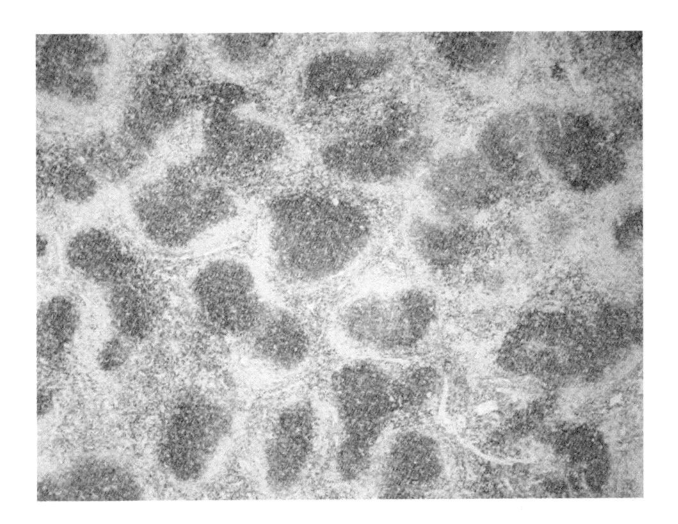

Figure 5.46 Grade 1 follicular lymphoma showing CD10 staining of follicle centre cells.

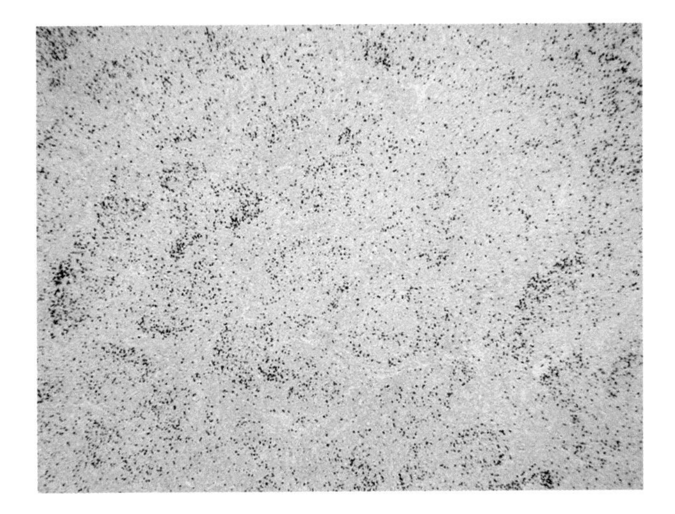

Figure 5.47 Grade 1 follicular lymphoma labelled for Ki67 showing a relatively low proliferation fraction. Note that many of the follicles show most labelling at the periphery.

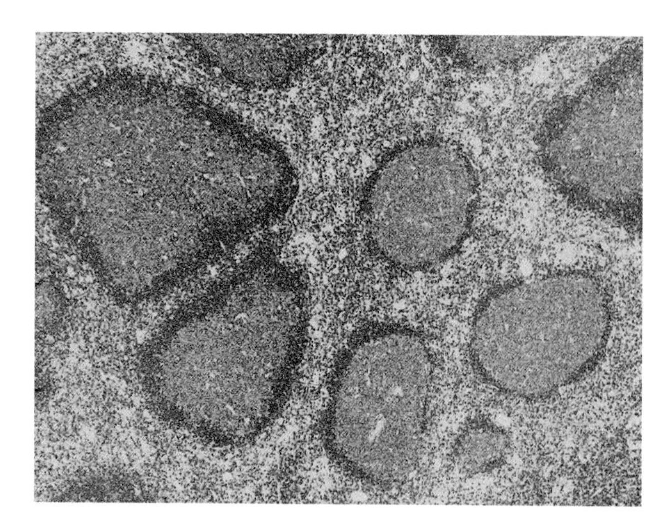

Figure 5.48 Follicular lymphoma stained for CD79a. The small dark cells are mantle cells that form attenuated mantles.

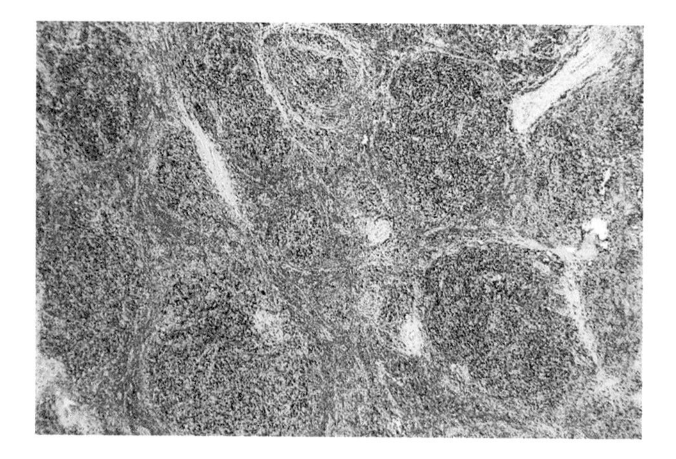

Figure 5.49 Follicular lymphoma stained for BCL-2 showing positivity of the follicle centres.

The predominant interfollicular cells in many follicular lymphomas are small B- or T-cells. However, tumour cells are often found in this situation, recognizable by their morphology or by their immunohistochemical staining.

In their cytomorphology, the tumour cells of many follicular lymphomas resemble normal centroblasts or centrocytes. The tumour cells in other cases, particularly of the higher grades, may show an atypical morphology. Centroblasts may show multilobated, serpiginous or multinucleated nuclei.

Grading of follicular lymphomas

In 1983, Mann and Berard proposed a method for the cytological subclassification (grading) of follicular lymphomas. This system is based on the number of centroblasts per high power field, with 0–5 being grade 1, 6–15 grade 2, and 16 or more grade 3. The prognostic value of the Mann and Berard grading system has been widely studied and debated and, despite its shortcomings, its use has been recommended in the World Health Organization (WHO) classification. It is recognized, however, that there is a progression between grades 1 and 2 and that the distinction between these two grades shows poor reproducibility. It is now common to include grades 1 and 2 in a low-grade group separate from grade 3. Grading may influence patient management; for example, patients with low-grade tumours may not be treated or may be offered only symptomatic treatment, whereas grade 3 tumours will usually be treated with chemotherapy. Patients with grade 3 follicular lymphoma have earlier relapses than grades 1 and 2, but similar overall survival. The poorer freedom from relapse of grade 3 disease may be overridden by chemotherapy.

The ability to recognize blasts depends on the fixation, thickness and quality of the section. Giemsa-stained sections in which the basophilia of the centroblast cytoplasm highlights these cells may be helpful. The grade is determined by counting ten fields within separate follicles. With the increasing use of core biopsies, however, the number of follicles available for analysis is likely to be fewer than ten.

Follicular lymphomas may have areas of diffuse growth pattern, and both the Revised European-American Lymphoma (REAL) classification and the WHO classification recommend that the degree of follicularity be reported as follows:

1—Predominantly follicular: >75 per cent follicular.
2—Follicular and diffuse: 25–75 per cent follicular.
3—Predominantly diffuse: <25 per cent follicular.

The significance of this subclassification with respect to treatment or survival is uncertain.

Gene expression studies have shown that different immune response signatures based on genes expressed by T-cells, macrophages and dendritic cells in the original diagnostic biopsy can be used to predict survival of patients with follicular lymphomas, separating 75 per cent of patients with indolent disease and survival of over 10 years from 25 per cent of patients with survival of less than 4 years. It is possible that these studies will eventually form the basis for immunophenotypic methods for predicting tumour behaviour (Dave *et al.* 2004).

In grade 3 tumours, diffuse areas represent areas of DLBCL. Such cases should not be reported as follicular and diffuse, but as follicular lymphoma grade 3 (*x* per cent) with transformation to DLBCL (*y* per cent). The presence of DLBCL usually dictates more aggressive therapy.

Grade 3 follicular lymphomas should be further subdivided into those with residual centrocytes (3A) and those without centrocytes (3B). A study of grade 3B follicular lymphoma has shown that cases can be separated into those with t(14;18) involving the BCL-2 gene, those with 3q27 abnormalities involving the BCL-6 gene, and those with neither of these but with other cytogenetic aberrations. Those with t(14;18) are probably related to the spectrum of follicular lymphomas, whereas those with 3q27 or other aberrations are more closely related to the majority of DLBCLs (Bosga-Bouwer *et al.* 2003).

Morphological variation within follicular lymphomas

Within the category of follicular lymphomas, there is considerable variation in histological appearances. Depending on the number of centroblasts and centrocytes in, and between, the follicles and the prominence, or otherwise, of the mantle zone, the follicles may be well defined or barely discernible. Follicularity may be highlighted by reticulin staining and is often accentuated by immunohistochemical staining for B- and T-cells. Other morphological variations are as follows.

SCLEROSIS

Sclerosis is a common feature of follicular lymphomas, particularly those involving the retroperitoneal lymph nodes. It frequently takes the form of fine or coarse fibrous bands surrounding groups of neoplastic follicles. Sometimes it surrounds and partially obliterates individual follicles. Less commonly, it is seen within follicles. Sclerosis has no prognostic significance.

AMORPHOUS, PERIODIC ACID–SCHIFF–POSITIVE, EXTRACELLULAR MATERIAL

Eosinophilic amorphous material may be found in reactive and in neoplastic follicles, where it may be so abundant that it obscures the tumour cells. Unlike collagen, this material is non-birefringent. Electron microscopic examination shows it to be composed of membranous material.

IMMUNOGLOBULIN INCLUSIONS

Immunoglobulin, most frequently of the IgM isotype, may accumulate in the endoplasmic reticulum or perinuclear space. In the former it appears as Russell bodies, and in the latter as intranuclear inclusions. The inclusions are usually strongly PAS positive.

SIGNET RING LYMPHOMA

Clear cytoplasmic inclusions that displace and compress the nucleus may be seen in a few or many tumour cells of a follicular lymphoma. These vacuoles do not stain with PAS. They have been reported to contain monoclonal IgG but it is often not possible to demonstrate immunoglobulin in these cases. Ultrastructurally, the vacuoles are seen to contain membranes and vesicles.

FOLLICULAR LYMPHOMA WITH PLASMACYTIC DIFFERENTIATION

A small number of follicle centre cell lymphomas show plasma cell differentiation with either intrafollicular or interfollicular monoclonal plasma cells.

FOLLICULAR LYMPHOMA WITH ATYPICAL NUCLEAR MORPHOLOGY

Rarely, follicular lymphomas show cells with cerebriform nuclei similar to those seen in the tumour cells of mycosis fungoides. The centroblasts within follicular lymphomas may show multilobated or serpiginous (centrocytoid) nuclei.

FLORAL VARIANT OF FOLLICULAR LYMPHOMA

In this uncommon variant of follicular lymphoma, the neoplastic follicle centres are broken up by ingrowths of small lymphocytes, giving an irregular flower-like outline to the follicles. The importance of recognizing this subtype is that it may be mistaken for progressive transformation of germinal centres or for nodular lymphocyte-predominant Hodgkin lymphoma (NLPHL).

FOLLICULAR LYMPHOMA WITH MARGINAL ZONE DIFFERENTIATION

Marginal zone differentiation in follicular lymphomas is more commonly seen in the spleen than in lymph nodes. Pale-staining collars of cells surround the neoplastic follicles. These cells are part of the neoplastic clone that has acquired more abundant clear or pale-staining cytoplasm. These cases may mimic marginal zone B-cell lymphoma morphologically. In marginal B-cell lymphoma the follicles are reactive and therefore will be BCL-2– and will not show light chain restriction.

FOLLICULAR LYMPHOMA WITH INFARCTION OF LYMPH NODE

Infarction of lymph nodes usually occurs in lymph nodes involved by lymphoma and, commonly, in those involved by follicular lymphoma. Multiple blocks should be examined to look for residual non-infarcted tissue. Reticulin staining may reveal underlying structural details that are not apparent in haematoxylin and eosin (H&E)– or Giemsa-stained sections. Some antigens, detectable by immunohistochemistry, survive for a considerable time after infarction, allowing the distribution and the phenotype of the tumour cells to be determined.

COMPOSITE LYMPHOMA

Follicular lymphomas may show areas of higher grade or diffuse growth pattern. These are not composite lymphomas but lymphomas showing tumour progression. Hodgkin lymphoma occasionally develops within a low-grade lymphoma, most commonly follicular lymphoma. In these cases, the Hodgkin lymphoma has the morphology and phenotype of one of the classical subtypes of the disease. The follicular lymphoma and the Hodgkin lymphoma may occur together in the same biopsy, or in separate synchronous or metachronous biopsies.

IN-SITU FOLLICULAR LYMPHOMA

Lymph nodes involved with *in-situ* follicular lymphoma show preservation of the normal architecture with follicles that have preserved mantles and germinal centres. The majority of the follicles are reactive; a few follicles have a more monomorphic morphology and show strong co-expression of CD10 and BCL-2. The neoplastic cells show t(14;18) by FISH analysis and the tumour cells are confined to the follicle centres. *In-situ* follicular lymphoma may be distinguished from partial nodal involvement by overt follicular lymphoma by the presence of architectural distortion in the latter condition (Carbone and Gloghini 2014).

IMMUNOHISTOCHEMISTRY

The tumour cells express surface, and often cytoplasmic, immunoglobulin showing light chain restriction. IgM is the commonest heavy chain isotype expressed. Surface immunoglobulin can be detected in frozen-section immunohistochemistry but is difficult to detect in paraffin sections because of the large amount of extracellular immunoglobulin present in most follicles. The detection of cytoplasmic immunoglobulin in paraffin sections requires good fixation and good technique.

The B-cell–associated antigens CD20 and CD79a are expressed. The tumour cells are negative for CD5, and CD43. CD23 may be expressed in follicular lymphoma, particularly follicular lymphoma involving inguinal lymph nodes (Olteanu *et al.* 2011). Many follicular lymphomas express CD10 and BCL-6, which serve to differentiate them from other low-grade B-cell lymphomas. Other germinal centre markers such as GCET1 or HGAL can be useful in the diagnosis of CD10/BCL-6 negative follicular lymphomas (Menter *et al.* 2015). Positive staining for CD45RA (MT2) may help to distinguish neoplastic from reactive follicles.

Approximately 90 per cent of low-grade follicular lymphomas express BCL-2 protein, whereas only 75 per cent of grade 3 tumours show positivity. This is a particularly valuable marker because in reactive follicle centre cells BCL-2 is not expressed. In interpreting BCL-2 staining, it should be noted that reactive T-cells, which are often abundant in neoplastic follicles, are BCL-2+. Ideally, a CD3 stain should be performed to determine the number and distribution of the T-cells, which should be distinguished from centrocytes in BCL-2–stained sections by their morphology. It should also be recognized that BCL-2 staining may be subtle, requiring a careful search for cytoplasmic positivity in perhaps only a proportion of the tumour cells. This contrasts with the total negativity of reactive follicle centre cells. BCL-2 negativity in low-grade follicular lymphoma may be the result of genuine absence of a BCL-2 gene mutation or, more usually, may be a false negative caused by the mutation masking the BCL-2 antibody binding site; in these cases positive expression may be seen using an alternative antibody clone. Low-grade follicular lymphomas with germline BCL-2 genes may show an aberrant phenotype expressing CD23 or lacking CD10 expression (Adam *et al.* 2013). Not all follicular lymphomas that express BCL-2 have the 14;18 translocation. Both neoplastic and reactive follicle centre cells express nuclear BCL-6 (Box 5.12).

BOX 5.12: Follicular lymphoma: Immunohistochemistry

- Surface/cytoplasmic immunoglobulin M (IgM)+
- Surface/cytoplasmic IgD > IgG > IgA may be positive
- CD45RA (MT2)+
- CD10+, CD20+, CD79a+
- BCL-6+
- BCL-2+; grade 1, 100 per cent; grade 3, 75 per cent
- CD21 and CD23 identify dendritic cell networks

GENETICS

Follicular lymphoma is characterized by the t(14;18)(q32;q21) translocation involving the immunoglobulin heavy chain gene on chromosome 14 and the BCL-2 gene on chromosome 18. This translocation occurs in greater than 80 per cent of follicle centre cell lymphomas, in 20 per cent of DLBCLs and in 1–2 per cent of B-CLL; t(14;18) has also been found in peripheral blood and reactive lymph node B-cells of individuals without lymphoma. Other genetic events, including mutations of p53, are associated with tumour progression.

Follicular lymphomas show clonal rearrangements of the immunoglobulin genes with a high level of ongoing mutations suggestive of antigen drive.

DIFFERENTIAL DIAGNOSIS

The main differential diagnosis of follicular lymphoma is with reactive follicular hyperplasia. The morphological features that aid the distinction between these two conditions are shown in Table 5.2. This differentiation is supported by immunohistochemical staining for BCL-2, CD45RA (MT2) and the proliferation marker Ki67. Apart from intrafollicular T-cells, reactive follicles are negative for BCL-2. Reactive follicles show a much higher proliferation fraction than most neoplastic follicles. Ki67 staining also highlights the zonal distribution of centroblasts in reactive follicles.

SLL with prominent proliferation centres may mimic follicular lymphoma. Morphologically the delicate nuclear structure of prolymphocytes and paraimmunoblasts can usually enable them to be differentiated from centroblasts and centrocytes. Immunohistochemistry shows the cells of SLL to be CD5+ and CD23+; follicle centre cells are negative for these two markers, although CD23 will label the dendritic cell network in the neoplastic follicles. Follicle centre cells are labelled by CD10 and show nuclear positivity for BCL-6; SLLs are negative for these markers.

Mantle cell lymphomas surround and colonize reactive germinal centres and often have a nodular structure. Residual germinal centre cells within these nodules may heighten the resemblance to follicular lymphoma although immunohistochemical staining will show them to be

Table 5.2 Morphological differentiation between reactive and neoplastic follicular proliferations

Reactive	Neoplastic
Follicles mainly distributed in cortex of node; sinus structure of node usually identifiable	Closely packed follicles distributed throughout node; sinus structure of node not usually identifiable
Follicles in perinodal tissues unusual	Follicles extending into perinodal and intranodal fat may be present
Mantle zone of small lymphocytes around follicles usually prominent	Mantle zone usually attenuated and may not be identifiable
Follicles of variable size; may show considerable variation in shape	Follicles of relatively uniform size and shape
Distribution of centroblasts and centrocytes zonal	Distribution of centroblasts and centrocytes random
Centroblasts form a large component of the follicle	Centroblasts usually form a minority population in the follicle and in Grade 1 follicular lymphoma are scanty
Tingible body macrophages containing apoptotic debris often prominent	Tingible body macrophages usually absent
Centroblasts and centrocytes usually have typical morphology	Serpiginous, multinucleated and multilobated centroblasts may be present in the follicles and in the interfollicular zones

BCL-2 negative. The tumour cells of mantle cell lymphoma show CD5 positivity and express nuclear cyclin D1, which is almost pathognomonic.

Nodal marginal zone lymphomas can be differentiated from follicular lymphomas on the basis that the follicles are reactive and have the morphological and immunohistochemical characteristics of reactive follicles. If they become totally colonized by tumour cells, in contrast to follicle centre cells, these will be CD10 and BCL-6 negative.

PAEDIATRIC-TYPE FOLLICULAR LYMPHOMA

Paediatric-type follicular lymphoma (PFL) is a rare variant of follicular lymphoma that generally occurs in children and adolescents but may be seen in adults and shows a marked male predominance. These lymphomas generally present with localized disease with a predilection for cervical nodes, Waldeyer's ring or testis. PFLs arising in lymph nodes or testis are indolent in contrast to those involving Waldeyer's ring, which behave in a more aggressive fashion. The nodal architecture is effaced by ill-defined, expanded, irregular follicles with attenuated mantle zones. The follicles show a 'starry sky' pattern and are composed of a monomorphic population of small to medium-sized blastoid cells with round to oval nuclei, fine chromatin and small nucleoli. Typical centroblasts and centrocytes are infrequent. All cases of PFL would be considered grade 3 based on the proportions of centroblasts and blastoid cells (Liu *et al.* 2013, Louissaint *et al.* 2012).

IMMUNOHISTOCHEMISTRY

The tumour cells express B-cell–associated antigens CD20 and CD79a. PFL shows strong expression of CD10 and BCL-6 and has a very high Ki67 proliferation fraction. However, in contrast to usual type follicular lymphomas, BCL-2 protein is usually negative in nodal PFL. PFLs involving Waldeyer's ring may lack CD10 expression and frequently show BCL-2 expression; all express MUM1.

GENETICS

PFLs have clonal immunoglobulin gene rearrangements but do not show the t(14;18)(q32;q21) translocation or rearrangements of the BCL-6 gene. A proportion of cases show IgH breaks by FISH analysis but specific translocations have not been identified. PFL involving Waldeyer's ring frequently has IRF4 breaks and may show an IGH-IRF fusion (Liu *et al.* 2013).

DIFFERENTIAL DIAGNOSIS

As for follicular lymphoma the main differential diagnosis of PFL is with reactive follicular hyperplasia. Because of the similar morphological features and absence of BCL-2 expression the diagnosis may be challenging. The presence of monotypic light chain expression or demonstration of clonal immunoglobulin rearrangements is essential for the diagnosis of this condition (Oschlies *et al.* 2010, Liu *et al.* 2013).

MANTLE CELL LYMPHOMA

Mantle cell lymphoma accounts for 3–10 per cent of non-Hodgkin lymphomas. It has a median age of 60 years and a 2:1 male predominance (Banks *et al.* 1992). It frequently involves extranodal tissues, particularly the oropharynx and gastrointestinal tract, where it may present as lymphomatous polyposis. Peripheral lymphadenopathy is, however, the commonest presentation. High-stage disease with infiltration of the spleen and involvement of the bone marrow is common. Tumour cells may be found in the peripheral blood of such patients. Orchard *et al.* (2003) studied 80 patients with mantle cell lymphoma, of whom 37 had blood and bone marrow involvement with no evidence of nodal disease. Six of the latter group were long-term survivors. Five of this group had mutated immunoglobulin genes. The average survival with mantle cell lymphoma is 3–5 years.

Mantle cell lymphomas frequently show a nodular growth pattern. This is because of their propensity to surround and replace pre-existing reactive germinal centres. Partially replaced germinal centres may be seen at the centre of tumour nodules. Their presence may be highlighted by immunohistochemistry, being BCL-2– and CD10+ in contrast with the tumour cells, and showing a high proliferation index in contrast with the usual low proliferation index of the tumour cells.

The tumour cells of mantle cell lymphoma typically have oval or angulated nuclei, which frequently exhibit small clefts. The nuclear chromatin is finely granular and nucleoli are small and insignificant. In some cases, owing to intrinsic features of the tumour cells or to poor fixation, the tumour cells have rounded nuclei and distinction from SLL may be difficult. The cytoplasm of the tumour cells is usually scanty and poorly defined. It is pale pink in H&E-stained sections and is weakly basophilic in Giemsa preparations.

The blood vessels in mantle cell lymphoma are often surrounded by hyaline basement membrane material. Reticulin stains highlight this material and may show spiky reticulin projections amongst the tumour cells. Another characteristic finding is the presence of histiocytes with large oval nuclei and abundant strongly eosinophilic cytoplasm.

There is a blastoid variant of mantle cell lymphoma in which the tumour cells have more open nuclear chromatin. These tumours may closely mimic lymphoblastic lymphomas. Rare pleomorphic (sarcomatoid) tumours occur. Proliferation indices, as measured by mitotic figures or immunohistochemistry, are variable in mantle cell lymphoma. The mitotic or Ki67 index is an important prognostic indicator and is high in the blastoid and sarcomatoid forms of the tumour (Boxes 5.13 and 5.14 and Figures 5.50 to 5.56).

IMMUNOHISTOCHEMISTRY

The tumour cells express surface immunoglobulin, usually IgM together with IgD. Cytoplasmic immunoglobulin and plasma cell differentiation are not seen. The B-cell antigens CD20 and CD79a are expressed, and CD5 is usually positive, staining less strongly than the reactive T-cells. Immunohistochemical detection of cyclin D1 in the nuclei of tumour cells provides an almost specific marker for the identification of mantle cell lymphoma. It must be stressed

BOX 5.13: Mantle cell lymphoma: Clinical features

- 3–10 per cent of all non-Hodgkin lymphomas
- Median age 60 years
- Male predominance, 2:1
- Usually present with stage III or IV disease
- Sites of involvement:
  - Lymph nodes
  - Bone marrow
  - Peripheral blood
  - Spleen
  - Waldeyer's ring
  - Gastrointestinal tract (lymphomatous polyposis)

BOX 5.14: Mantle cell lymphoma: Morphology

- Loss of normal lymph node architecture
- Tumour cells surround and infiltrate reactive germinal centres giving:
  - Mantle zone growth pattern
  - Nodular growth pattern
  - Diffuse growth pattern
- Small to medium-sized cells with angulated nuclei; may be more rounded in some cases
- Nucleoli inconspicuous
- No paraimmunoblasts or neoplastic centroblasts
- Do not show plasma cell differentiation
- Blood vessels surrounded by hyaline basement membrane
- Characteristic histiocytes in some cases
- Blastoid variant (uncommon)

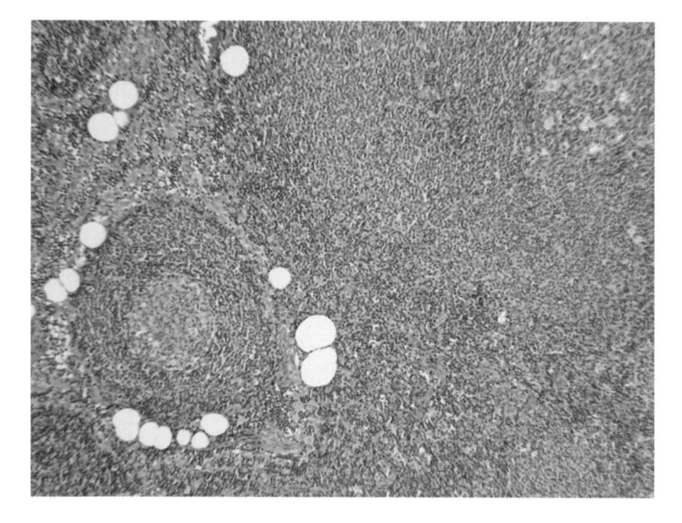

Figure 5.50 Mantle cell lymphoma showing a reactive follicle surrounded by tumour cells (lower left); a second follicle engulfed and infiltrated by tumour is visible at upper right.

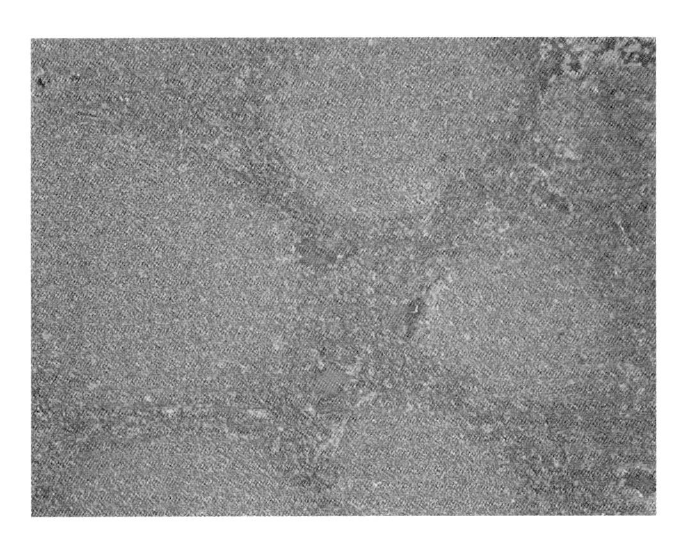

Figure 5.51 Mantle cell lymphoma showing a nodular growth pattern.

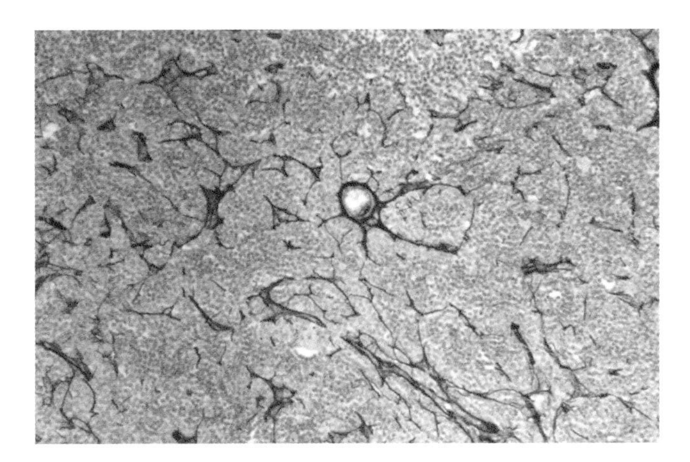

Figure 5.52 Mantle cell lymphoma stained for reticulin. Note condensation of reticulin around blood vessel with short radiating spikes.

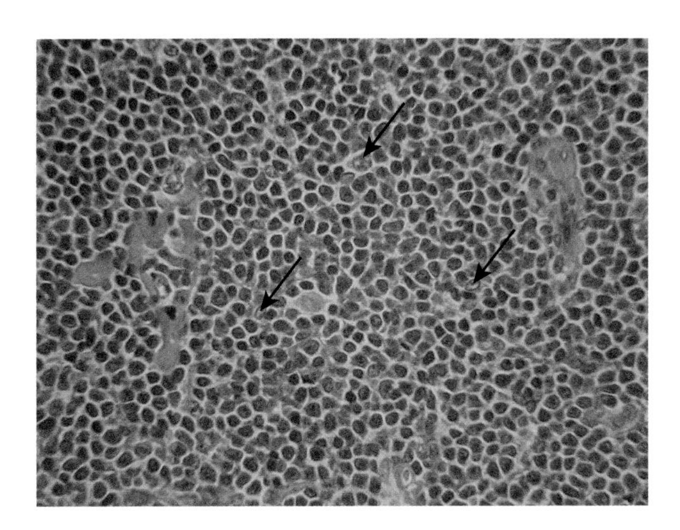

Figure 5.53 Mantle cell lymphoma showing characteristic perivascular hyaline material. Arrows indicate the nuclei of dispersed follicular dendritic cells.

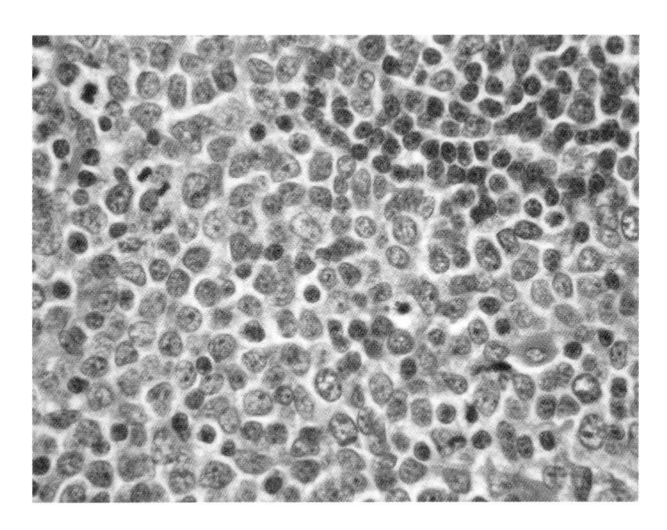

Figure 5.54 Blastic variant of mantle cell lymphoma. The tumour cells contrast in size with those of the usual type, visible at upper right. The nuclear size, fine chromatin and visible (but small) nucleoli mimic the appearance of lymphoblastic lymphoma.

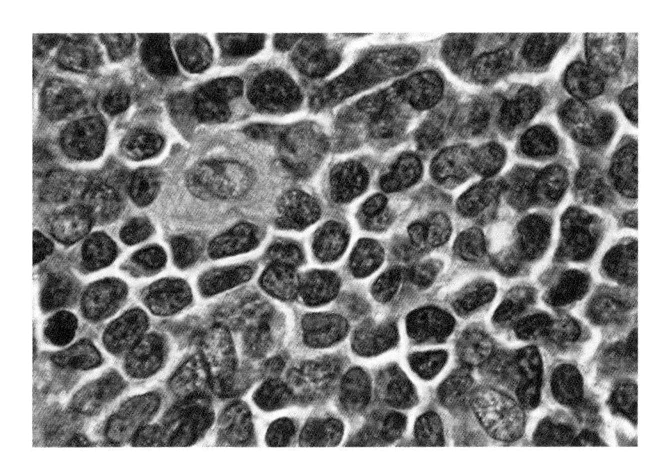

Figure 5.55 Mantle cell lymphoma showing characteristic cytomorphology. Note the characteristic histiocyte.

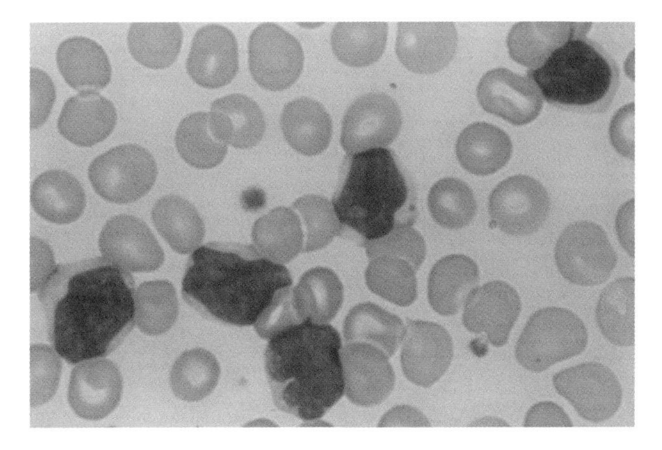

Figure 5.56 Blood film from a patient with mantle cell lymphoma. The circulating tumour cells show characteristic cleft nuclei with visible nucleoli and pale staining cytoplasm.

that the expression of cyclin D1 must be in the nucleus, not the cytoplasm or nucleus and cytoplasm, which indicate artefactual staining. Endothelial cells provide a positive internal control for cyclin D1 staining. Rare mantle cell lymphomas do not express cyclin D1 but stain strongly for cyclin D2 or D3 (Gesk *et al.* 2006). SOX11 is expressed in the nuclei of the majority of cases of mantle cell lymphoma and not by other small B-cell lymphomas; this is a useful diagnostic aid in cases of cyclin D1–negative mantle cell lymphoma. Lack of SOX11 expression in cyclin D1–positive mantle cell lymphoma is associated with a more indolent clinical course (Lu *et al.* 2013). Staining for CD21 or CD23 usually shows an expanded and dispersed follicular dendritic cell network within the tumour (Box 5.15 and Figures 5.57 to 5.64).

GENETICS

Mantle cell lymphomas are characterized by the translocation t(11;14)(q13;q32), which brings the cyclin D1 gene on chromosome 11 under promoter influence of the immunoglobulin heavy chain gene on chromosome 14. Cyclin D1, together with cyclin-dependent kinase (CDK4), phosphorylates retinoblastoma protein (pRB), releasing transcription factors that allow transition from the G1 to the S phase of

BOX 5.15: Mantle cell lymphoma: Immunohistochemistry

- Surface immunoglobulin M (IgM)+
- Surface IgD (usually)+
- CD5+, CD20+, CD79a+
- Nuclear cyclin D1 (BCL-1)+
- Nuclear SOX11 (usually)+
- CD10−
- BCL-6−
- CD21 and CD23 show expanded dendritic cell network

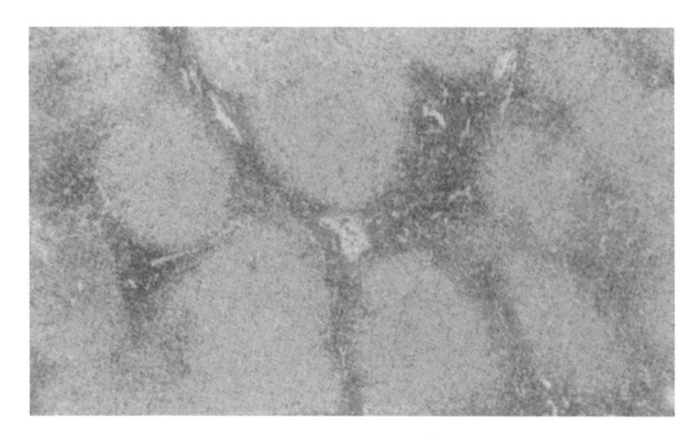

Figure 5.57 Nodular growth pattern of a mantle cell lymphoma highlighted by CD3+ T-cells.

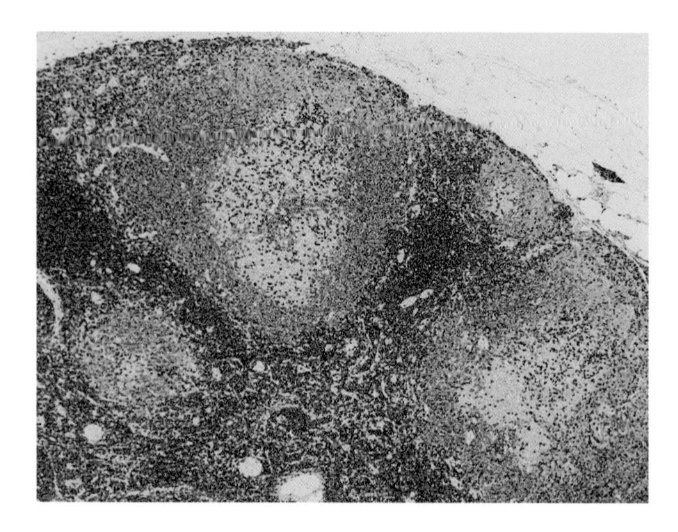

Figure 5.58 CD5+ mantle cell lymphoma. The dark cells are reactive T-lymphocytes. The unstained islands are residual reactive germinal centres surrounded by lymphoma.

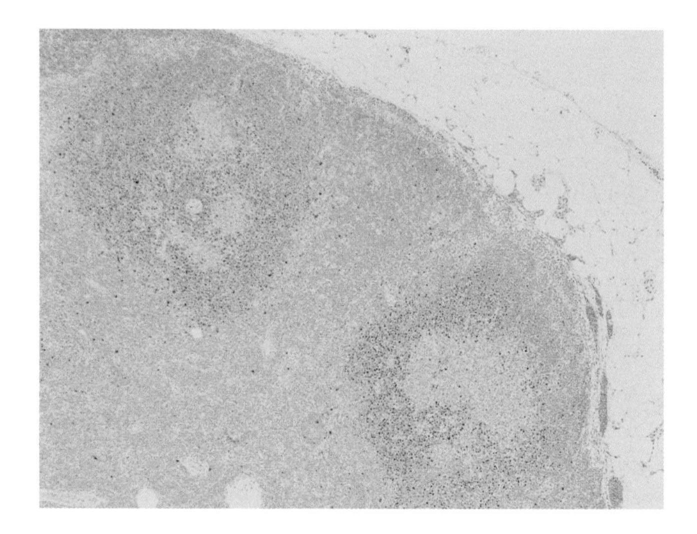

Figure 5.59 Mantle cell lymphoma stained for cyclin D1, same field as in Figure 5.58. Note that the staining is nuclear and varies in intensity.

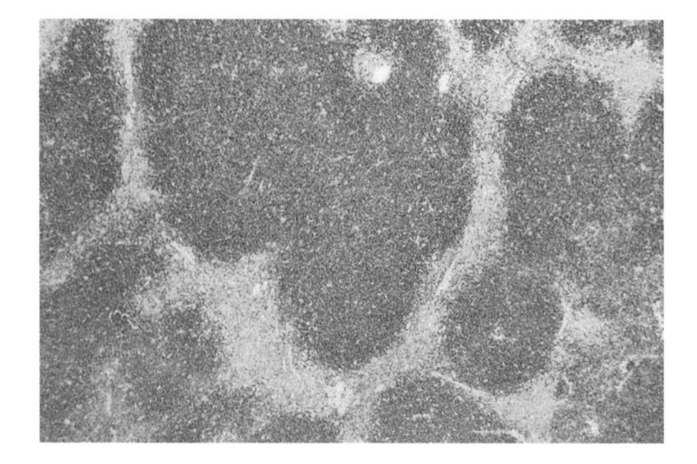

Figure 5.60 Mantle cell lymphoma stained for CD20 showing a nodular growth pattern.

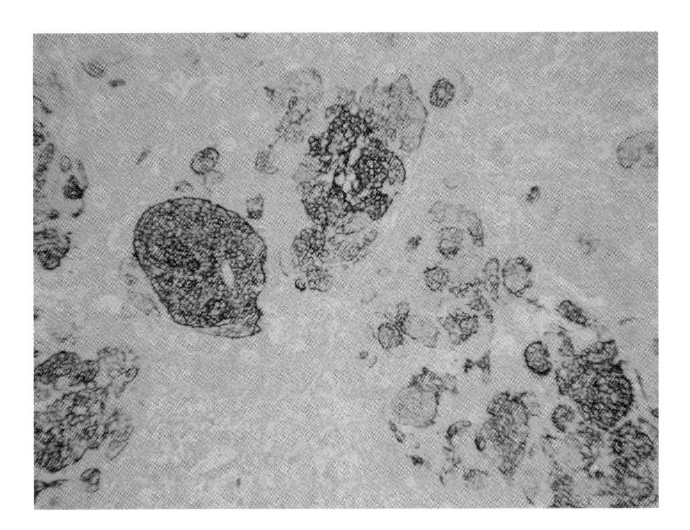

Figure 5.61 Mantle cell lymphoma stained for CD23 showing the characteristic expansion and dispersal of follicular dendritic cells.

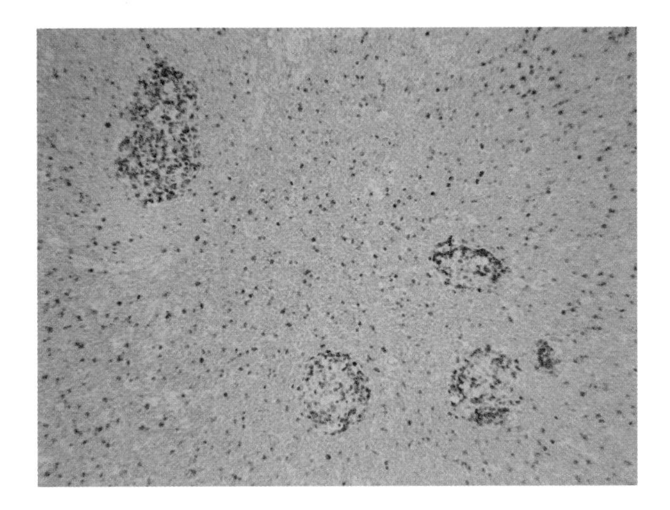

Figure 5.62 Mantle cell lymphoma stained for Ki67, showing a low proliferation fraction. The areas of intense labelling are residual reactive germinal centres.

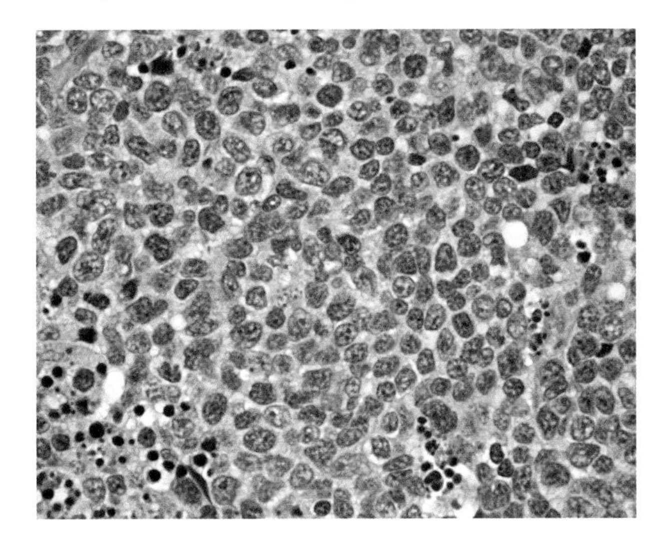

Figure 5.63 Blastoid mantle cell lymphoma showing numerous apoptotic bodies.

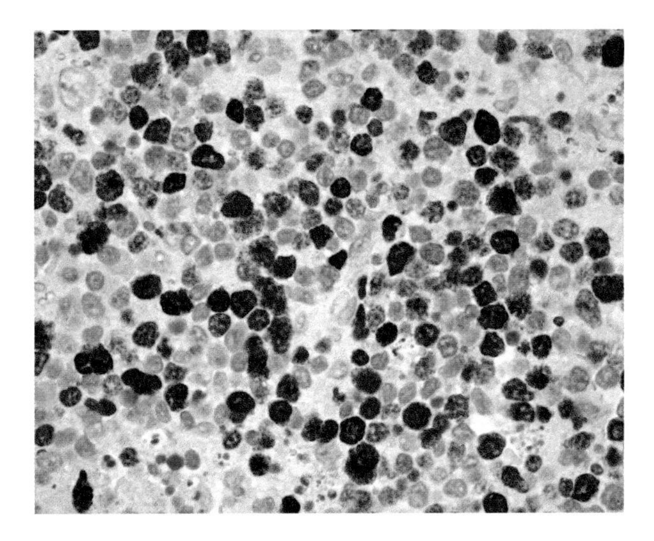

Figure 5.64 Same case as in Figure 5.63 stained for Ki67. A high proportion of the tumour cells can be seen to be in cycle.

the cell cycle. In the normal cell, p16/CDKN2 suppresses the cyclin D1/CDK4 phosphorylation of pRB. Deletions of the gene for p16/CDKN2 are associated with progression of mantle cell lymphoma and are positively correlated with proliferative activity of the tumour. The tumour is also associated with a large number of non-random secondary chromosomal aberrations. Mantle cell lymphomas may show mutated or unmutated immunoglobulin genes.

DIFFERENTIAL DIAGNOSIS

The differentiation between mantle cell lymphoma and B-lymphocytic lymphoma can be difficult. This is an important differentiation because mantle cell lymphomas have a poor prognosis in comparison with all other small B-cell lymphomas. Morphologically, the identification of paraimmunoblasts, either singly or in proliferation centres, identifies lymphocytic lymphoma. Proliferation centres must be differentiated from the remains of reactive germinal centres that may be found in mantle cell lymphomas. Conclusive separation may be obtained by immunohistochemistry, with CD23 positivity identifying lymphocytic lymphoma and cyclin D1 and/or SOX11 reactivity identifying mantle cell lymphoma.

Mantle cell lymphomas that have a nodular growth pattern may be mistaken for follicular lymphomas, especially if they have residual reactive centroblasts within the nodules. Mantle cell lymphoma cells (CD5+, cyclin D1+) can be distinguished from follicle centre cells (CD10+, BCL-6+) using immunohistochemistry.

The distinction between the blastic variant of mantle cell lymphoma and lymphoblastic lymphoma on morphological criteria can be difficult. Immunophenotypically, the distinction is clear-cut with lymphoblastic lymphomas displaying an immature T- or B-cell phenotype and usually expressing terminal deoxynucleotidyl transferase (TdT) but not cyclin D1.

DIFFUSE LARGE B-CELL LYMPHOMA

DLBCL encompasses all B-cell lymphomas with a diffuse growth pattern and cells more than twice the size of small lymphocytes. It accounts for approximately 40 per cent of adult B-cell non-Hodgkin lymphomas and has a wide age range with a median in the seventh decade. Involvement of extranodal sites occurs in 40 per cent of cases, frequently as primary tumours. DLBCL usually appear to arise de novo but may follow transformation of a pre-existing low-grade B-cell lymphoma.

DLBCL is not a homogeneous group of tumours. Apart from the fact that some appear to be primary and others evolve from pre-existing low-grade B-cell lymphomas, they can be divided on the basis of morphology, molecular characteristics and immunophenotype. The WHO classification recognizes four subtypes of DLBCL and eight other lymphomas of large B-cells. The classification also includes two borderline categories between DLBCL and Burkitt lymphoma and DLBCL and Hodgkin lymphoma (Table 5.3, Boxes 5.16 and 5.17, and Figures 5.65 to 5.74).

CENTROBLASTIC VARIANT

Centroblasts have rounded nuclei, open nuclear chromatin and two to five nucleoli that often appear to be attached to the nuclear membrane. They have basophilic cytoplasm. Lennert and co-workers (Hui *et al.* 1988) recognized four morphological variants of centroblastic lymphoma cells: monomorphic, multilobated, centrocytoid and polymorphic. Although the centrocytoid variant described has subsequently been interpreted as the blastic variant of mantle cell lymphoma, it may be worth retaining the term 'centrocytoid' for the large and small serpiginous cells often seen in follicular lymphomas and DLBCL. This morphological variation is not known to have any biological or clinical significance, although primary large B-cell lymphomas of bone are reported to be composed predominantly of the multilobated subtype. Histopathologists should, however, be familiar with these subtypes, since their recognition can aid the histological diagnosis of centroblastic lymphoma and its separation from other neoplasms.

Monomorphic centroblastic lymphoma is composed of uniform cells as described earlier. Multilobated centroblasts exhibit lobated nuclei, fine nuclear chromatin and

Table 5.3 Diffuse large B-cell lymphoma (DLBCL): variants, subgroups and subtypes

DLBCL, not otherwise specified (NOS)	Common morphological variants: Centroblastic Immunoblastic Anaplastic
	Molecular subgroups: Germinal centre B-cell like (GCB) Activated B-cell like (ABC)
	Immunohistochemical subgroups: CD5-positive DLBCL Germinal centre B-cell-like (GCB) Activated B-cell like (ABC)
DLBCL subtypes	T-cell/histiocyte rich large B-cell lymphoma EBV-positive DLBCL of the elderly *Primary DLBCL of the CNS* *Primary cutaneous DLBCL, leg type*
Other lymphomas of large B-cells	Primary mediastinal (thymic) large B-cell lymphoma (PMBL) ALK-positive LBCL *Intravascular large B-cell lymphomas* *DLBCL associated with chronic inflammation* *Lymphomatoid granulomatosis* *Plasmablastic lymphoma* *Large B-cell lymphoma arising in HHV-8–associated multicentric Castleman disease* *Primary effusion lymphoma*
Borderline cases	B-cell lymphoma unclassifiable with features intermediate between DLBCL and Burkitt lymphoma B-cell lymphoma unclassifiable with features intermediate between DLBCL and classical Hodgkin lymphoma

ALK, anaplastic lymphoma kinase; HHV-8, human herpesvirus 8.
This table is based on the WHO classification Table 10.14 (Stein *et al.* 2008).
The distinction between 'DLBCL subtypes' and 'other lymphomas of large B-cells' is presumably because of the uncertainties concerning the histogenesis of the latter group.
Those entities shown in italics are uncommon or rarely involve lymph nodes and are briefly discussed at the end of this section.

BOX 5.16: Diffuse large B-cell lymphoma: Clinical features

- 30–40 per cent of adult non-Hodgkin lymphomas
- May occur at all ages; median age seventh decade
- Slight male predominance
- 60 per cent nodal; 40 per cent extranodal
- May be primary or may represent transformation of a low-grade B-cell lymphoma

BOX 5.17: Diffuse large B-cell lymphoma: Morphology

- Tumour cells at least twice the size of small lymphocytes
- Various morphological forms: immunoblasts, centroblasts (monomorphic, multilobated, centrocytoid, polymorphic) in varying proportions
- Partial or complete effacement of nodal architecture
- Sinusoidal pattern of infiltration uncommon
- Variable admixed T-cells and/or histiocytes
- Variable sclerosis

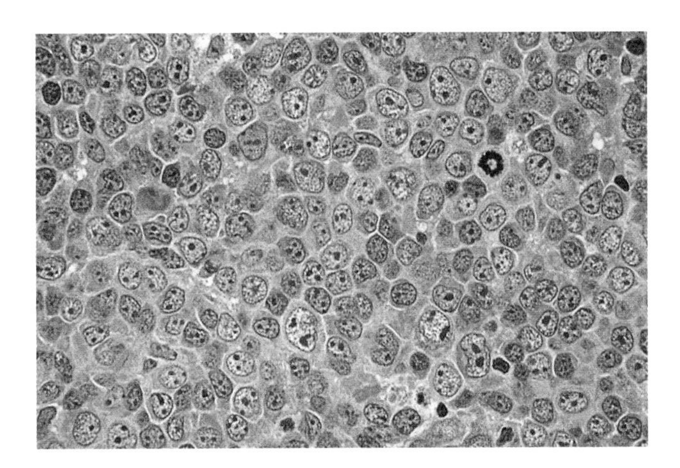

Figure 5.65 Diffuse large B-cell lymphoma (centroblastic monomorphic).

inconspicuous nucleoli. Centrocytoid centroblasts have elongated, sometimes serpiginous nuclei within which nucleoli are visible. The polymorphic subtype may contain all of these variants as well as immunoblasts and more pleomorphic, often multinucleated, cells.

IMMUNOBLASTIC VARIANT

Immunoblasts have round or oval nuclei with a prominent central nucleolus. Their cytoplasm is deeply basophilic. Tumours in which greater than 90 per cent of the cells have this morphology are categorized as immunoblastic variants.

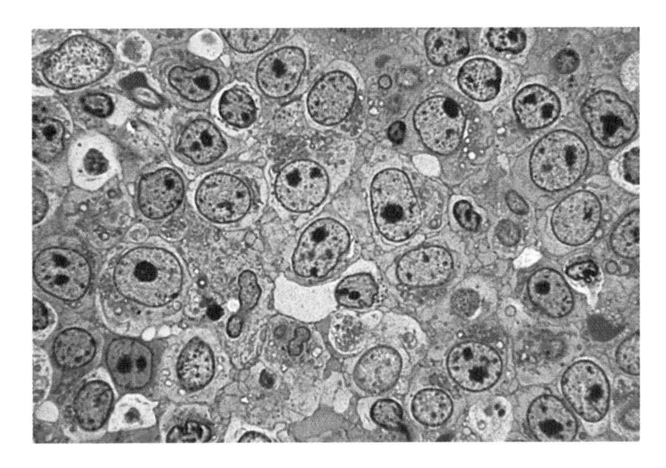

Figure 5.66 Diffuse large B-cell lymphoma (centroblastic monomorphic). Plastic-embedded section.

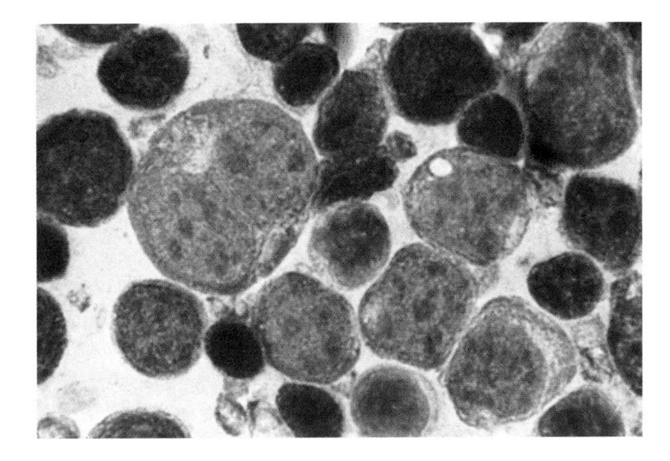

Figure 5.67 Diffuse large B-cell lymphoma (centroblastic). Imprint preparation showing moderate pleomorphism.

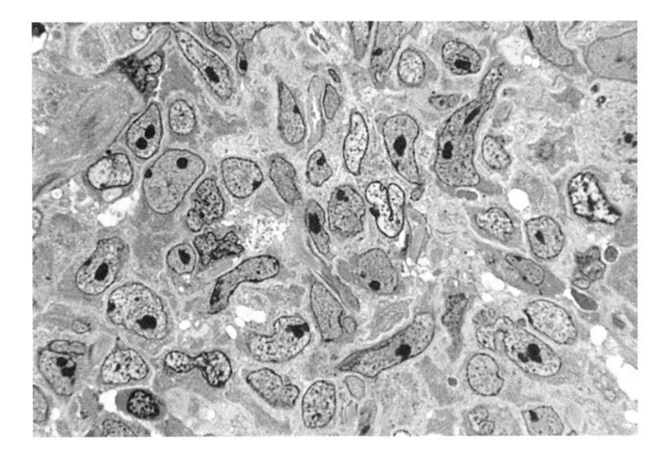

Figure 5.68 Diffuse large B-cell lymphoma (centroblastic centrocytoid). Plastic-embedded section showing large centrocytoid cells with prominent nucleoli.

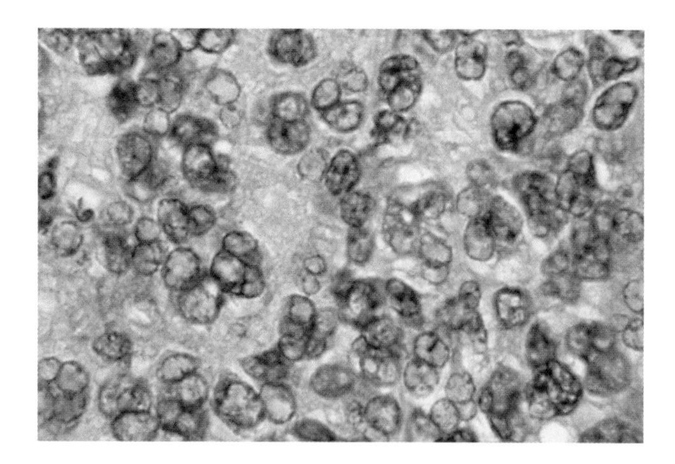

Figure 5.69 Diffuse large B-cell lymphoma (centroblastic multilobated).

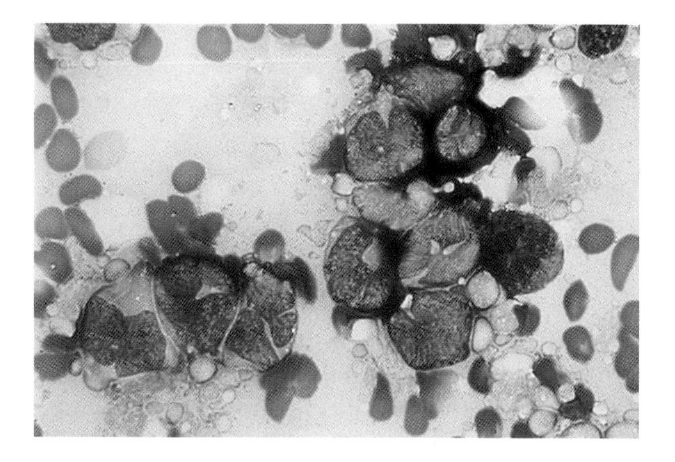

Figure 5.70 Diffuse large B-cell lymphoma (centroblastic multilobated). Imprint preparation.

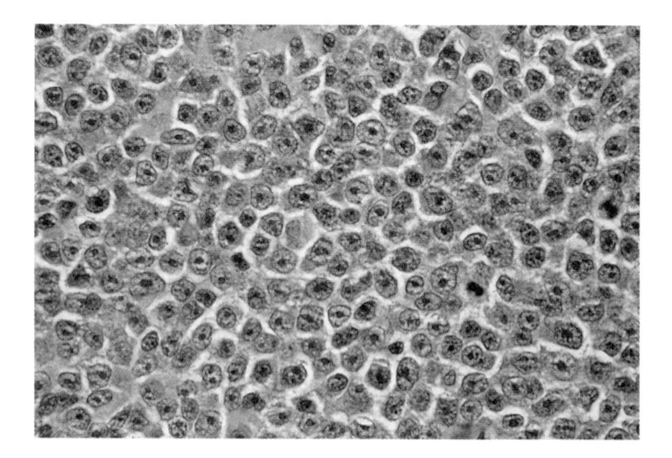

Figure 5.71 Diffuse large B-cell lymphoma (immunoblastic).

ANAPLASTIC VARIANT

This is an uncommon variant of DLBCL in which the tumour cells are large with abundant cytoplasm and atypical nuclei. They may have multinucleated nuclei and resemble RS cells or the pleomorphic cells of anaplastic large cell lymphoma.

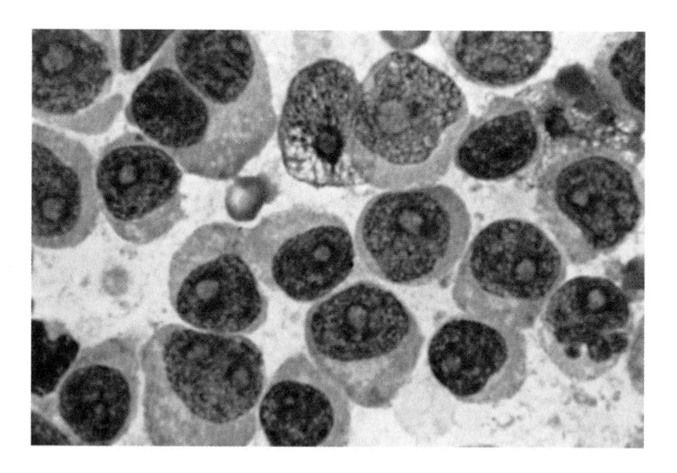

Figure 5.72 Diffuse large B-cell lymphoma (immunoblastic). Imprint preparation.

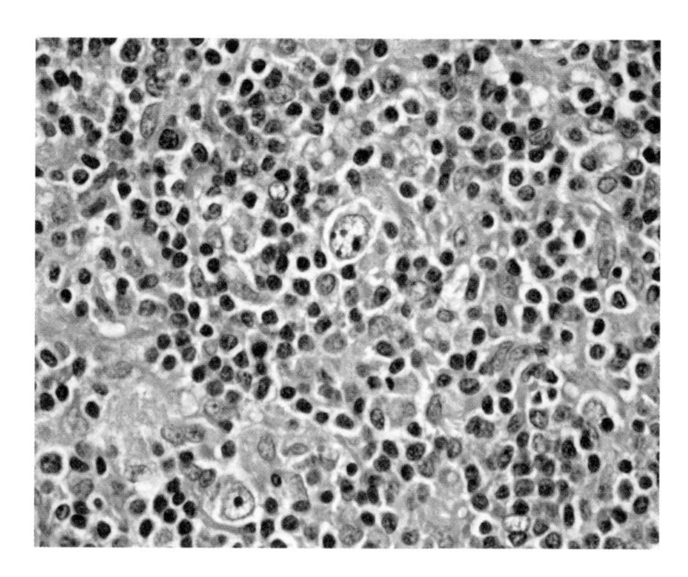

Figure 5.73 T-cell/histiocyte-rich B-cell lymphoma showing a tumour cell with centroblast morphology. The surrounding cells are predominantly T-cells and histiocytes.

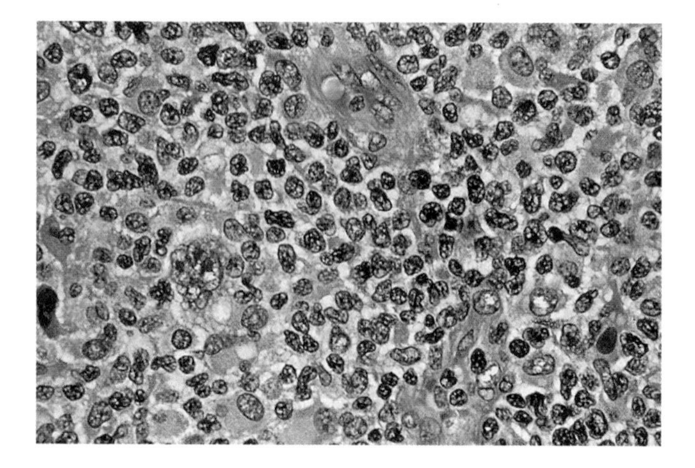

Figure 5.74 T-cell/histiocyte-rich diffuse large B-cell lymphoma with a tumour cell resembling a Reed–Sternberg cell.

These cells are often CD30+ but CD15–. The tumour cells may show sinusoidal permeation and resemble metastatic carcinoma. Differentiation from carcinoma and anaplastic large cell lymphoma of T-cell or null-cell type can be made by immunohistochemical staining (Haralambieva *et al.* 2000).

MOLECULAR SUBGROUPS

Gene expression profiling identifies two subgroups of DLBCL (Alizadeh *et al.* 2000). One subgroup has the gene expression profile of germinal centre B-cells (GCB group) and one the profile of activated B-cells (ABC group). The two groups are associated with different chromosomal aberrations and have different mechanisms for blocking tumour cell apoptosis (Davis *et al.* 2010). With CHOP-like chemotherapy the 5-year overall survival rate for the GCB group was 60 per cent and for the ABC group 30 per cent (Wright *et al.* 2003).

The addition of rituximab to CHOP (R-CHOP) has improved the survival of patients with DLBCL. Lenz *et al.* (2008) studied the stromal gene signatures in DLBCL and correlated these with survival in patients treated with CHOP or R-CHOP. They identified two stromal gene signatures that were prognostically significant in both groups of patients. The prognostically favourable signature reflected extracellular matrix deposition and histiocyte infiltration. The prognostically unfavourable signature reflected blood vessel density in the tumours.

IMMUNOHISTOCHEMISTRY

The cells of DLBCL express CD45 and B-cell lineage markers CD19, CD20, CD22 and CD79a. Most tumours express surface and/or cytoplasmic immunoglobulin, although this will be difficult to interpret in poorly fixed specimens. Immunoglobulin inclusions may be seen. Scattered cells may express CD30; this is most commonly seen in the anaplastic variant. Approximately 10 per cent of DLBCLs express CD5, but this does not imply a relationship to B-CLL/SLL or mantle cell lymphoma. Cyclin D1 is negative in these cases (Box 5.18 and Figures 5.75 and 5.76).

BOX 5.18: Diffuse large B-cell lymphoma: Immunohistochemistry

- CD20+ and CD79a+
- Surface and/or cytoplasmic immunoglobulin IgM > IgG > IgA
- Approximately 10 per cent of cases are CD5+
- 25–50 per cent of cases are CD10+
- 30–50 per cent of cases are BCL-2+
- Many cases are BCL-6+
- CD30+ in anaplastic variant, occasional cells positive in many cases
- Ki67 labelling fraction 40–90 per cent

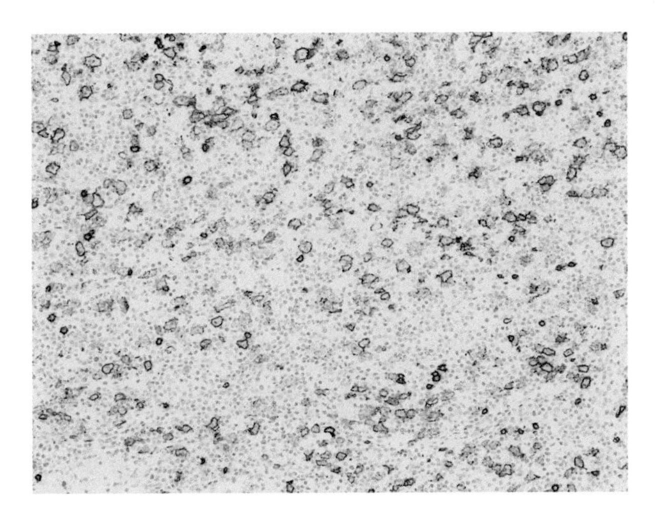

Figure 5.75 Low-power view of T-cell/histiocyte-rich diffuse large B-cell lymphoma stained for CD20 showing the sparse tumour cells.

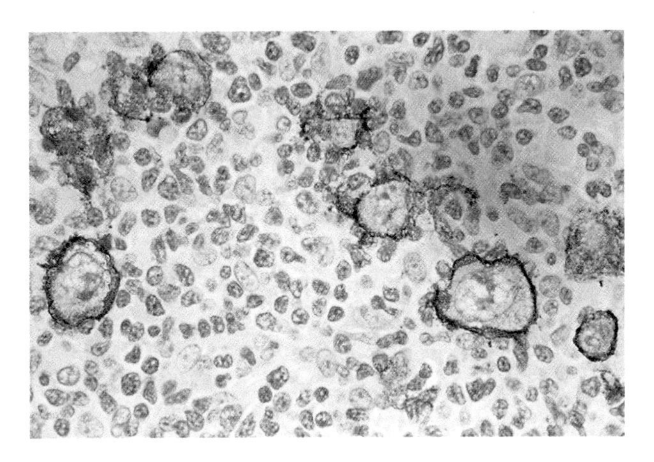

Figure 5.76 High-power view of T-cell/histiocyte-rich diffuse large B-cell lymphoma stained for CD20 showing strong membrane staining of the tumour cells.

Three immunochemical markers—CD10, BCL-6 and MUM1—may be used as surrogates for molecular analysis to determine whether the tumours are of the GCB or non-GCB (ABC) type (Hans *et al.* 2005); the algorithm is shown in Box 5.19.

The correlation between the GCB and ABC subtypes as determined by gene expression profiling and as determined by immunohistochemistry is good but not exact (Hwang *et al.* 2014).

GENETICS

Most DLBCLs have a post-germinal centre genotype with rearrangement of the immunoglobulin genes and mutated v-region genes. Approximately one quarter of cases show t(14;18), probably indicating evolution from follicle centre cell lymphoma. Another third show translocations and other abnormalities of 3q27 involving the BCL-6 gene. MYC rearrangements are found in up to 10 per cent of

BOX 5.19: Diffuse large B-cell lymphoma: Decision tree to determine GCB or non-GCB type

For a marker to be classified as positive 30% or more tumour cells must show expression (based on Hans *et al.* 2005).

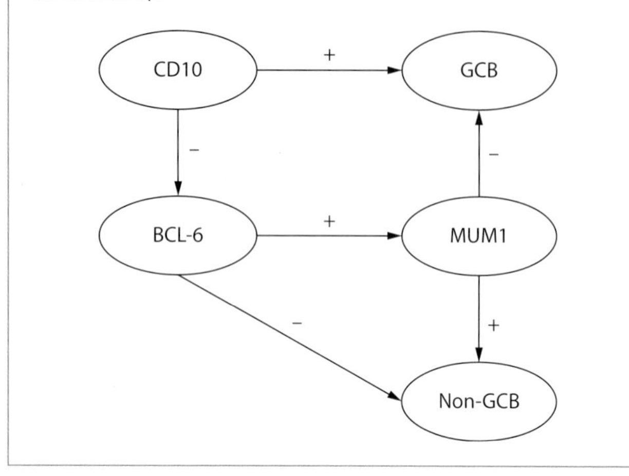

DLBCLs; these are the tumours that often have histological features of Burkitt lymphoma. Double-hit or triple-hit DLBCLs have an MYC rearrangement and either a BCL-2 or a BCL-6 rearrangement or both; these have an aggressive clinical course and poor prognosis (Aukema *et al.* 2011). Many cases show a complex genotype with multiple abnormalities.

T-CELL/HISTIOCYTE-RICH LARGE B-CELL LYMPHOMA

T-cell/histiocyte-rich large B-cell lymphoma (THRLBCL) is categorized as a morphological subtype of DLBCL in the WHO classification. The tumour shows a male predominance with a wide age range. Patients present with fever, malaise and lymphadenopathy. Splenomegaly, hepatomegaly and bone marrow involvement are common, most patients having high-stage disease. The disease usually follows an aggressive course (Achten *et al.* 2002).

The neoplastic large B-cells constitute 10 per cent or less of the tumour cell population. They occur singly and are surrounded by large numbers of small T-cells and histiocytes. After recurrence, the number of large B-cells may increase and these cells may form confluent aggregates. Eosinophils and plasma cells are not found in this lymphoma. The tumour cells usually have the appearance of centroblasts but less commonly have the appearance of LP cells of NLPHL or resemble H/RS cells of classical Hodgkin lymphoma (Lim *et al.* 2002).

The large B-cells express B-cell lineage markers; the expression of CD20 usually is particularly strong. They also express BCL-6 but not CD10. BCL-2 is negative and there is an absence of t(14;18). The T-cells are CD3+, CD5+, CD8+ and TIA-1+ but granzyme B–. CD68 immunostaining usually reveals larger numbers of histiocytes than would be expected from the H&E appearance. Lymphomas with THRLBCL morphology and EBV expression should be classified as EBV-positive DLBCL rather than THRLBCL (Menon *et al.* 2012).

This lymphoma should be distinguished from DLBCL containing large numbers of T-cells and histiocytes in which the large B-cells occur in aggregates or sheets. The main differential diagnosis of THRLBCL is with NLPHL. This is not a problem with the nodular, predominantly small B-cell variant of NLPHL but it can be with the variants that have a more diffuse growth pattern and are rich in T-cells (Fan *et al.* 2003). These two diseases, which have very different clinical and pathological characteristics, may show overlapping features, and both lymphomas may occur concurrently or subsequently in the same patient (Pittaluga and Jaffe 2010). A study of the two diseases using gene expression profiling found that NLPHL has a microenvironment similar to a lymph node showing follicular hyperplasia, whereas THRLBCL has a stromal signature that points towards a tumour tolerogenic immune response (Van Loo *et al.* 2010). Unfortunately the study did not include borderline cases. Features that histopathologists can use to differentiate between these two entities are shown in Table 5.4.

Table 5.4 Differential diagnosis between T-cell/histiocyte-rich large B-cell lymphoma (THRLBCL) and nodular lymphocyte-predominant Hodgkin lymphoma (NLPHL)

	THRLBCL	NLPHL
Growth pattern	Diffuse	Nodular or nodular and diffuse
Tumour cell morphology	Centroblasts, less commonly LP cells or Reed–Sternberg cells	LP cells
T-cell rosetting of tumour cells	No	Yes
CD30	Some+	Tumour cells–
Background reactive cells	Mostly CD8+ T-cells and histiocytes	Variable small B-cells. T-cells: CD3+, CD4+, CD57+, PD1-positive+
Nodular meshwork of dendritic cells	Absent	Present

'LP cells' is the term applied to the large neoplastic B-cells (popcorn cells) in NLPHL.

EPSTEIN–BARR VIRUS–POSITIVE DIFFUSE LARGE B-CELL LYMPHOMA OF THE ELDERLY

The identification of this subtype of DLBCL is based mainly on reports from Japan (Oyama *et al.* 2007) and Korea (Park *et al.* 2007), where it accounts for 8–10 per cent of all DLBCL. It is defined as EBV-positive clonal B-cell lymphoid proliferation that occurs in patients older than 50 years and without any known immunodeficiency or previous lymphoma. It is thought that this is an EBV-driven lymphoma related to reduced immunological surveillance (senescence of immunity) with increasing age. Defined entities that may be EBV positive, such as plasmablastic lymphoma, primary effusion lymphoma and DLBCL associated with chronic inflammation, are excluded from this category.

The median age of EBV-positive DLBCL is 71 years, with 20–25 per cent of patients being older than 90 years. Nodal disease alone occurs in 30 per cent of patients, 50 per cent have nodal and extranodal disease and 20 per cent have extranodal disease alone. The sites of the extranodal tumours differ in frequency from those seen in EBV-negative DLBCL (Oyama *et al.* 2007).

The International Prognostic Index (IPI) is high in over 50 per cent of patients and the overall prognosis is unfavourable compared with EBV-negative DLBCL. Age >70 years and B-symptoms show the strongest correlation with survival. Positivity for both of these carries a mean survival of 8.5 months, one factor alone 25.2 months and neither factor 56.3 months.

The tumours are often pleomorphic and may have a polymorphic background containing many small lymphocytes and histiocytes. Giant cells, which may resemble RS cells, are common. Areas of necrosis are frequently present. The tumour cells are usually CD20+ and/or CD79a+. Large cells and RS-like cells are often CD30+ but CD15–. These tumours more frequently have the activated B-cell (71 per cent) than the germinal centre (29 per cent) phenotype (Park *et al.* 2007). Epstein–Barr virus–encoded RNAs (EBERs) are positive in all the tumour cells and by molecular techniques can be shown to be clonal. Latent membrane protein 1 (LMP1) is expressed by 94 per cent and Epstein-Barr nuclear antigen 2 (EBNA2) by 28 per cent of the tumours (Box 5.20 and Figures 5.77 and 5.78).

BOX 5.20: Epstein–Barr virus (EBV)–positive diffuse large B-cell lymphoma (DLBCL) of the elderly

- In Asian countries accounts for 8–10 per cent of DLBCL
- 20–25 per cent of patients aged over 90 years (mean age 71 years)
- 30 per cent nodal, 50 per cent nodal and extranodal; 20 per cent extranodal only
- Extranodal sites most commonly skin, lung, stomach and tonsil
- High International Prognostic Index and poor survival compared with EBV-negative DLBCL
- May have pleomorphic or polymorphic histology or a mixture of both; RS-like cells may be present
- Tumour cells usually CD20+ and/or CD79a+; large cells and RS-like cells often CD30+ but CD15–
- All tumour cells express EBERs, LMP1+ in 94 per cent, EBNA2+ in 28 per cent

EBERs, Epstein–Barr virus–encoded RNAs; EBNA2, Epstein-Barr nuclear antigen 2; LMP1, latent membrane protein 1.

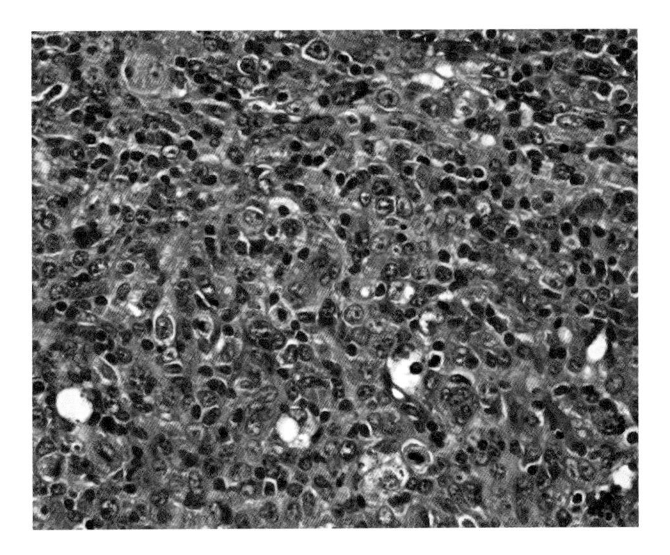

Figure 5.77 Diffuse large B-cell lymphoma in an elderly man. The tumour cells show moderate pleomorphism and there is a background of small lymphocytes and histiocytes.

PRIMARY MEDIASTINAL (THYMIC) LARGE B-CELL LYMPHOMA

Primary mediastinal large B-cell lymphoma (PMBL) is thought to arise from thymic B-cells and to be distinct from other DLBCLs. Its highest incidence is in young adults and it affects twice as many females as males. It usually presents with symptoms and signs of a mediastinal mass. Direct tumour spread may occur into other intrathoracic organs, and lymphatic spread into cervical and supraclavicular lymph nodes. Distant dissemination may involve the kidneys, adrenals, liver and central nervous system. Bone marrow involvement is uncommon. The 5-year survival of PMBL (64 per cent) is better than that of DLBCL (46 per cent).

The morphology of PMBL is distinctive with fine compartmentalizing sclerosis and medium to large tumour cells with clear cytoplasm. The tumour may show borderline features with classical Hodgkin lymphoma, and composite tumours of these two lymphomas rarely occur.

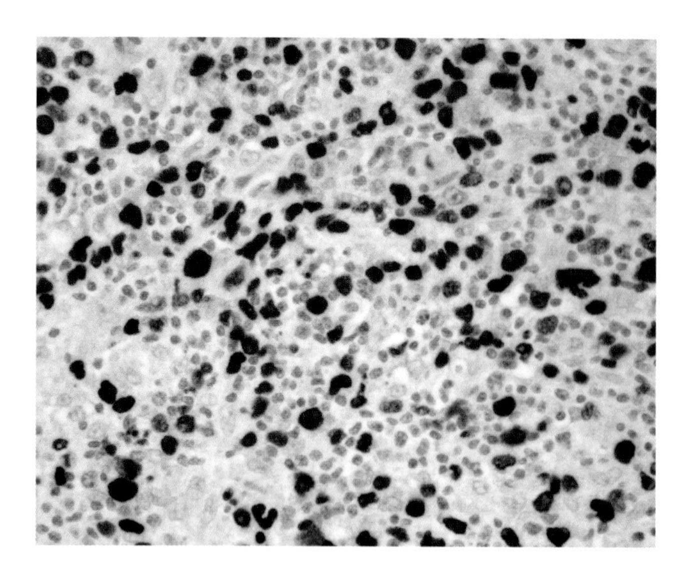

Figure 5.78 The same case as shown in Figure 5.77 stained to show Epstein–Barr virus–encoded RNAs (EBERs). All of the tumour cells appear to be positive, revealing the unstained background of small lymphocytes and histiocytes.

IMMUNOHISTOCHEMISTRY

The tumour cells express B-cell lineage markers but lack immunoglobulin synthesis. CD30 is expressed in 80 per cent of cases but it is usually weak. CD15 is sometimes present. Positive staining for MUM1 occurs in 75 per cent of cases and for CD23 in 70 per cent. Tumours are usually negative for CD5 and CD10. The tumour cells are positive for the MAL antigen.

GENETICS

Immunoglobulin genes are rearranged and show a high load of somatic mutations. Genetic abnormalities that characterize other types of DLBCL are absent. Abnormal expression of the REL gene and the MAL gene and gains of JAK2 have been reported in a large number of cases (Steidl and Gascoyne 2011). Recently translocations involving the CIITA gene, a major histocompatibility complex (MHC) class II transactivator, have been reported in 38% of cases of PMBL (Harris 2013). The gene expression profile of PMBL differs from that of other DLBCL and has features in common with classical Hodgkin lymphoma (Savage *et al.* 2003). Immunohistochemical detection of nuclear phosphorylated STAT6, a feature shared by PMBL and classical Hodgkin lymphoma, may aid the differentiation between PMBL and other subtypes of DLBCL (Guiter *et al.* 2004). PDL2, which codes a regulator of T-cell activation, discriminates PMBL from other DLBCL and is also expressed in classical Hodgkin lymphoma cell lines. PDL2 is encoded in the 9p chromosomal region, which shows gains and amplification in both PMBL and classical Hodgkin lymphoma (Rosenwald *et al.* 2003).

DIFFERENTIAL DIAGNOSIS

The differential diagnosis between PMBL and other types of DLBCL in a mediastinal biopsy may be difficult. The presence of thymic remnants, clear cells and/or sclerosis will favour PMBL. In the absence of these features, immunophenotypic and genetic characteristics may aid the differentiation, in particular detection of MAL protein or STAT6. The differentiation between PMBL and classical Hodgkin lymphoma is discussed in the section on borderline tumours between DLBCL and classical Hodgkin lymphoma.

ALK-POSITIVE LARGE B-CELL LYMPHOMA

ALK-positive large B-cell lymphoma is a rare variant of DLBCL first described by Delsol *et al.* (1997). The review of 38 patients by Laurent *et al.* (2009) included the majority of patients reported to date. It is possible, however, that the tumour is more common than this number would suggest since cases may be interpreted as immunoblastic or plasmablastic tumours without investigation of their ALK-1 status. The median age of the patients reported by Laurent *et al.* was 43 years. A few paediatric cases have been reported (Onciu *et al.* 2003). Involvement of lymph nodes with or without extralymphatic tumour occurs in most patients. Over half the patients evaluated have had stage III or IV disease. Most have been treated with CHOP or CHOP-derived regimens and have a poor prognosis in comparison with DLBCL not otherwise specified (NOS).

The majority of ALK-positive large B-cell lymphomas have been shown to have a t(2;17)(p23;q23) translocation involving the ALK gene on chromosome 2 band p23 and the clathrin gene on chromosome 17 band q23 (De Paepe *et al.* 2003, Gascoyne *et al.* 2003). The clathrin gene encodes a protein involved in receptor-mediated endocytosis in coated vesicles. The consequent disturbance of this protein may be responsible for the granular cytoplasmic staining for ALK-1 seen in these cases. Laurent *et al.* (2009) suggest that granular cytoplasmic staining for ALK-1 may be used as a surrogate marker for t(2;17). A few cases expressing the t(2;5) NPM–ALK fusion transcript have been reported. These cases show smooth cytoplasmic and nuclear staining for ALK-1. Additional fusion partner genes for ALK including SQSTM1, SEC31A and RANBP2 have recently been described. SQSTM1 results in a diffuse cytoplasmic pattern of ALK with ill-demarcated spots, and RANBP2 produces a membrane staining pattern (Lee *et al.* 2014).

Dot-like paranuclear staining for cytokeratin has been described in a small number of patients with ALK-positive large B-cell lymphoma. This feature, together with EMA positivity and in some cases CD45 negativity, might lead to a misdiagnosis of anaplastic carcinoma.

The poor prognosis of ALK-positive DLBCL treated with CHOP-based regimens highlights the need to identify these lymphomas accurately and to target therapy, including the use of ALK inhibitors (Li 2009, Wass *et al.* 2013). The sinusoidal growth pattern of this tumour together with the unusual histochemical profile should indicate the need for ALK-1 staining (Boxes 5.21 to 5.23 and Figures 5.79 to 5.84).

OTHER LARGE B-CELL LYMPHOMAS

In addition to the aforementioned categories of DLBCL, the WHO classification includes a number of subtypes that are not usually associated with lymphadenopathy.

PRIMARY LARGE B-CELL LYMPHOMA OF THE CENTRAL NERVOUS SYSTEM

This tumour rarely disseminates to extraneural sites and is not discussed further.

BOX 5.21: ALK-positive diffuse large B-cell lymphoma: Clinical features

- Age range 14–85 years (median age 43 years)
- Ratio of males to females 5:1
- Recorded at a number of extranodal sites but almost all have nodal disease
- 60 per cent of patients have advanced stage III or IV disease
- 25 per cent have bone marrow infiltration
- Aggressive clinical course with 25 per cent 5-year survival (worse for stage III or IV disease than stage I or II)

ALK, anaplastic lymphoma kinase.

BOX 5.22: ALK-positive diffuse large B-cell lymphoma: Morphology

- Tumour shows sinusoidal growth pattern in lymph nodes
- Tumour cells show immunoblastic/plasmablastic features with round nuclei, a prominent central nucleolus and abundant basophilic cytoplasm; multinucleated tumour cells may be seen
- Necrosis may be present

ALK, anaplastic lymphoma kinase.

BOX 5.23: ALK-positive diffuse large B-cell lymphoma: Immunohistochemistry

- Marker percentage of patients with positive result (based on Laurent *et al.* 2009):
  - ALK-1 — 100 per cent
  - CD45 — 82 per cent
  - CD30 — 5 per cent[a]
  - EMA — 100 per cent
  - CD20 — 9 per cent[a]
  - CD79a — 9 per cent[a]
  - CD138 — 10 per cent
  - IgA — 88 per cent
  - CD4 — 41 per cent
  - Cytokeratin — 14 per cent

ALK, anaplastic lymphoma kinase.
[a] Rare weakly positive cells.

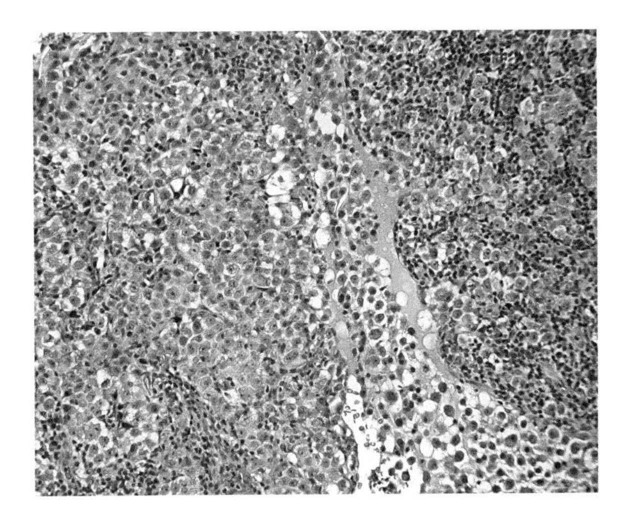

Figure 5.79 Lymph node containing ALK-positive large B-cell lymphoma showing characteristic sinus infiltration.

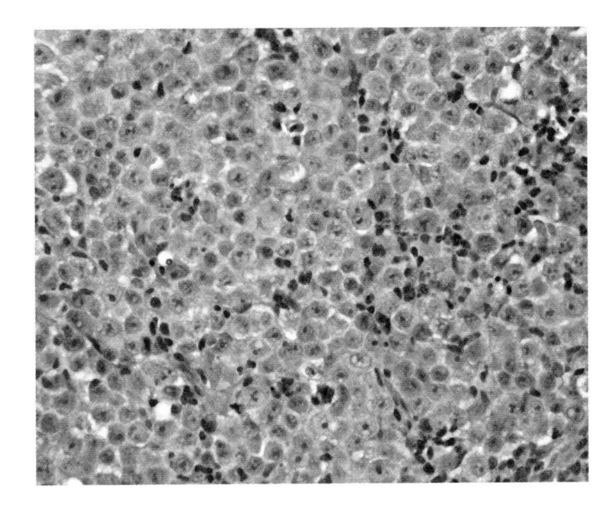

Figure 5.80 Higher-power view of ALK-positive large B-cell lymphoma showing the regular rounded immunoblastic cells with prominent single central nucleoli.

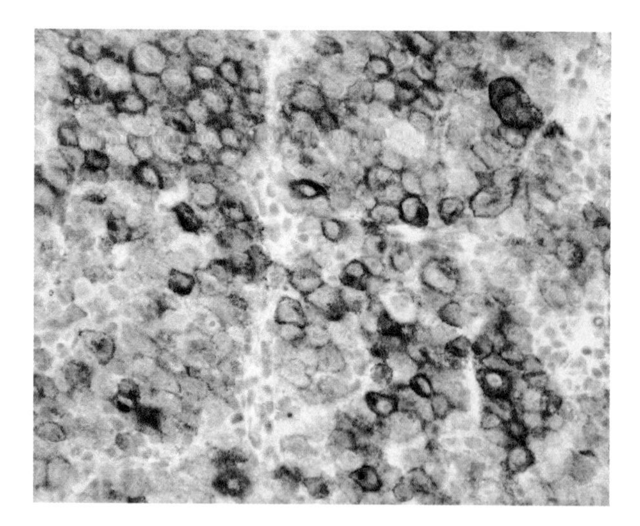

Figure 5.81 ALK-positive large B-cell lymphoma stained for CD10. The tumour cells show variable positivity. Note that this membrane antigen is showing a granular cytoplasmic distribution, possibly caused by altered clathrin function.

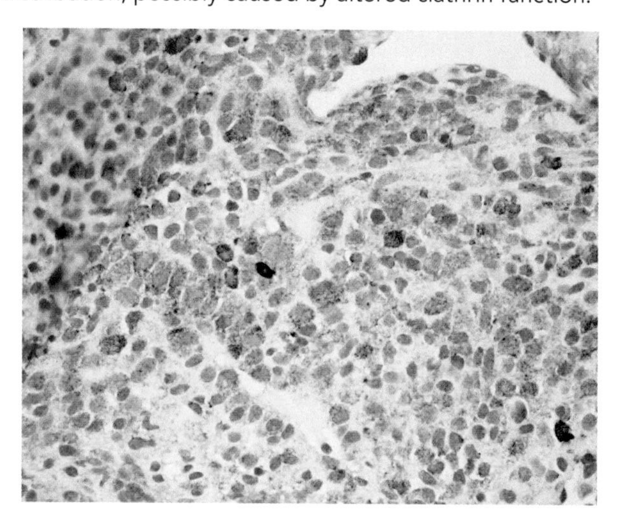

Figure 5.82 ALK-positive large B-cell lymphoma stained for CD79a. The tumour cells are focally positive and show granular cytoplasmic staining.

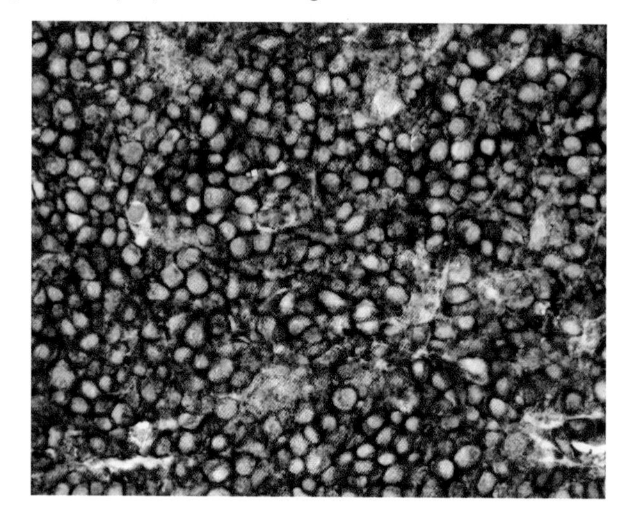

Figure 5.83 ALK-positive large B-cell lymphoma stained for CD138, showing strong positive staining of the tumour cells.

Figure 5.84 ALK-positive large B-cell lymphoma stained for ALK-1. Note the granular positivity characteristic of the ALK-1/clathrin gene translocation.

PRIMARY CUTANEOUS DLBCL, LEG TYPE

These tumours primarily involve the skin and are not discussed further.

DLBCL ASSOCIATED WITH CHRONIC INFLAMMATION

Pyothorax-associated DLBCL is the prototypic form of this lymphoma but it may occur at other sites of chronic inflammation, and lymph node involvement is rare (Gurbaxani *et al.* 2009). The tumours usually express CD20 and CD79a but may lose these markers and express the plasma cell–associated antigens CD138 and MUM1. The tumours usually express EBER, EBNA2 and LMP1 (latency type III). They are thought to be derived from EBV-infected germinal centre or post–germinal centre B-cells.

LYMPHOMATOID GRANULOMATOSIS

This entity has been moved from the immunodeficiency-associated lymphoproliferative disorders to the mature B-cell lymphoma section of the WHO classification (Pittaluga *et al.* 2008). The disease occurs in patients with overt immunodeficiency or underlying reduced immune function, and is EBV related. Most patients have pulmonary involvement. Brain, kidney, liver and skin are frequently involved but lymph nodes and spleen are rarely affected.

Lymphomatoid granulomatosis manifests as an angiocentric and angiodestructive polymorphous lymphoid/inflammatory infiltrate. It is graded from grade I, in which there is a predominantly small-cell lymphoid infiltrate, through to grade III in which there are many large, often pleomorphic B-cells. At this stage the tumour shows clonal immunoglobulin genes. The large B-cells are EBV positive. Patients with grade III disease should be regarded as having DLBCL.

INTRAVASCULAR LARGE B-CELL LYMPHOMA

This is a rare type of extranodal DLBCL that predominantly manifests as neurological or cutaneous disease. An Asian variant that presents with cytopenias, haemophagocytic syndrome and multiorgan failure has been described (Gurbaxani *et al.* 2009). The tumour cells are confined to the lumina of small and medium-sized vessels, probably because of a lack of homing receptors.

PLASMABLASTIC LYMPHOMA

This neoplasm occurs most frequently in HIV-infected individuals but may be seen in other immunodeficiency states including old age. It was originally described as an oral tumour but may involve other extranodal sites. Morphologically the tumours are composed of immunoblasts that may show varying degrees of plasma cell differentiation. The tumour cells usually show loss of CD45 and CD20, express the plasma cell–associated antigens CD38 and CD138 and are EBER positive. The clinical course is aggressive with poor survival.

LARGE B-CELL LYMPHOMA ARISING IN HUMAN HERPESVIRUS-8–ASSOCIATED MULTICENTRIC CASTLEMAN DISEASE

This rare lymphoma occurs in HIV-positive individuals who have developed human herpesvirus 8 (HHV-8)–positive multicentric Castleman disease (MCD) (see Chapter 3). This may occur in HIV-negative individuals from areas where HHV-8 is endemic. Although the tumour cells have the morphology of plasmablasts they have unmutated immunoglobulin genes and correspond to naive B-cells.

The tumour cells form small aggregates within the mantle zones around the involuted follicles of Castleman disease. These microlymphomas coalesce to form tumour cell sheets that efface the lymph node architecture. The tumour cells express HHV-8 nuclear antigen 1. The plasmablasts seen in MCD may show immunoglobulin monotypia but can be shown to be polyclonal by molecular studies. The tumour cells in HHV-8 plasmablastic lymphoma are IgM positive and are monoclonal. The prognosis of this disease is very poor.

PRIMARY EFFUSION LYMPHOMA

This large B-cell lymphoma usually presents as an effusion without solid tumour. It occurs in patients with immunodeficiency, usually acquired immune deficiency syndrome (AIDS). Half of the patients have, or develop, Kaposi sarcoma.

Immunophenotypically the tumour cells usually express CD45 but lack B-cell lineage markers. Non-lineage markers such as CD38, CD138 and EMA are often present. HHV-8 latent nuclear antigen 1 can be demonstrated in all cases and EBERs are expressed in most. The prognosis of this neoplasm is very poor.

BURKITT LYMPHOMA

Denis Burkitt, while working as a surgeon in Uganda, described a childhood tumour syndrome with characteristic clinico-anatomical features. This tumour appeared to be limited to the warm wet tropics of Africa and Papua New Guinea. A meeting of haematopathologists held in 1968 under the auspices of the WHO proposed that this tumour should be defined by its cytology and histology. Using this definition, it became apparent that Burkitt lymphoma was not confined to the wet tropics but occurred sporadically throughout the world. In later years, a proportion of high-grade B-cell lymphomas occurring in patients with AIDS were found to have the cytological and histological features of Burkitt lymphoma.

Burkitt lymphoma may therefor, be divided into three subtypes—endemic Burkitt lymphoma, sporadic Burkitt lymphoma and immunodeficiency-associated Burkitt lymphoma—with identical cytological and histological features but different clinical and gross anatomical characteristics. The common cytomorphology of these tumours is related to the presence in all three subtypes of translocations between the *c-myc* oncogene and immunoglobulin genes that result in unrestrained cell proliferation without differentiation. It must be emphasized that the term 'endemic Burkitt lymphoma' should be used for tumours with the characteristics described by Denis Burkitt and not for all tumours with the microscopic features of Burkitt lymphoma occurring in the wet tropics. HIV infection is prevalent in the wet tropics and a proportion of cases occurring in these regions will be AIDS associated, and rare cases may be of the sporadic type (Box 5.24 and Figures 5.85 to 5.90).

BOX 5.24: Burkitt lymphoma: Clinical features

ENDEMIC BURKITT LYMPHOMA

- Childhood predominance (median age 7 years)
- Male predominance 2:1
- Sites of involvement: jaws, gastrointestinal tract, kidneys, liver, pancreas, retroperitoneum, gonads, breast, endocrine organs, brain
- Leukaemic manifestations rare

SPORADIC BURKITT LYMPHOMA

- Children and young adults
- Male predominance 5:1
- Sites of involvement: ileocaecal region, abdomen and pelvis, Waldeyer's ring, peripheral lymph nodes
- Leukaemic manifestations uncommon

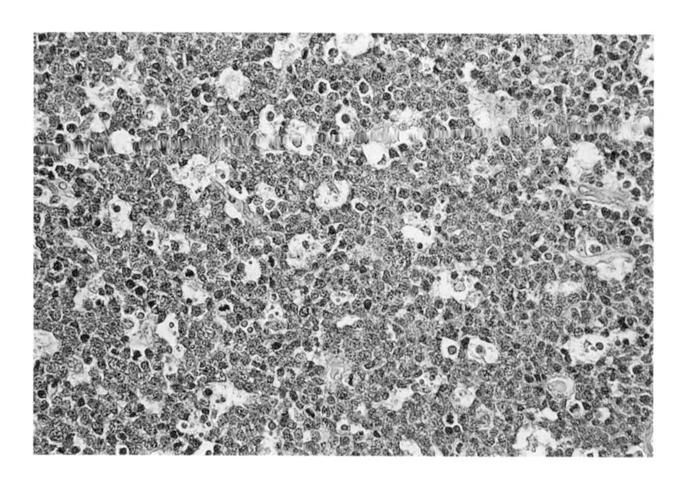

Figure 5.85 Burkitt lymphoma showing a 'starry-sky' pattern.

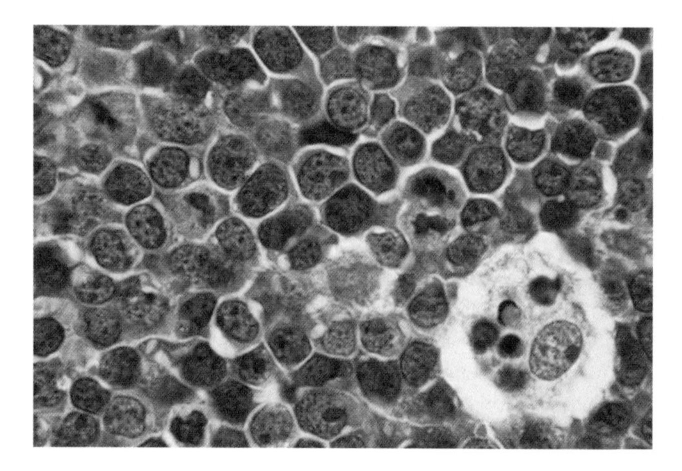

Figure 5.86 High-power view of Burkitt lymphoma showing characteristic morphology of the tumour cells and apoptotic debris in a 'starry-sky' macrophage.

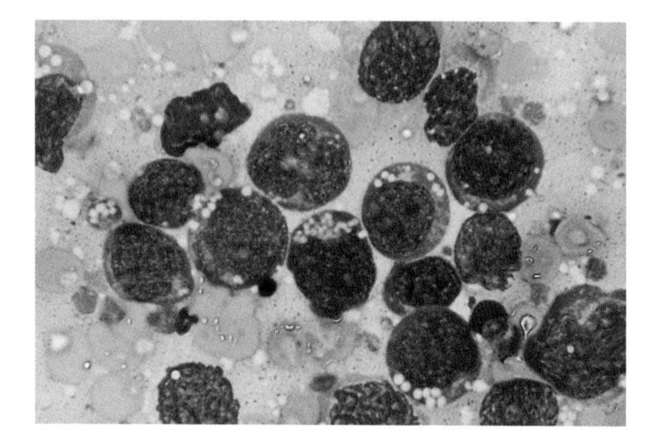

Figure 5.87 Endemic Burkitt lymphoma. Imprint showing characteristic tumour cells.

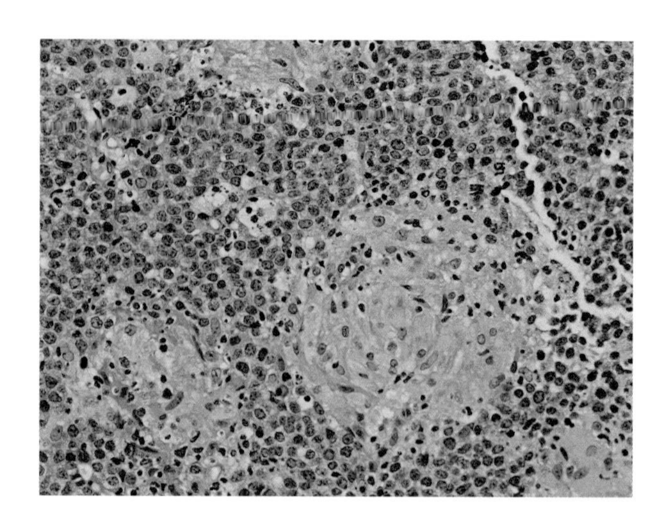

Figure 5.88 Sporadic Burkitt lymphoma with a florid granulomatous response.

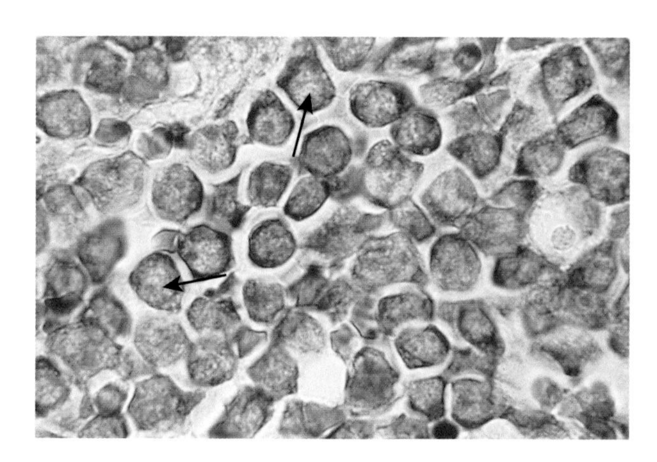

Figure 5.89 Burkitt lymphoma, stained with Giemsa, showing lipid vacuoles in deeply basophilic cytoplasm (arrows).

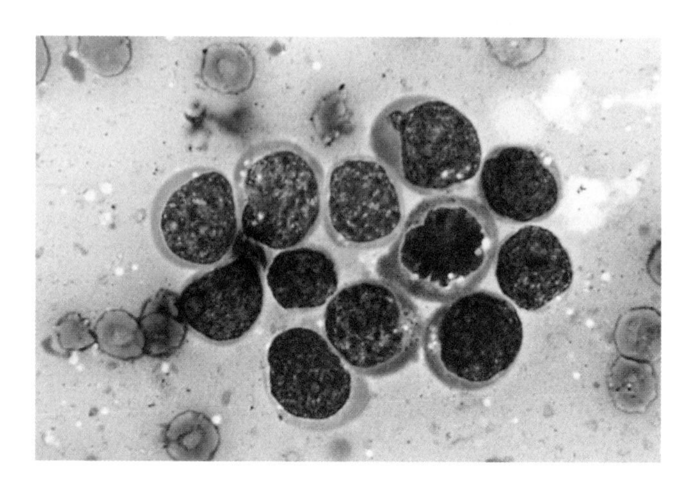

Figure 5.90 Sporadic Burkitt lymphoma. Imprint showing features similar to those of the endemic tumour.

TYPES OF BURKITT LYMPHOMA

Endemic Burkitt lymphoma

Endemic Burkitt lymphoma is rare before the age of 2 years and has a peak incidence at age 7 years; the incidence after the age of 15 years is low. It has a male to female sex incidence of approximately 2:1. The most characteristic and best-known feature of this tumour is the occurrence of one or, more commonly, multiple jaw tumours. This feature is age dependent and is almost certainly related to dental development and to the presence of cellular marrow in the jaws. All Burkitt lymphoma patients aged 3 years in Uganda were found to have jaw tumours, the proportion falling thereafter to 10 per cent at the age of 15 years. This accounts for the observation that the overall incidence of jaw tumours is in the region of 50 per cent, a proportion that will vary with the age of the patient population studied. Other common sites of involvement are brain (18 per cent), heart (32 per cent), stomach (26 per cent), small intestine (28 per cent), liver (37 per cent), pancreas (43 per cent), kidney (77 per cent), ovaries (82 per cent of females), testis (12 per cent of males), thyroid (37 per cent) and adrenals (58 per cent). Intra-abdominal lymph nodes, often within retroperitoneal masses, are frequently involved; peripheral lymphadenopathy is uncommon. Massive bilateral breast involvement may occur in females who are pregnant or lactating.

There is strong evidence that the aetiology of endemic Burkitt lymphoma is related to the interaction of holoendemic *Plasmodium falciparum* malaria with EBV (Thorley-Lawson and Duca 2007).

Sporadic Burkitt lymphoma

Burkitt lymphoma is the commonest non-Hodgkin lymphoma of children under the age of 15 years in the developed world. The tumour is also seen in young adults and, less commonly, in older age groups. It has a male predominance of 5:1 or more. It presents most commonly with intra-abdominal tumour, which often appears to have originated in the ileocaecal region. Involvement of the lymphoid tissue of Waldeyer's ring is also seen. These anatomical sites are rarely involved in endemic Burkitt lymphoma. Peripheral lymphadenopathy is seen more frequently than in endemic Burkitt lymphoma. Sporadic Burkitt lymphoma may also involve the kidneys, ovaries and breasts. Bone marrow involvement may be associated with leukaemic overspill into the peripheral blood. In rare sporadic cases in young children, the disease may closely resemble endemic Burkitt lymphoma in its distribution, with jaw tumours and multiple visceral involvement.

Immunodeficiency-associated Burkitt lymphoma

Burkitt lymphoma may present as nodal or extranodal disease in patients with AIDS and is often the initial manifestation of AIDS in these patients. It occurs in the early stages of the disease before CD4 counts fall and is often the presenting manifestation. It is an uncommon tumour in other forms of immunodeficiency.

MORPHOLOGY

Burkitt lymphoma is most easily diagnosed on imprint or other cytological preparations. Nuclei are rounded and may show clefts. The nuclear chromatin is finely granular and there are up to five visible, but not usually conspicuous, nucleoli. The cytoplasm is deeply basophilic with a paler area corresponding to the position of the Golgi body. Cytoplasmic vacuolation caused by neutral fat droplets is characteristic but not always present.

Histologically, the tumour cells have rounded and relatively uniform nuclei that approximate the size of the nuclei of starry-sky (tingible body) macrophages within the tumour. The nuclear chromatin appears granular. Two to five nucleoli are usually discernible, but are not usually prominent. The cytoplasm is best seen at the edge of the section where the cells tend to separate. It forms a well-defined rim around the nucleus, and appears amphophilic in H&E-stained sections and deeply basophilic in Giemsa-stained preparations. High-power microscopy will often reveal small vacuoles within this cytoplasmic rim. These fat droplets can be demonstrated in frozen sections of formalin-fixed tissue with neutral fat stains. They are characteristic but not specific for Burkitt lymphoma.

Burkitt lymphoma always shows a high mitotic index, and apoptotic nuclei and nuclear fragments are common. Many of these are engulfed within macrophages that give a characteristic starry-sky pattern to the tumour.

A small number of cases of sporadic Burkitt lymphoma showing a florid granulomatous reaction have been reported (Hollingsworth *et al.* 1993, Haralambieva *et al.* 2004, Schrager *et al.* 2005). The patients ranged in age from 4 to 52 years, all were EBER positive and none of those tested were HIV positive. It is suggested that a human leukocyte antigen (HLA)–restricted T-cell response to the EBNA1 antigen may be the cause of this reaction. The diagnostic importance of this feature is that the granulomatous reaction may overshadow the underlying lymphoma. All the reported cases have had a good clinical outcome (Box 5.25).

BOX 5.25: Burkitt lymphoma: Morphology

- Monomorphic small blast cells
- Defined basophilic (amphophilic) cytoplasm
- Cytoplasmic lipid droplets/vacuoles
- Proliferation index 100 per cent
- Numerous apoptotic bodies
- 'Starry-sky' macrophages

BOX 5.26: Burkitt lymphoma: Immunohistochemistry

- Membrane immunoglobulin M (IgM) with light chain restriction
- Do not express cytoplasmic immunoglobulin
- CD20+ and CD79a+
- CD10+
- BCL-6+
- BCL-2–
- TdT–

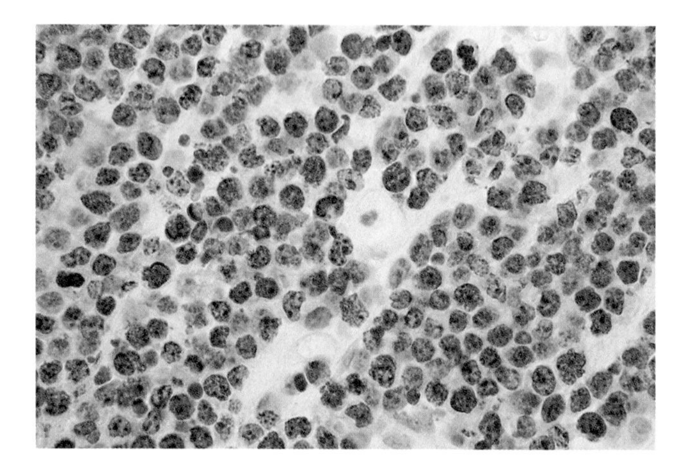

Figure 5.91 Burkitt lymphoma stained for Ki67, showing labelling of 100 per cent of the tumour cells.

IMMUNOHISTOCHEMISTRY

The tumour cells express the B-cell antigens CD20 and CD79a. They also express CD10 and BCL-6, consistent with germinal centre cell origin, but are BCL-2–. Staining for Ki67 is diagnostically helpful because all Burkitt lymphoma cells are in cycle as a consequence of *c-myc* deregulation; therefore the labelling index of the viable tumour cells will be 100 per cent. Burkitt leukaemia, previously categorized as a type of lymphoblastic leukaemia, has a mature B-cell phenotype and is TdT–, in contrast to precursor B-cell leukaemia/lymphoma (Box 5.26 and Figure 5.91).

GENETICS

All types of Burkitt lymphoma show translocations between the *c-myc* oncogene on chromosome 8 and one of the immunoglobulin genes: usually the heavy chain gene on chromosome 14, less commonly the κ light chain gene on chromosome 2 or the λ light chain gene on chromosome 22. There are subtle differences in the breakpoint region on chromosome 14 between endemic (joining region) and sporadic (switch region) Burkitt lymphoma.

EBV genomes can be demonstrated in all cases of endemic Burkitt lymphoma. In sporadic Burkitt lymphoma, the incidence shows geographical variation, being approximately 30 per cent in the developed world, but 70–80 per cent in North Africa and South America. EBV genomes are exhibited by 25–40 per cent of immunodeficiency-associated Burkitt lymphoma. Burkitt lymphoma shows latency pattern 1, expressing only EBNA1 and EBER.

DIFFERENTIAL DIAGNOSIS

The main differential diagnosis of Burkitt lymphoma is with DLBCL. This is discussed under the new WHO category 'B-cell lymphoma, unclassifiable, with features intermediate between DLBCL and Burkitt lymphoma'.

B-CELL LYMPHOMA, UNCLASSIFIABLE, WITH FEATURES INTERMEDIATE BETWEEN DLBCL AND BURKITT LYMPHOMA

This category comprises tumours that have morphological features that are intermediate between Burkitt lymphoma and DLBCL or have an immunophenotype that is not typical for Burkitt lymphoma (usually showing BCL-2 positivity). They have a high proliferation fraction and many apoptotic nuclei and may show a starry sky pattern. These tumours are often widespread at presentation, involving both nodal and extranodal sites. Bone marrow infiltration with leukaemic manifestations may occur. These tumours usually follow an aggressive course (Kluin *et al.* 2008).

GENETICS

Most tumours in this category show MYC translocations. In contrast to Burkitt lymphoma in which the translocation partner is an immunoglobulin gene, the translocation in the intermediate group often involves nonimmunoglobulin genes. In addition, in this group tumours often show multiple cytogenetic abnormalities including translocations involving genes such as BCL-2 and BCL-6 (so-called double-hit or triple-hit lymphomas).

B-CELL LYMPHOMA, UNCLASSIFIABLE, WITH FEATURES INTERMEDIATE BETWEEN DIFFUSE LARGE B-CELL LYMPHOMA AND CLASSICAL HODGKIN LYMPHOMA

The majority of cases of classical Hodgkin lymphoma are thought to be derived from B-cells. Rarely, B-cell lymphomas transform into classical Hodgkin lymphoma, and in some cases clonal identity between the lymphoma B-cells and the H/RS cells has been established. Among B-cell lymphomas, primary mediastinal large B-cell lymphoma (PMBL) has unique molecular features, and gene

Table 5.5 Morphological, immunophenotypic and genetic features that distinguish Burkitt lymphoma (BL) from diffuse large B-cell lymphoma (DLBCL)

Characteristic	BL	Intermediate BL/DLBCL	DLBCL
Morphology			
Only small or medium-sized cells	Yes	Common	No
Only large cells	No	No	Common
Mixture	No	Sometimes	Rare
Proliferation			
>90%	Yes	Common	Rare
<90%	No	Sometimes	Common
BCL-2 expression			
Negative or weak	Yes	Sometimes	Sometimes
Strong	No	Sometimes	Sometimes
Genetic features			
Ig-MYC	Yes	Sometimes	Rare
Non Ig-MYC	No	Sometimes	Rare
Double hit	No	Sometimes	Rare
MYC—simple karyotypes[a]	Yes	Rare	Rare
MYC—complex karyotypes[b]	No	Common	Rare

[a] No, or few, chromosomal abnormalities other than MYC translocation.
[b] Multiple chromosomal abnormalities in addition to MYC translocation.

expression profiling has shown a relationship between this tumour and nodular sclerosis classical Hodgkin lymphoma (NSCHL). This relationship is also revealed by the occurrence of PMBL and NSCHL as composite or metachronous tumours. Traverse-Glehen *et al.* (2005) reported 21 cases that they described as 'mediastinal gray zone lymphoma'. In this report they also included six cases of synchronous NSCHL and mediastinal large B-cell lymphoma and nine cases in which these two tumours occurred sequentially.

All of the 21 patients with grey zone lymphomas presented with a mediastinal mass and a third had cervical or supraclavicular lymphadenopathy. Morphologically, 11 tumours were categorized as NSCHL but showed a large number of mononuclear variants and a diminished inflammatory background. The ten PMBL-like cases had the morphology of PMBL but with admixed H/RS cells and lacunar cells. Immunohistochemistry in these tumours gave paradoxical results. Those cases with the morphology of PMBL showed variable CD20 expression, strong CD30 expression and in half the cases CD15 expression. Those cases with NSCHL morphology stained strongly for CD20. Irrespective of morphology, seven of nine cases tested were positive for MAL protein. The B-cell transcription factors PAX5, Oct-2 and BOB.1 were usually expressed but surface and cytoplasmic immunoglobulin was absent.

It is suggested that epigenetic factors are responsible for the transitions between morphology and phenotype in these grey zone lymphomas (Hertel *et al.* 2002). Oncologists in general do not welcome grey zone diagnoses, particularly in this instance since treatments for PMBL and NSCHL differ. Some preliminary results suggest that CHOP together with rituximab for CD20-expressing tumours may provide effective therapy for this lymphoma (Table 5.5).

REFERENCES

Achten R, Verhoef G, Vanuytsel L, De Wolf-Peeters C 2002 T-cell/histiocyte rich large B-cell lymphoma: A distinct clinicopathological entity. *Journal of Clinical Oncology* **20:** 1269–1277.

Adam P, Baumann R, Schmidt J, *et al.* 2013 The BCL2 E17 and SP66 antibodies discriminate 2 immunophenotypically and genetically distinct subgroups of conventionally BCL2-"negative" grade 1/2 follicular lymphomas. *Human Pathology* **44:** 1817–1826.

Alizadeh AA, Eisen MB, Davis RE, *et al.* 2000 Distinct types of diffuse large B-cell lymphoma identified by gene expression profiling. *Nature* **403:** 503–511.

Andriko J-AW, Swerdlow SH, Aguilera NI, Abbondanzo SL 2001 Is lymphoplasmacytic lymphoma/immunocytoma a distinct entity? A clinicopathological study of 20 cases. *American Journal of Surgical Pathology* **25:** 742–751.

Andrulis M, Penzel R, Weichert W, *et al.* 2012 Application of a BRAF V600E mutation–specific antibody for the diagnosis of hairy cell leukemia. *American Journal of Surgical Pathology* **36:** 1796–1800.

Asplund SL, McKenna RW, Howard M, Croft SH 2002 Immunophenotype does not correlate with lymph node histology in chronic lymphocytic leukemia/small lymphocytic lymphoma. *American Journal of Surgical Pathology* **26:** 624–629.

Attygalle AD, Liu H, Shirali S, *et al.* 2004 Atypical marginal zone hyperplasia of mucosa-associated lymphoid tissue: A reactive condition of childhood showing immunoglobulin lambda light-chain restriction. *Blood* **97:** 3343–3348.

Aukema SM, Siebert R, Schuuring E *et al.* 2011 Double-hit B-cell lymphomas. *Blood* **117:** 2319–2331.

Ballesteros E, Osborne BM, Matushima AY 1998 CD5+ low grade marginal zone B-cell lymphomas with localized presentation. *American Journal of Surgical Pathology* **22:** 201–207.

Banks PM, Chan J, Cleary ML, *et al.* 1992 Mantle cell lymphoma. A proposal for unification of morphologic, immunologic and molecular data. *American Journal of Surgical Pathology* **16:** 637–640.

Bosga-Bouwer AG, van Imhoff GW, Boonstra R, *et al.* 2003 Follicular lymphoma grade 3B includes 3 cytogenetically defined subgroups with primary t(14;18), 3q27, or other translocations: t(14;18) and 3q27 are mutually exclusive. *Blood* **101:** 1149–1154.

Camacho FI, Algara P, Mollejo M, *et al.* 2003 Nodal marginal zone lymphoma: A heterogeneous tumor. *American Journal of Surgical Pathology* **27:** 762–771.

Campo E, Miquel R, Krenacs L, *et al.* 1999 Primary nodal marginal zone lymphomas of splenic and MALT type. *American Journal of Surgical Pathology* **23:** 59–68.

Carbone A, Gloghini A 2014 Emerging issues after recognition of in situ follicular lymphoma. *Leukemia and Lymphoma* **55:** 482–490.

Chang CC, McClintock S, Cleveland RP, *et al.* 2004 Immunohistochemical expression patterns of germinal center and activation B-cell markers correlate with prognosis in diffuse large B-cell lymphoma. *American Journal of Surgical Pathology* **28:** 464–470.

Dave SS, Wright G, Tan B, *et al.* 2004 Prediction of survival in follicular lymphoma based on molecular features of tumor-infiltrating immune cells. *New England Journal of Medicine* **351:** 2159–2169.

Davis RE, Ngo VN, Lenz G, *et al.* 2010 Chronic active B-cell signalling in diffuse large B-cell lymphoma. *Nature* **463:** 88–92.

De Paepe P, Baens M, van Krieken H, *et al.* 2003 ALK activation by the CLTC-ALK fusion is a recurrent event in large B-cell lymphoma. *Blood* **102:** 2638–2641.

Delsol G, Lamant L, Mariame B, *et al.* 1997 A new subtype of large B-cell lymphoma expressing the ALK kinase and lacking the 2;5 translocation. *Blood* **89:** 1483–1490.

Fan Z, Natkunam Y, Bair E, *et al.* 2003 Characterization of variant patterns of nodular lymphocyte predominant Hodgkin lymphoma with immunohistologic and clinical correlation. *American Journal of Surgical Pathology* **27:** 1346–1356.

Gascoyne RD, Lamant L, Martin-Subero JI, *et al.* 2003 ALK-positive diffuse large B-cell lymphoma is associated with Clathrin–ALK rearrangements: Report of 6 cases. *Blood* **102:** 2568–2573.

Gesk S, Klapper W, Martin-Subero JI, *et al.* 2006 A chromosomal translocation in cyclin D1 negative/cyclin D2 positive mantle cell lymphoma fuses the CCND2 gene to the IgK locus. *Blood* **108:** 1109–1110.

Guiter C, Dusanter-Fourt I, Copie-Bergman C, *et al.* 2004 Constitutive STAT6 activation in primary mediastinal large B-cell lymphoma. *Blood* **104:** 543–549.

Gurbaxani S, Anastasi J, Hyjek E. 2009 Diffuse large B-cell lymphoma—more than a diffuse collection of large B cells. An entity in search of a meaningful classification. *Archives of Pathology and Laboratory Medicine* **133:** 1121–1134.

Hans CP, Weisenberger DD, Greiner TC, *et al.* 2005 Confirmation of the molecular classification of large B-cell lymphoma by immunohistochemistry using a tissue microarray. *Blood* **103:** 275–282.

Haralambieva E, Pulford KA, Lamant L, *et al.* 2000 Anaplastic large-cell lymphomas of B-cell phenotype are anaplastic lymphoma kinase (ALK) negative and belong to the spectrum of diffuse large B-cell lymphomas. *British Journal of Haematology* **109:** 584–591.

Haralambieva E, Rosati S, van Noesel C, *et al.* 2004 Florid granulomatous reaction in Epstein-Barr virus positive nonendemic Burkitt lymphomas. *American Journal of Surgical Pathology* **28:** 379–383.

Harris N 2013 Shades of gray between large B-cell lymphomas and Hodgkin lymphomas: Differential diagnosis and biological implications. *Modern Pathology* **26:** S57–S70.

Hertel CB, Zhou XG, Hamilton-Dutoit SJ, Junker S 2002 Loss of B-cell identity correlates with loss of B-cell specific transcription factors in Hodgkin/Reed-Sternberg cells of classical Hodgkin lymphoma. *Oncogene* **21:** 4908–4920.

Hisada M, Chan BE, Jaffe ES, Travis LB 2007 Second cancer incidence and cause specific mortality among 3104 patients with hairy cell leukemia: A population based study. *Journal of the National Cancer Institute* **99:** 215–222.

Hollingsworth HC, Longo DL, Jaffe ES 1993 Small noncleaved cell lymphoma associated with florid epithelioid granulomatous response: A clinico-pathologic study of seven patients. *American Journal of Surgical Pathology* **17:** 51–59.

Hui PK, Feller AC, Lennert K 1988 High-grade non-Hodgkin's lymphoma of B-cell type I. *Histopathology* **12:** 127–143.

Hwang HS, Park CS, Yoon DH, Suh C, Huh J 2014 High concordance of gene expression profiling-correlated immunohistochemistry algorithms in diffuse large B-cell lymphoma not otherwise specified. *American Journal of Surgical Pathology* **38:** 1046–1057.

Kienle D, Benner A, Laufle C, *et al.* 2010 Gene expression factors as predictors of genetic risk and survival in chronic lymphocytic leukemia. *Haematologica* **95:** 102–109.

Kluin PM, Harris NL, Stein H, *et al.* 2008 B-cell lymphoma, unclassifiable, with features intermediate between diffuse large B-cell lymphoma and Burkitt lymphoma. In Swerdlow SH, Campo E, Harris NL, *et al.* (Eds) *WHO Classification of Tumours of Haematopoietic and Lymphoid Tissues.* Lyon: IACR Press.

Kremer M, Ott G, Nathrath M, *et al.* 2005 Primary extramedullary plasmacytoma and multiple myeloma: Phenotypic differences revealed by immunohistochemical analysis. *Journal of Pathology* **205:** 92–101.

Landgren O, Tageja N 2014 MYD88 and beyond: Novel opportunities for diagnosis, prognosis and treatment in Waldenstrom's macroglobulinemia. *Leukemia* **28**: 1799–1803.

Laurent C, Do C, Gascoyne RD, *et al.* 2009 Anaplastic lymphoma kinase-positive diffuse large B-cell lymphoma: A rare clinicopathologic entity with poor prognosis. *Journal of Clinical Oncology* **27:** 4211–4216.

Lee SE, Kang SY, Takeuchi K, Ko YH 2014 Identification of RANBP2–ALK fusion in ALK positive diffuse large B-cell lymphoma. *Hematological Oncology* **32:** 221–224.

Lenz G, Wright G, Dave SS, *et al.* 2008 Stromal gene signatures in large B-cell lymphomas. *New England Journal of Medicine* **359:** 2313–2323.

Li S 2009 Anaplastic lymphoma kinase-positive large B-cell lymphoma: A distinct clinicopathological entity. *International Journal of Clinical and Experimental Pathology* **2:** 508–518.

Lim MS, Beaty M, Sorbara L, *et al.* 2002 T-cell/histiocyte-rich large B-cell lymphoma. A heterogeneous entity with derivation from germinal center B-cells. *American Journal of Surgical Pathology* **26:** 1458–1466.

Lin P, Mansoor A, Bueso-Ramos C, *et al.* 2003 Diffuse large B-cell lymphoma occurring in patients with lymphoplasmacytic lymphoma/Waldenstrom macroglobulinemia. Clinicopathological features of 12 cases. *American Journal of Clinical Pathology* **120:** 246–253.

Liu Q, Salaverria I, Pittaluga S, *et al.* 2013 Follicular lymphomas in children and young adults: A comparison of the pediatric variant with the usual follicular lymphoma. *American Journal of Surgical Pathology* **37:** 333–343.

Louissaint A, Ackerman AM, Dias-Santagata D, *et al.* 2012 Pediatric-type nodal follicular lymphoma: An indolent clonal proliferation in children and adults with high proliferation index and no BCL2 rearrangement. *Blood* **120:** 2395–2402.

Lu TX, Li J-, Xu W 2013 The role of SOX11 in mantle cell lymphoma. *Leukemia Research* **37:** 1412–1419.

Mann RB, Berard CW 1983 Criteria for the cytologic subclassification of follicular lymphomas: A proposed alternative method. *Hematological Oncology* **1:** 187–192.

Menon MP, Pittaluga S, Jaffe E 2012 The histological and biological spectrum of diffuse large B-cell lymphoma in the WHO classification. *Cancer Journal* **18:** 411–420.

Menter T, Gasser A, Juskevicius D, *et al.* 2015 A diagnostic utility of the germinal center–associated markers GCET1, HGAL and LMO2 in haematolymphoid neoplasms. *Applied Immunohistochemistry and Molecular Morphology* **23:** 491–498.

Olteanu H, Fenske TS, Harrington AM, *et al.* 2011 CD23 Expression in follicular lymphoma. *American Journal of Clinical Pathology* **135:** 46–53.

Onciu M, Behm FG, Downing JR, *et al.* 2003 ALK-positive plasmablastic B-cell lymphoma with expression of the NPM-ALK fusion transcript: Report of 2 cases. *Blood* **102:** 2642–2644.

Orchard J, Garand R, Davis Z, *et al.* 2003 A subset of t(11;14) lymphoma with mantle cell features displays mutated IgVh genes and includes patients with good prognosis, non-nodal disease. *Blood* **101:** 4975–4981.

Oschlies I, Salaverria I, Mahn F, *et al.* 2010 Pediatric follicular lymphoma—a clinico-pathological study of a population-based series of patients within the non-Hodgkin's Lymphoma—Berlin-Frankfurt-Munster (NHL-BFM) multicenter trials. *Haematologica* **95:** 253–259.

Oyama T, Yamamoto K, Asano N, *et al.* 2007 Age-related EBV-associated B-cell lymphoproliferative disorders constitute a distinct clinicopathological group: A study of 96 patients. *Clinical Cancer Research* **13:** 5124–5132.

Park S, Lee J, Ko YH, *et al.* 2007 The impact of Epstein-Barr virus status on clinical outcome in diffuse large B-cell lymphoma. *Blood* **110:** 972–978.

Pittaluga S, Jaffe ES 2010 T-cell/histiocyte-rich large B-cell lymphoma. *Haematologica* **95:** 352–356.

Pittaluga S, Wilson WH, Jaffe ES 2008 Lymphomatoid granulomatosis. In Swerdlow SH, Campo E, Harris NL, *et al.* (Eds) *WHO Classification of Tumours of Haematopoietic and Lymphoid Tissues.* Lyon: IACR Press.

Rosenwald A, Wright G, Leroy K, *et al.* 2003 Molecular diagnosis of primary mediastinal B-cell lymphoma identifies a clinically favorable subgroup of diffuse large B-cell lymphoma related to Hodgkin lymphoma. *Journal of Experimental Medicine* **198:** 851–862.

Sanchez-Beato M, Sanchez-Aguilera A, Piris MA 2003 Cell cycle deregulation in B-cell lymphomas. *Blood* **101:** 1220–1235.

Savage KJ, Monti S, Kutok JL, *et al.* 2003 The molecular signature of mediastinal large B-cell lymphoma differs from that of other diffuse large B-cell lymphomas and shares features with classical Hodgkin lymphoma. *Blood* **102:** 3871–3879.

Schrager JA, Pittaluga S, Raffeld M, Jaffe ES 2005 Granulomatous reaction in Burkitt lymphoma: Correlation with EBV positivity and clinical outcome. *American Journal of Surgical Pathology* **29:** 1115–1116.

Sheibani K, Burke JS, Swartz WG, *et al.* 1988 Monocytoid B-cell lymphoma: Clinicopathologic study of 21 cases of a unique type of low-grade lymphoma. *Cancer* **62:** 1531–1538.

Sheibani K, Sohn CC, Burke JS, *et al.* 1986 Monocytoid B-cell lymphoma: A novel B-cell neoplasm. *American Journal of Pathology* **124:** 310–318.

Sherman MJ, Hanson CA, Hoyer JD 2011 An assessment of the usefulness of immunohistochemical stains in the diagnosis of hairy cell leukemia. *American Journal of Clinical Pathology* **136:** 390–399.

Steidl C, Gascoyne RD 2011 The molecular pathogenesis of primary mediastinal large B-cell lymphoma. *Blood* **118:** 2659–2669.

Stein H, Warnke RA, Chan WC, *et al.* 2008. Diffuse large B-cell lymphoma, not otherwise specified. In Swerdlow SH, Campo E, Harris NL, *et al.* (Eds) *WHOCclassification of Tumours of Haematopoietic and Lymphoid Tissues*. Lyon: IACR Press.

Taddesse-Heath L, Pittaluga S, Sorbara L, *et al.* 2003 Marginal zone B-cell lymphoma in children and young adults. *American Journal of Surgical Pathology* **27:** 522–531.

Thieblemont C, Nasser V, Felman P, *et al.* 2004 Small lymphocytic lymphoma, marginal zone B-cell lymphoma and mantle cell lymphoma exhibit distinct gene expression profiles allowing molecular diagnosis. *Blood* **103:** 2727–2737.

Thorley-Lawson DA, Duca KA 2007 Pathogens cooperate in lymphomagenesis. *Nature Medicine* **13:** 906–907.

Traverse-Glehen A, Pittaluga S, Gaulard P, *et al.* 2005 Mediastinal gray zone lymphoma. The missing link between classical Hodgkin's lymphoma and mediastinal large B-cell lymphoma. *American Journal of Surgical Pathology* **29:** 1411–1421.

van den Brand M, van Krieken JH 2013 Recognizing nodal marginal zone lymphoma: Recent advances and pitfalls. A systematic review. *Haematologica* **98:** 1003–1013.

Van Loo P, Tousseyn T, Vanhentenrijk V, *et al.* 2010 T-cell/histiocyte rich large B-cell lymphoma shows transcriptional features suggestive of a tolerogenic host immune response. *Haematologica* **95:** 440–448.

Wass M, Behlendorf T, Schädlich B, *et al.* 2013 Crizotinib in refractory ALK-positive diffuse large B-cell lymphoma: A case report with a short term response. *European Journal of Haematology* **92:** 268–270.

Weistner A, Rosenwald A, Barry TS, *et al.* 2003 ZAP-70 expression identifies a chronic lymphocytic leukemia subtype with unmutated immunoglobulin genes, inferior clinical outcome, and distinct gene expression profile. *Blood* **101:** 4944–4951.

Wright G, Tan B, Rosenwald A, *et al.* 2003 A gene expression based method to diagnose clinically distinct subgroups of diffuse large B-cell lymphoma. *Proceedings of the National Academy of Sciences USA* **100:** 9991–9996.

Yu SC, Chen SU, Lu W, *et al.* 2011 Expression of CD19 and lack of miR-223 distinguish extramedullary plasmacytoma from multiple myeloma. *Histopathology* **58:** 896–905.

Zhang XM, Aguilera N 2014 New immunohistochemistry for B-cell lymphoma and Hodgkin lymphoma. *Archives Pathology and Laboratory Medicine* **138:** 1666–1672.

6

T-cell lymphomas

OVERVIEW

T-cell neoplasms are less common than B-cell lymphomas and their incidence shows both geographical and racial variations. In the International Non-Hodgkin Lymphoma Classification Study of cases from the United States, Europe, Asia and South Africa, peripheral T-cell lymphomas accounted for 7.6 per cent of the total, with anaplastic large cell lymphoma (ALCL) forming an additional 2.4 per cent (Non-Hodgkin's Lymphoma Classification Project 1997). T-cell lymphomas are more common in Asia and there is a high incidence of nasal-type natural killer (NK)/T-cell lymphomas in Asian races. In parts of Japan and the Caribbean, the increased incidence of adult T-cell leukaemia/lymphoma (ATLL) is related to the high prevalence of infection by the human T-cell lymphotrophic virus 1 (HTLV-1).

For a number of reasons, the classification of lymphomas derived from post-thymic or mature T-cells is less satisfactory than the classification of B-cell neoplasms. With the exception of diffuse large B-cell lymphoma (DLBCL), which remains a heterogeneous group, the current World Health Organization (WHO) classification of B-cell lymphomas reflects reasonably well their histogenesis and provides a guide to treatment. By contrast, neoplastic T-cells show marked morphological variation. They may be difficult to characterize because of variable or aberrant expression of the T-cell antigens, and their relationship to the normal sequence of T-cell maturation and differentiation is far from clear. Some lymphomas exhibit combined features of T-cells and the closely related NK cells and it is usual to consider T-cell and NK-cell neoplasms together.

Classifications of T-cell lymphomas have been an uncomfortable mix of purely descriptive terms and fairly well-defined entities. The Working Formulation made no attempt to separate lymphomas according to phenotype, and T-cell lymphomas were hidden in descriptive terminology designed primarily for B-cell lymphomas. The Kiel classification improved on this by recognizing T-cell neoplasms and separating them into low-grade and high-grade categories. The least satisfactory area concerned nodal

T-cell lymphomas. With the exception of ALCL and angioimmunoblastic lymphoma, both of which are now fairly well characterized, node-based peripheral T-cell lymphomas remained a poorly defined group, which could not be subclassified with any degree of consistency.

The Revised European-American Lymphoma (REAL) classification (Harris *et al.* 1994) and, more recently, the WHO classification (Jaffe *et al.* 2001, Swerdlow *et al.* 2008) attempted an objective reassessment and accepted only those entities that can be reliably distinguished on clinical, pathological or genetic grounds. They recognize a number of well-defined T-cell lymphomas, some of which are primarily extranodal or leukaemic, and they leave largely intact a group of node-based peripheral T-cell lymphomas under the umbrella term 'not otherwise specified' (NOS), on the basis that further subclassification is subject to considerable interobserver variation and, at present, has little clinical validity.

The anatomic localization of neoplastic T-cells and NK cells parallels in part their proposed normal cellular counterparts and functions. T-cells of the adaptive immune system are mainly based in lymph nodes and peripheral blood, whereas lymphomas derived from T-cells and NK cells of the innate immune system are mainly extranodal (Bajor-Dattilo *et al.* 2013). Leukaemias and extranodal lymphomas fall outside the scope of a book concerned with lymph node pathology; however, some show secondary involvement of lymph nodes and it is important, therefore, to recognize their features. The modification of the WHO classification shown in Box 6.1 indicates which tumours are primarily nodal and which involve nodes secondarily.

BOX 6.1: Lymph node involvement by T-cell lymphomas

T-CELL LYMPHOMAS WITH PRIMARY NODAL INVOLVEMENT

Angioimmunoblastic T-cell lymphoma
Anaplastic large cell lymphoma
Peripheral T-cell lymphoma not otherwise specified

T-CELL LEUKAEMIA/LYMPHOMAS WITH NODAL INVOLVEMENT

T-lymphoblastic leukaemia/lymphoma (see Chapter 4)
Adult T-cell leukaemia/lymphoma
T-cell prolymphocytic leukaemia

EXTRANODAL T-CELL LYMPHOMAS WITH SECONDARY NODAL INVOLVEMENT

Mycosis fungoides/Sézary syndrome
Enteropathy-associated T-cell lymphoma
Primary cutaneous CD30+ T-cell anaplastic large cell lymphoma

EXTRANODAL T-CELL LYMPHOMAS RARELY SHOWING NODAL INVOLVEMENT

Extranodal NK/T-cell lymphoma, nasal type
Hepatosplenic T-cell lymphoma
Subcutaneous panniculitis-like T-cell lymphoma
Rare subtypes of primary cutaneous peripheral T-cell lymphoma

T-CELL LEUKAEMIAS AND RELATED SYSTEMIC DISORDERS

Without nodal involvement

T-cell large granular lymphocytic leukaemia
Aggressive NK-cell leukaemia

With nodal involvement

Chronic lymphoproliferative disorders of NK cells
Epstein–Barr virus–positive T-cell lymphoproliferative disorders of childhood

ANGIOIMMUNOBLASTIC LYMPHOMA

In the mid-1970s, the first descriptions of angioimmunoblastic T-cell lymphoma (AITL) as 'immunoblastic lymphadenopathy' or 'angioimmunoblastic lymphadenopathy with dysproteinaemia' regarded it as an essentially reactive systemic process but recognized that it was at least premalignant and carried a poor prognosis, with a proportion of patients developing frank lymphoma. Not until monoclonal T-cell populations could be demonstrated by molecular techniques did it become clear that, at least in the majority of cases, AITL is a T-cell neoplasm from the outset (Dogan 2003).

As with most lymphomas, the aetiology of AITL is unknown. In early series, many of the patients had a history of medication, particularly with antibiotics, but it is likely that these were given because systemic symptoms in the early stages of the disease can mimic an infectious process. Search for a viral cause has shown an interesting association with Epstein–Barr virus (EBV), which is further discussed later, but no other convincing associations.

AITL accounts for about 15–20 per cent of T-cell lymphomas and 1–2 per cent of all non-Hodgkin lymphomas. Most patients are aged 50–70 years, with a median age of 59–64 years and equal sex incidence. They typically present with systemic symptoms, generalized lymphadenopathy and hepatosplenomegaly. Some also have a skin rash with pruritus. Other features include ascites, pleural effusions and arthritis. Investigations show a polyclonal hypergammaglobulinaemia, elevated erythrocyte sedimentation rate (ESR) and lactate dehydrogenase (LDH) and anaemia, often of Coombs-positive autoimmune type. Other autoantibodies, such as rheumatoid factor, thyroid autoantibodies and anti–smooth muscle antibodies may be present.

Despite intensive chemotherapy, the prognosis remains poor, with a 5-year survival of only 20–30 per cent. This is

BOX 6.2: Angioimmunoblastic T-cell lymphoma: Clinical features

- Systemic symptoms
- Widespread lymphadenopathy and hepatosplenomegaly
- Rash
- Hypergammaglobulinaemia
- Haemolytic anaemia
- Other autoantibodies
- Immune deficiency
- Fatal infections

partly because of the immune deficiency associated with the disease; many patients develop fatal infectious complications (Box 6.2).

HISTOLOGY

Primary diagnosis of AITL usually depends on lymph node biopsy. Biopsies from involved extranodal sites, such as skin, spleen, liver, bone marrow and lung, are less reliable because the diagnosis is dependent on a combination of architectural and cytological features.

Typically, the lymph node is moderately enlarged and replaced by a polymorphous cellular infiltrate. This may extend into extracapsular tissues, but a useful low-power clue to the diagnosis is provided by the subcapsular sinus, which is usually at least partially preserved, and often focally dilated (Ottaviani *et al.* 2004). The characteristic vascular proliferation that gives the disease its name takes the form of prominent, branching or arborizing post-capillary or high endothelial venules, present in groups. These are highlighted in sections stained for reticulin. Periodic acid–Schiff (PAS) staining shows a thick layer of PAS-positive basement membrane material around the vessels, but also demonstrates deposition of amorphous, eosinophilic, PAS-positive stromal material in the vicinity of the proliferating vessels. Much of the infiltrate appears reactive and includes small lymphocytes, plasma cells, eosinophils and histiocytes. The neoplastic T-cell component may be difficult to identify. It is best seen in the areas of vascular proliferation as loose clusters of intermediate-sized cells with clear cytoplasm and atypical hyperchromatic nuclei, usually without particularly prominent nucleoli. The numbers of clearly atypical cells and their degree of atypia show considerable variation. In addition to the neoplastic cells, a further consistent feature is the presence of scattered, large, blast-like cells that form a conspicuous feature because of their large nuclei and their occasional resemblance to Hodgkin/Reed–Sternberg (H/RS) cells. The majority of AITL biopsies also include hypocellular, eosinophilic areas corresponding to areas of burnt-out follicles in which there is dendritic cell proliferation. Although this remains a useful feature, it is now recognized that cases with hyperplastic follicles may occur. This is further discussed later in the section on differential diagnosis (Ree *et al.* 1998, Rodriguez-Justo *et al.* 2009) (Box 6.3 and Figures 6.1 to 6.5).

BOX 6.3: Angioimmunoblastic T-cell lymphoma: Morphology

- Complete or partial effacement of node
- Extension into perinodal tissues, often leaving subcapsular sinus patent
- Mixed population of reactive and neoplastic cells
- Scattered B-blasts—often EBV+
- Prominent, branching high endothelial venules surrounded by periodic acid–Schiff (PAS)–positive basement membrane
- Expanded follicular dendritic cell network
- Clusters of medium-sized T-cells with clear cytoplasm

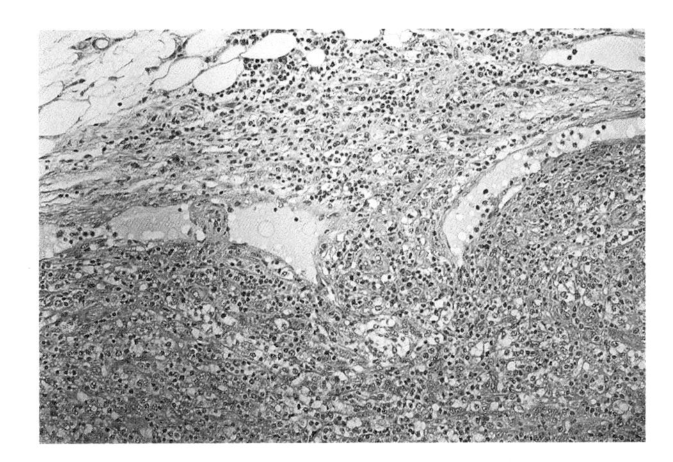

Figure 6.1 Angioimmunoblastic lymphoma. Even when the infiltrate extends beyond the capsule of the node into surrounding tissue, there is a tendency for the peripheral sinus to remain widely patent.

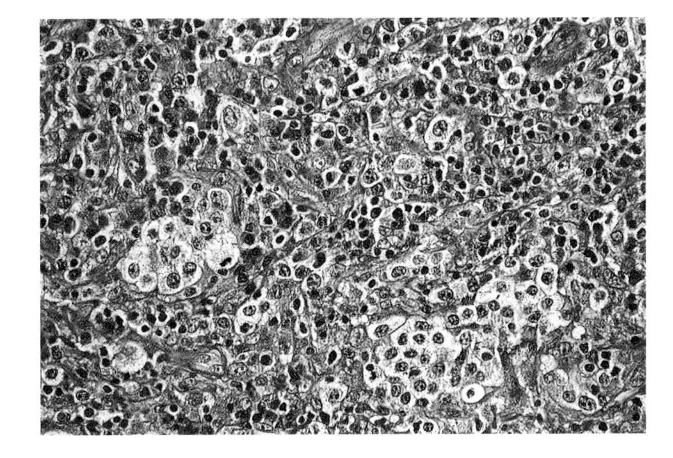

Figure 6.2 The typical polymorphous infiltrate in angioimmunoblastic lymphoma includes plasma cells, large blasts, eosinophils and clear cells.

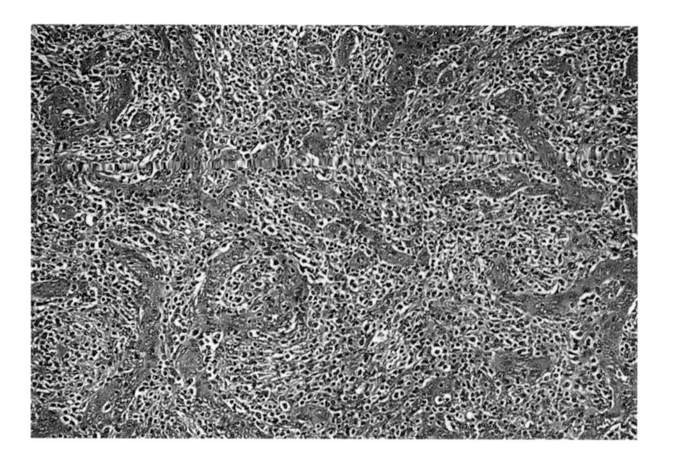

Figure 6.3 The characteristic vascular proliferation in angioimmunoblastic lymphoma is caused by 'arborizing' high-endothelial venules.

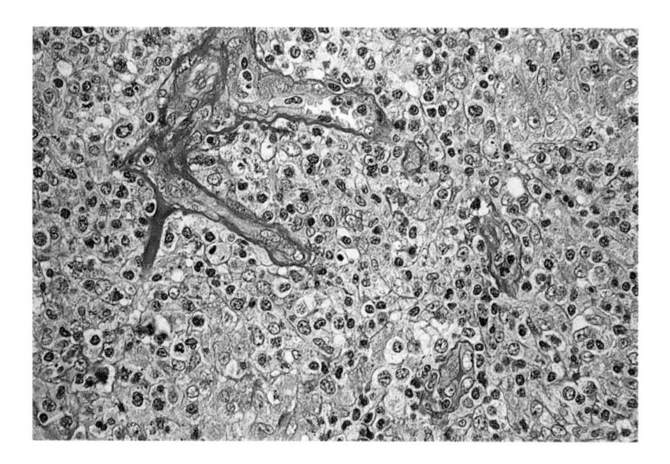

Figure 6.4 Angioimmunoblastic lymphoma. The vessels are well seen in periodic acid–Schiff stains owing to staining of the basement membranes.

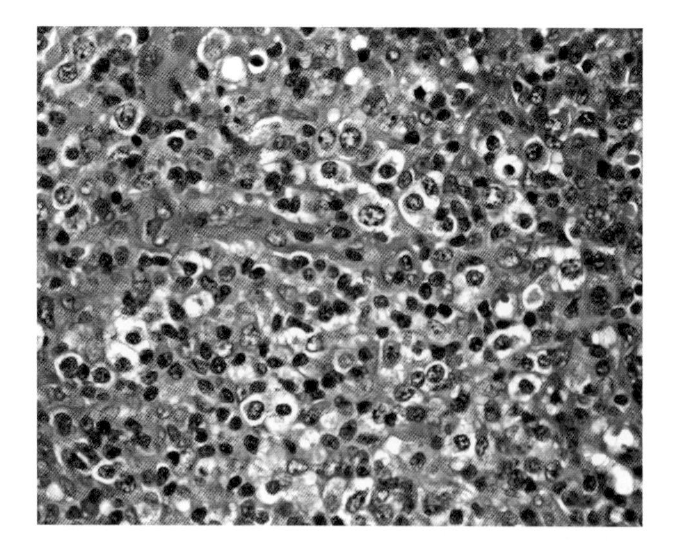

Figure 6.5 Clear cells. The neoplastic cells in angioimmunoblastic lymphoma tend to cluster around vessels, and have clear cytoplasm with nuclei of intermediate size and variable pleomorphism.

IMMUNOHISTOCHEMISTRY

Pan-T-cell markers such as CD3 and CD5 show that the majority of cells in the infiltrate, including the atypical clear cells, are T-cells. The neoplastic T-cells are typically CD4+, although this may be difficult to appreciate because background, reactive CD8+ cells are present in variable numbers (Attygalle *et al.* 2002). The neoplastic cells are thought to originate from an effector subset of T-helper cells normally found in the germinal centre (T-follicular helper [TFH] cells) and, as such, display a characteristic immunophenotype (Rodriguez-Justo *et al.* 2009). They are typically CD10+ and BCL-6+, although staining for these markers is inconsistent and variable, and also show reactivity for two novel markers, the chemokine CXCL13 and programmed cell death protein-1 (PD-1 or CD279).

The B-cell markers CD20 and CD79a indicate that the scattered large blasts are of B-cell origin. These are prominent because the normal B-cell areas are depleted, with residual aggregates of small B-cells confined to the subcapsular areas at the periphery of the node. The large B-blasts are usually positive for CD30, and even occasionally CD15, and the presence of latent EBV infection can usually be demonstrated with latent membrane protein 1 (LMP1) or, more reliably, using *in-situ* hybridization for Epstein–Barr virus–encoded RNAs (EBERs). The markers for follicular dendritic cells (FDCs), CD21, CD23 and CD35, show a very distinctive proliferation of these cells, producing an expanded dendritic cell network in the region of the proliferating vessels, corresponding to residual disrupted follicles and to the eosinophilic material seen in haematoxylin and eosin (H&E) staining (Jones *et al.* 1998). CD23 immunostaining may be better than CD21 in demonstrating FDC expansion (Attygalle *et al.* 2014). Similar dendritic cell networks and related CD10+ cells may also be identified in extranodal deposits (Attygalle *et al.* 2004) (Box 6.4 and Figures 6.6 to 6.12).

GENETICS

Clonal rearrangement of T-cell receptor (TCR) genes is found in AITL; however, immunoglobulin gene rearrangement may also be identified in a significant number of cases.

BOX 6.4: Angioimmunoblastic T-cell lymphoma: Immunohistochemistry

- CD3+, CD5+, CD4+ T-cells, often CD10+, PD-1+
- Residual B-cell areas (CD20+ and CD79a+) pushed to periphery of node
- CD20+, CD79a+ blasts, often CD30+
- B-blasts may express Epstein–Barr virus (EBV)–encoded RNAs (EBERs) and EBV-latent antigens
- Expanded network of dendritic cells demonstrated by CD21, CD23, CD35

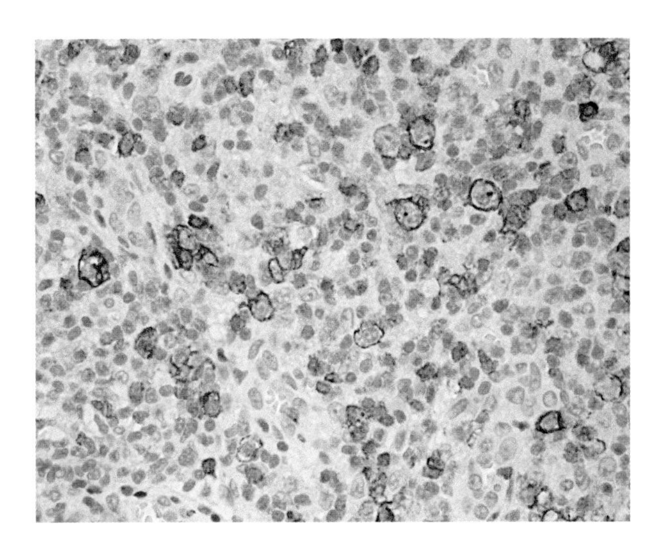

Figure 6.6 Angioimmunoblastic T-cell lymphoma. The neoplastic T-cells express PD-1 with a membranous pattern of staining.

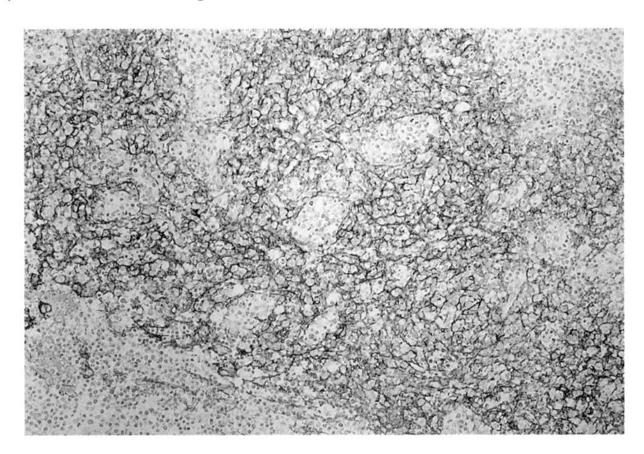

Figure 6.7 Angioimmunoblastic lymphoma stained with CD21. Both CD21 and CD23 can be used to demonstrate the complex network of follicular dendritic cells associated with the vascular proliferation.

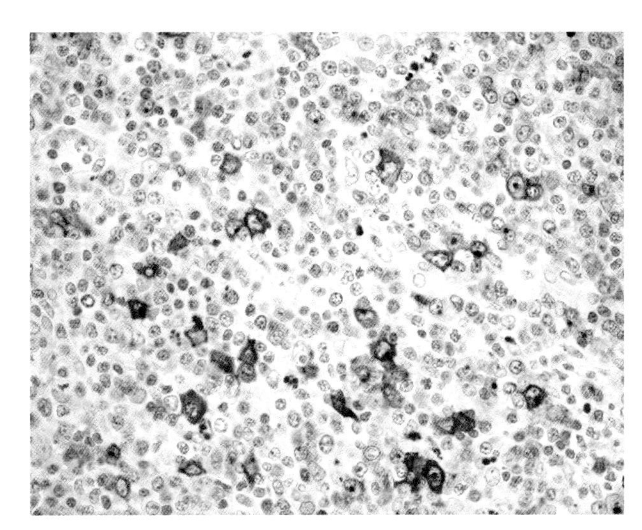

Figure 6.8 Angioimmunoblastic lymphoma stained with CD30. Scattered large blasts express B-cell markers and are frequently CD30+.

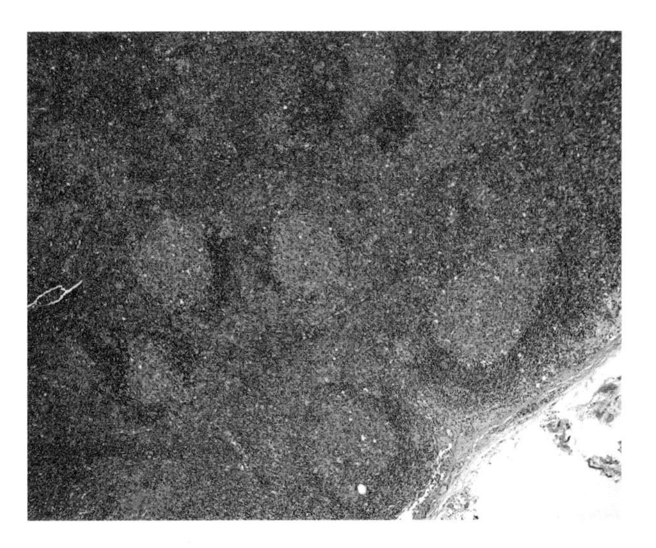

Figure 6.9 Early or incomplete patterns of node involvement by angioimmunoblastic T-cell lymphoma. Follicles remain in parts of the node but may be poorly defined with thin or incomplete mantle zones.

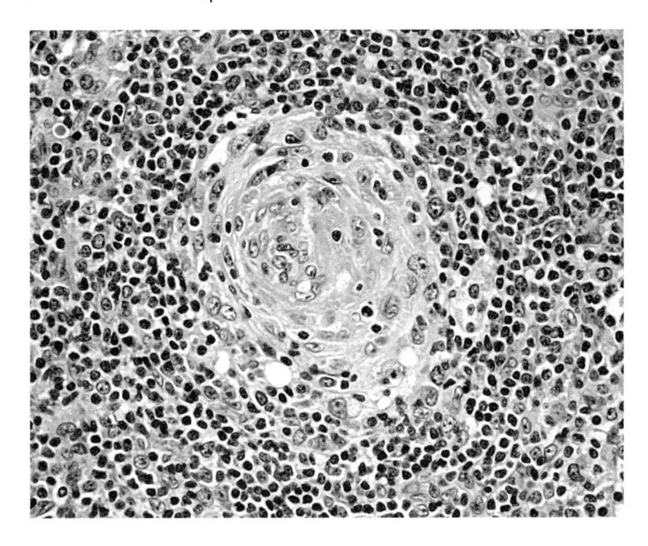

Figure 6.10 Some residual follicles in angioimmunoblastic T-cell lymphoma resemble those seen in Castleman disease, consisting largely of follicular dendritic cells.

It has been suggested that these B-cell clones are related to the proliferation of EBV-driven cells. This suggests a possible role for EBV in the pathogenesis of AITL (Swerdlow *et al.* 2008), or the clones may emerge owing to the immunodeficiency associated with AITL, a situation analogous to post-transplant lymphoproliferative disease. Gene expression analysis of AITL indicates derivation from TFH cells (Pittaluga *et al.* 2007a) and suggests that the tumour microenvironment significantly contributes to the prognosis of AITL (Iqbal *et al.* 2014). These studies have also shown that isocitrate dehydrogenase 2 gene (IDH2) mutations are mostly confined to AITL and can differentiate these cases from peripheral T-cell lymphoma not otherwise specified (PTCL-NOS).

Cytogenetic studies reveal both clonal and non-clonal abnormalities, the most consistent being trisomy 3, trisomy 5 and an additional X chromosome.

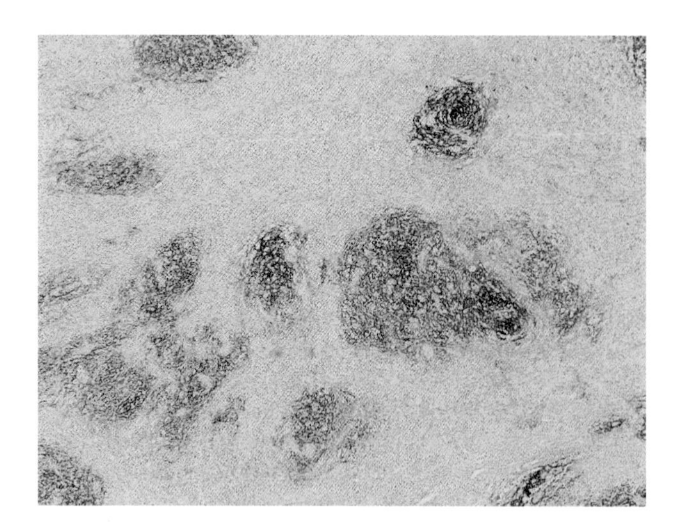

Figure 6.11 Angioimmunoblastic T-cell lymphoma stained for CD21 showing some normal follicular dendritic cell networks, and others show irregular expansion and appear frayed.

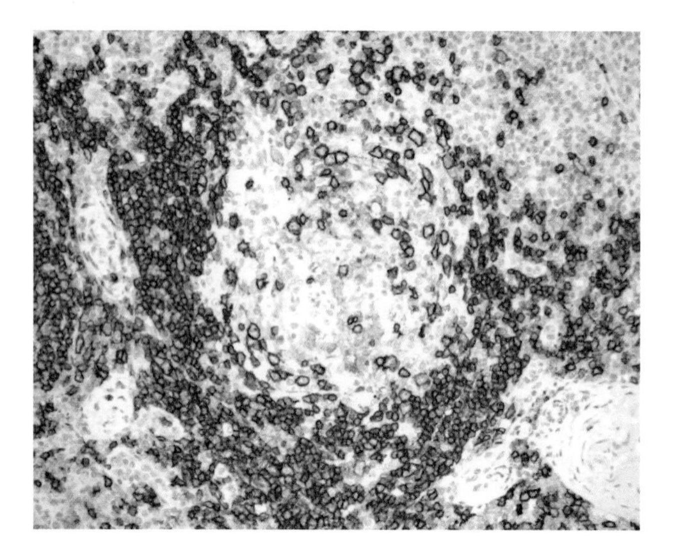

Figure 6.12 Angioimmunoblastic T-cell lymphoma stained for CD20 showing abnormal follicles almost devoid of B-cells, leaving scattered larger B-cells centrally and small B-cells in the surrounding node.

DIFFERENTIAL DIAGNOSIS

The diagnosis of AITL presents little difficulty when the changes are typical. However, a spectrum or sequence of changes can sometimes be identified in simultaneous or consecutive multiple lymph node biopsies. Early or partial involvement may be difficult to distinguish from a reactive process, and at the other end of the spectrum there may be some features of AITL in a node that would otherwise be regarded as PTCL-NOS. Interpretation of the changes is made more difficult in those cases of AITL that are complicated by the development of a DLBCL. Recognition of the evolution of the changes of AITL usually enables a confident diagnosis to be made. The histological features have been separated into three patterns (Attygalle *et al.* 2002). In pattern I, the lymph node structure is partially preserved with hyperplastic B-cell follicles. These are usually poorly defined and have narrow, irregular mantle zones. They merge into an expanded paracortex with a polymorphous infiltrate and prominent vessels. Follicles are less prominent in pattern II. Those remaining appear depleted, leaving groups of FDCs, sometimes with a concentric pattern reminiscent of the follicles in Castleman disease of the hyaline-vascular type. Pattern III refers to the fully developed, typical changes described earlier, where follicles are no longer seen and the whole node is replaced by a polymorphous infiltrate. In patterns I and II atypical clear cells are not easily identified and diagnosis may hinge on interpretation of more subtle features, helped by an appropriate clinical presentation. It should be emphasized that the three patterns are arbitrary and overlap, with different areas within the same node or different nodes showing different combinations.

In patterns I and II, when the nodal architecture may show little disruption, the main differential diagnosis is reactive hyperplasia of the paracortical type, most commonly associated with viral infections. Immunohistochemistry will often help to identify those follicles showing the disorganization associated with AITL: depletion of the normal B-cell population and loss of the normal tight FDC networks, with abnormally spread or frayed dendritic cell meshworks. The interfollicular areas show many of the features associated with reactive hyperplasia, including proliferation of high endothelial venules and scattered blasts in a mixed background. Detection of abnormally situated TFH cells in the perifollicular areas and around vessels in the interfollicular areas is facilitated by staining for CD10, PD-1 and CXCL13.

When the neoplastic cells are clearly atypical and form the majority population, the distinction from other forms of T-cell lymphoma, particularly the NOS type, may be difficult. Some examples have a similar polymorphic population to AITL but lack some of the typical features, particularly cells expressing CD10, PD-1 and CXCL13 (Grogg *et al.* 2006). Whether these should be grouped with AITL is debatable and not at present of great clinical significance. The distinction can usually be made using a number of features characteristic of AITL (Table 6.1). The presence of widespread disease and systemic symptoms should be taken into consideration if the histological appearances are equivocal.

The presence of CD30+, Reed–Sternberg-like cells may suggest classical Hodgkin lymphoma of mixed cellularity type. In AITL the H/RS-like cells usually lack typical morphological features and are usually, although not invariably, CD15−. Unlike H/RS cells they consistently express CD45 and the B-cell markers CD20 and CD79a. Attention to the background will demonstrate the other features of AITL. The presence of large blast-like B-cells in a mixed background with large numbers of T-cells may

Table 6.1 Differential diagnosis of angioimmunoblastic T-cell lymphoma (AITL)

	Reactive follicles	Vascular proliferation	Atypical clear cells	Large CD30+, EBV+ blasts	Hodgkin/ Reed–Sternberg cells	Follicular dendritic cell networks	Clonality by PCR
AITL pattern III	Usually absent	Conspicuous	CD3+, CD5+, CD4+ variably CD10+, CXCL13+, PD-1+	Scattered	Hodgkin-like cells may be present	Prominent expanded networks	T-cell clone
AITL patterns I and II	Present but poorly defined, some Castleman-like	Present	Not easily seen but CXCL13+, PD-1+	Occasional	Absent	Occasional expanded network	T-cell clone
Reactive hyperplasia	Present with well-defined mantle zones	Present	Absent	Absent	Absent	Normal cohesive networks	Polyclonal
Classical Hodgkin lymphoma	May be present	Absent	Absent	Present	Present CD30+, CD15+, CD45–, CD20+/–	Absent or normal	Not demonstrable
PTCL-NOS	May be present	May be present	Absent	Absent	Hodgkin-like T-cells may be present	Absent	T-cell clone
Combined AITL and DLBCL	Absent	Present	Present	Abundant and forming sheets	Absent	Present in places	B- and T-cell clones
T-cell/ histiocyte-rich B-cell lymphoma	Absent	Absent	Absent	CD20++, CD79a+, CD45+, CD30– and EBV–	Hodgkin-like cells may be present	Absent	B-cell clone, polyclonal T-cells

DLBCL, diffuse large B-cell lymphoma; EBV, Epstein-Barr virus; PCR, polymerase chain reaction; PTCL-NOS, peripheral T-cell lymphoma not otherwise specified.

also suggest large B-cell lymphoma of T-cell/histiocyte-rich type (THRLBCL). The typical vascular pattern, the mix and distribution of the cells, and the pattern of FDC networks all help to make the distinction from AITL. In THRLBCL the background T-cells are small and uniform without atypical features. The neoplastic B-cells are positive for CD45, showing strong membrane staining for CD20 and variable staining for CD79a. They are negative for CD30 and EBV.

The B-cell proliferation in AITL may be extensive and some patients develop secondary EBV-positive DLBCLs. These are comparable to those of DLBCLs arising in immunodeficient patients. In AITL the number of large B-blasts varies, but they usually remain scattered throughout the node. DLBCL should be considered when they form more cohesive sheets in parts of the node. If they become the dominant population, the underlying AITL is difficult to identify and clonality studies may be necessary to confirm the presence of both B- and T-cell neoplasms.

The TFH phenotype is also a feature of the follicular variant of peripheral T-cell lymphoma (PTCL-F), which has a nodular architecture (de Leval *et al.* 2001). In contradistinction to AITL, it does not have prominent vascular proliferation or extrafollicular expansion of FDC networks (see PTCL section).

ANAPLASTIC LARGE CELL LYMPHOMA

ALCL was first described as a specific entity by Stein and his colleagues in 1985 (Stein *et al.* 1985, 2000, Jaffe 2001). This followed identification of the antigen that was originally called Ki-1 but was subsequently designated CD30. The CD30 antigen is expressed on the H/RS cells of classic Hodgkin lymphoma and in normal lymph nodes by activated lymphoid blasts typically scattered around reactive B-cell follicles. The large lymphoma cells of ALCL uniformly

BOX 6.5: (T/null) anaplastic large cell lymphoma: Clinical features

- Most common in young people, with peak in second decade
- Male predominance
- Widespread disease with B-symptoms, often extranodal
- ALK+ tumours have good prognosis
- ALK− tumours have a worse prognosis, more comparable to T-cell lymphoma not otherwise specified

and strongly express CD30 on the cell membrane and in the Golgi region. It is often weakly expressed or not expressed on the small cells of the small cell and lymphohistiocytic variants of ALCL, although it is more consistently positive on the large cells found in these lymphomas.

It was subsequently found that many examples of ALCL had a t(2;5) chromosomal translocation, causing the NPM (nucleophosmin) gene located on 5q35 to fuse with a gene on 2p23 encoding the tyrosine kinase receptor anaplastic lymphoma kinase (ALK).

ALCL constitutes 2 per cent of all lymphomas but forms 10 per cent or more of childhood lymphomas. ALK-positive cases have a bimodal age distribution with the major peak in the second decade and a small peak in later life. It is more common in males, particularly in the younger age group. Most patients have advanced disease with systemic symptoms at the time of presentation. The majority have lymphadenopathy but extranodal disease is frequent and, in a small number of patients, ALCL is exclusively extranodal. Skin, bone, soft tissues, bone marrow, lung and liver are the most frequently involved sites. Leukaemic peripheral blood involvement is uncommon, except in the small cell variant where it is a frequent finding, and is an indicator of poor prognosis (Onciu *et al.* 2003, Summers and Moncor 2010).

ALK-negative tumours occur in an older age group, have a lower male to female ratio and tend to present at an earlier stage. Untreated ALCL pursues an aggressive course but patients with ALK-positive tumours have a good response to treatment with excellent (>75 per cent) overall survival. ALK-negative tumours behave similarly to other types of peripheral T-cell lymphoma with a relatively poor prognosis (Box 6.5).

HISTOLOGY

The histology of ALCL is variable but all variants contain at least occasional hallmark cells. These are large cells with reniform, horseshoe- or fetus-shaped nuclei, and abundant eosinophilic or amphophilic cytoplasm. The nuclei usually contain several nucleoli that may appear basophilic or eosinophilic. Multinucleated cells in which the nuclei form a wreath-like circle are sometimes seen. Binucleate forms may resemble RS cells, although they do not usually show the prominent eosinophilic nucleoli of the latter. Cytoplasmic inclusions within nuclei may produce a clear central area, giving an appearance resembling a doughnut. Cells of ALCL often appear cohesive and, in lymph node biopsies, may be seen within preserved sinuses, giving an appearance suggestive of metastatic carcinoma or melanoma. ALCL cells also have a propensity to accumulate around blood vessels.

Anaplastic large cell lymphoma common variant

This variant accounts for 70–80 per cent of cases. The tumour cells are large and have abundant cytoplasm. The nuclear morphology varies from pleomorphic to predominantly rounded and monomorphic. Hallmark cells can usually be identified, with the morphological variations described earlier.

Anaplastic large cell lymphoma lymphohistiocytic variant

This variant accounts for approximately 10 per cent of cases. Since histiocytes are the dominant cell and may obscure the neoplastic cells, this subtype needs to be recognized. The lymphoma cells, which may be relatively small, are highlighted by immunostaining for CD30 or ALK, and may be clustered around blood vessels. The non-neoplastic histiocytes may show erythrophagocytosis.

Anaplastic large cell lymphoma small cell variant

This is an uncommon variant that frequently leads to a mistaken diagnosis of PTCL-NOS. The majority of the cells have small to medium-sized nuclei, which are typically hyperchromatic and irregular in outline. Nucleoli are therefore usually inconspicuous. The cytoplasm is pale-staining or clear, giving the cells a 'fried egg' appearance. Hallmark cells may be difficult to find and are often clustered around blood vessels. They are highlighted by staining for CD30, which is expressed more strongly on these cells than on the small cells. Although the small cell variant of ALCL is ALK positive and, like the common variant, occurs in the younger age group, it tends to take a more aggressive course, often with a leukaemic presentation, and responds less well to treatment (Youd *et al.* 2009).

Other variants

In addition to the aforementioned variants recognized in the WHO classification, other histological patterns have been reported. A hypocellular variant associated with a myxoid stroma mimicking an inflammatory process has been reported (Cheuk *et al.* 2000). Spindle cell sarcomatoid, signet ring and giant-cell–rich subtypes have also been

Table 6.2 Differential diagnosis of anaplastic large cell lymphoma: immunophenotype

	ALK+ anaplastic large cell lymphoma	ALK− anaplastic large cell lymphoma	Classical Hodgkin lymphoma H/RS cells	Diffuse large B-cell lymphoma	ALK+ diffuse large B-cell lymphoma	Peripheral T-cell lymphoma NOS
CD30	Positive membrane and Golgi	Positive membrane and Golgi	Positive membrane and Golgi	Variably positive; may show anaplastic morphology	Negative	Often positive but variable
ALK	Positive; usually nuclear and cytoplasmic	Negative	Negative	Negative	Positive; granular cytoplasmic and Golgi	Negative
CD20	Negative	Negative	Variable; approximately 20 per cent weakly positive	Positive	Negative	Negative
T-cell antigens	Variable CD3, CD4, CD5 or 'null'	Variable CD3, CD4, CD5 or 'null'	Rarely weakly positive	Negative, except for CD5+ subgroup	Negative	Variable—usually abnormal expression
EMA	Positive membrane and Golgi	Variable and weak	Negative	Negative	Positive membrane, variable Golgi	Rarely positive
CD45	Positive	Variable and weak	Negative	Positive	Positive	Positive
CD15	Negative	Negative	Positive in approximately 75 per cent	Negative	Negative	Occasionally positive
Cytotoxic T-cell markers	Positive (TIA-1, granzyme, perforin)	Variable	Usually negative	Negative	Negative	Variably positive
PAX5/BSAP	Negative	Negative	Weakly positive	Positive	Positive	Negative

ALK, anaplastic lymphoma kinase; EMA, epithelial membrane antigen; H/RS, Hodgkin/Reed–Sternberg; NOS, not otherwise specified; TIA, T-cell intracellular antigen.

reported (Benharroch *et al.* 1998). Different subtypes may be seen in the same patient at different sites and at different times. Occasionally, H/RS-like cells are present in sufficient numbers to mimic classical Hodgkin lymphoma: the 'Hodgkin-like' variant of ALCL. Misdiagnosis is most likely to occur when the atypical cells resemble lacunar cells and are accompanied by sclerosis. Examples of nodular sclerosing Hodgkin lymphoma with unusual features, such as complete lack of B-cell markers and CD15 expression, should be subjected to additional immunohistochemistry (Table 6.2).

IMMUNOHISTOCHEMISTRY

In the classical variant of ALCL, CD30 is strongly and uniformly expressed on the cell membrane and in the Golgi region. CD30 expression in the small cell variant may be weaker or more variable, with strongest expression on perivascular large cells. ALK immunohistochemistry is recommended in all T-cell lymphomas with expression of CD30 since cases with variant morphology might otherwise be misclassified. ALK protein is expressed in 60–90 per cent of cases, mainly those occurring in children and young adults, with ALK-negative ALCL occurring in older patients. The most common pattern of expression, associated with t(2;5), shows positive nuclear and cytoplasmic staining. Other, less common translocations give rise to membrane or cytoplasmic staining. Epithelial membrane antigen (EMA) expression is seen in all ALK-positive lymphomas but is less commonly expressed in ALK-negative cases. CD45 and CD45RO are variably expressed in ALCL. Expression of T-cell antigens is variable: most examples express one or more T-cell markers, but some are of the 'null cell' phenotype. CD4 is most often positive, CD3 is expressed by less than 25 per cent of tumours, CD5 and CD7 are frequently negative, and CD8 is usually negative. Both CD43 and CD25 are frequently positive. Most lymphomas are positive for the cytotoxic granule markers T-cell intracellular antigen (TIA-1), granzyme and perforin, and some express CD56 (Box 6.6 and Figures 6.13 to 6.18).

ALK-negative anaplastic large cell lymphoma

Evidence is accumulating that ALK-negative ALCL is distinct from both ALK-positive ALCL and PTCL-NOS. Although sharing many features with ALK-positive ALCL, ALK-negative lymphomas show a peak incidence later in life, occurring in the fifth and sixth decades, and

BOX 6.6: (T/null) anaplastic large cell lymphoma: Morphology

- May have sinusoidal growth pattern with cohesive cells
- Hallmark cells found in all types but typical of 'common' type
- Variants include small cell, lymphohistiocytic, giant cell, monomorphic and sarcomatoid
- Pattern may be mixed in same site or different nodes

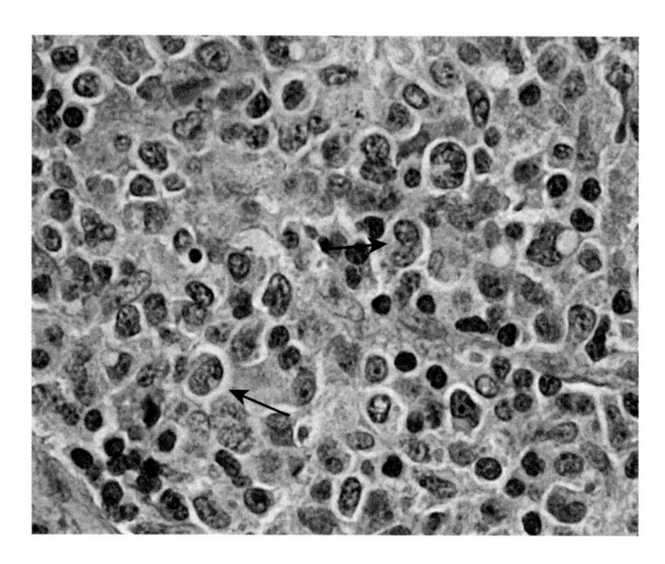

Figure 6.13 In the 'common' variant of anaplastic large cell lymphoma, cells show the features of large blasts, often with fairly abundant cytoplasm and forming cohesive sheets. Scattered hallmark cells are present (arrows).

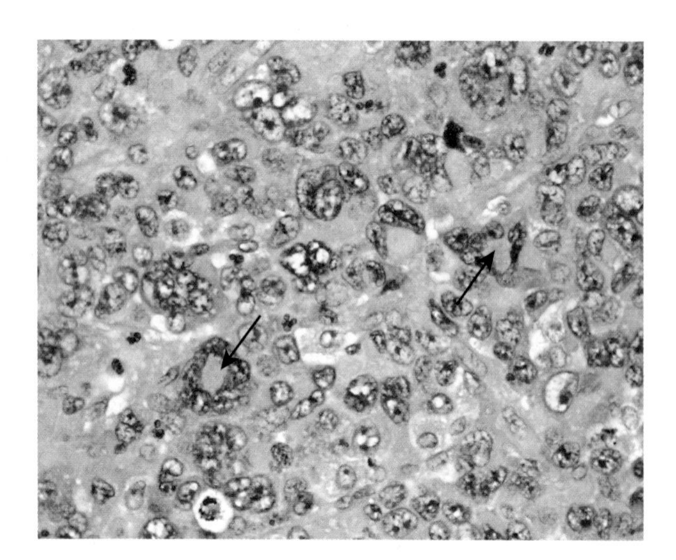

Figure 6.14 Anaplastic large cell lymphoma. Other examples show more marked pleomorphism with atypical mitoses. Nuclei are open and lobated and include the hallmark cells, in which the lobes form horseshoe or ring shapes (arrows).

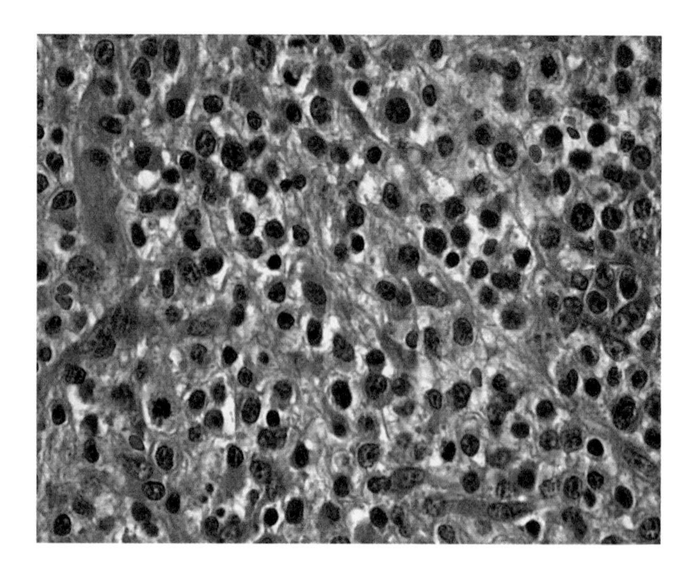

Figure 6.15 Anaplastic large cell lymphoma—small cell variant. Nuclei are smaller and irregular and many lack prominent nucleoli. The cells have abundant pale-staining or clear cytoplasm.

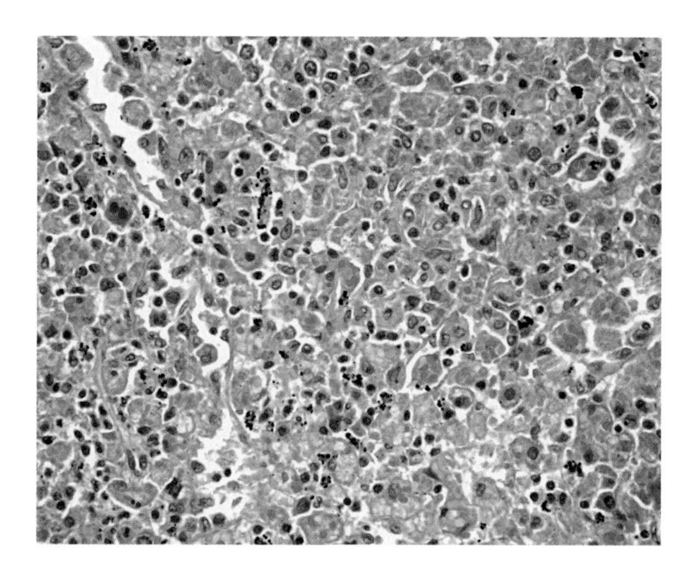

Figure 6.16 Anaplastic large cell lymphoma—lymphohistiocytic variant. Neoplastic cells are obscured by numerous histiocytes, some of which show erythrophagocytosis.

extranodal sites are less often involved. Response to treatment and survival rates are poor in comparison to ALK-positive cases, but recent studies suggest that they are better than PTCL-NOS. The morphological features of ALK-negative ALCL are similar to those of the common type of ALK-positive ALCL, including hallmark cells. The cells are cohesive and show a tendency to infiltrate lymph node sinuses. By definition, CD30 is strongly expressed, with a membranous and Golgi pattern of staining. The profile of staining for T-cell markers is similar to ALK-positive ALCL with loss of some T-cell antigens. However, many lymphomas express CD4 and CD43 and a significant number express CD3. In common with ALK-positive ALCL,

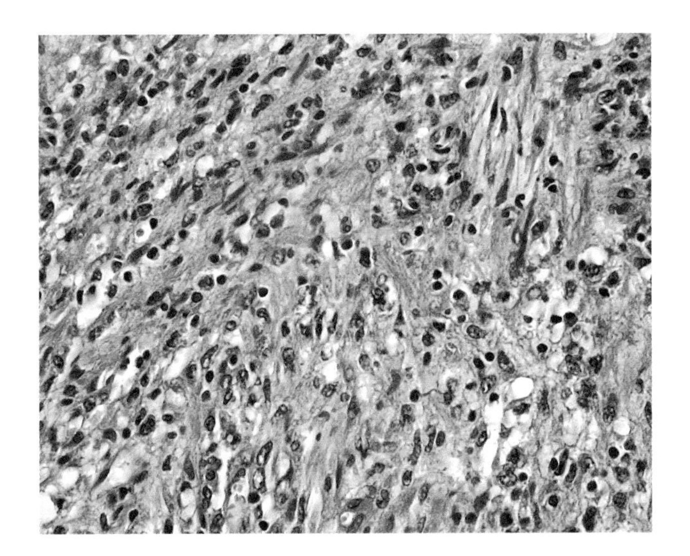

Figure 6.17 Anaplastic large cell lymphoma—sarcomatoid variant. Many of the neoplastic cells have oval or spindle morphology and mimic a connective tissue sarcoma.

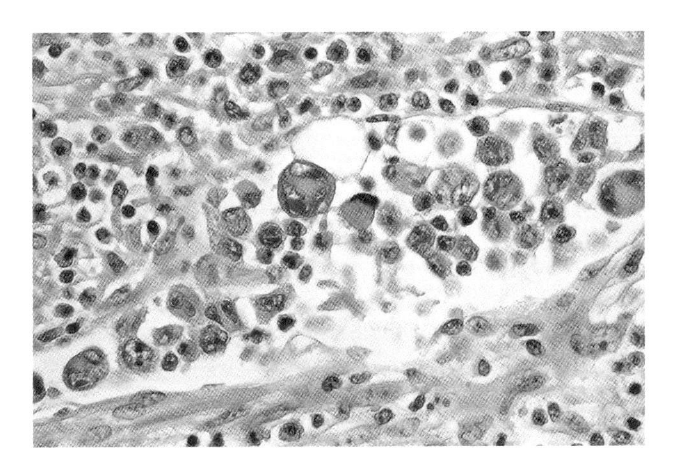

Figure 6.18 Anaplastic large cell lymphoma. A sinus pattern of spread is typical of early nodal involvement.

cytotoxic granule markers are variably expressed, but only a small minority show staining for EMA and CD25.

Breast implant–associated anaplastic large cell lymphoma

Breast implant–associated anaplastic large-cell lymphoma is an extremely rare but increasingly recognised entity. Although this is a predominantly extranodal disease, rare cases present with involvement of axillary lymph nodes. The average patient is 50 years of age and the tumour develops on average at 10 years post-surgery. The lymphoma usually arises within the fibrous capsule surrounding the implant and is often associated with an effusion; less frequently the tumour presents as a mass or with nodal disease (Alobeid *et al.* 2009, Aladily *et al.* 2012, Ye *et al.* 2014). The majority of cases are ALK negative and are associated with textured surface silicone-coated implants. Cytological examination of the seroma fluid shows large, pleomorphic neoplastic cells with basophilic cytoplasm. In histological sections, the appearances are similar to those seen in other cases of ALCL and sinusoidal distribution has been described in the involved lymph nodes. Immunohistochemistry demonstrates that the cells are strongly CD30 positive and negative for ALK, CD15, CD20 and EBER. The cells express T-cell associated markers; CD4 expression is reported to be seen most frequently, in up to 84 per cent of cases in one study, with the same study demonstrating CD3 expressed in fewer than 50 per cent of cases (Taylor *et al.* 2013). CD43 is frequently positive and there is variable expression of other T-cell associated antigens, cytotoxic granule-associated proteins, EMA, and clusterin. Most cases show clonal rearrangement of the TCR genes.

The majority of cases of breast implant–associated ALCL appear to have a good prognosis and patients who present with an effusion around the implant, without a tumour mass or nodal involvement, may achieve complete remission with removal of the implant and capsulectomy alone. However, those with a tumour mass or nodal involvement are more likely to have clinically aggressive disease requiring surgery and systemic therapy (Alobeid *et al.* 2009, Miranda *et al.* 2014) (Boxes 6.7 and 6.8 and Figures 6.19 to 6.21).

BOX 6.7: (T/null) anaplastic large cell lymphoma: Immunophenotype

- CD30 expression essential
- ALK expression provides important division into prognostic groups
- EMA typically positive in ALK+ tumours
- CD3, CD5, CD7, CD8 usually negative, CD4 more often positive
- TIA-1, granzyme, perforin usually positive in ALK+ tumours
- CD56 often positive
- Neoplastic cells often aggregate around blood vessels

BOX 6.8: CD30-positive tumours with anaplastic large cell morphology

- ALK+ anaplastic large cell lymphoma
- ALK− anaplastic large cell lymphoma
- Anaplastic variants of diffuse large B-cell lymphoma
- Primary cutaneous anaplastic large cell lymphoma
- Intestinal T-cell lymphoma (enteropathy-associated T-cell lymphoma)
- Adult T-cell lymphoma/leukaemia
- Peripheral T-cell lymphoma not otherwise specified

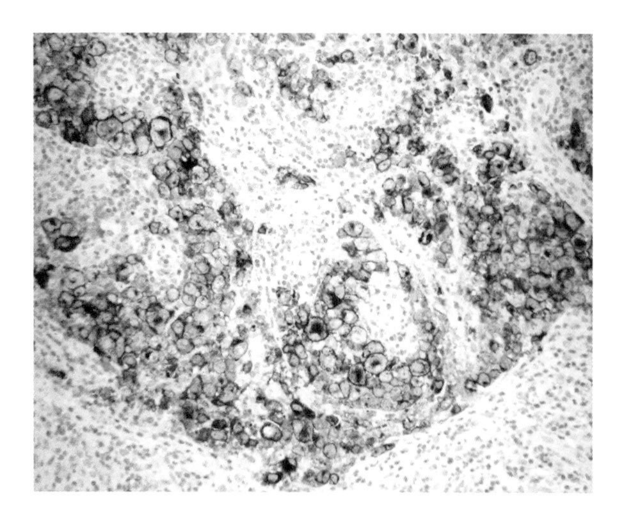

Figure 6.19 Anaplastic large cell lymphoma. CD30 staining typically shows membrane and Golgi positivity similar to that seen in Hodgkin lymphoma. CD30 is useful in identifying cells in the sinuses.

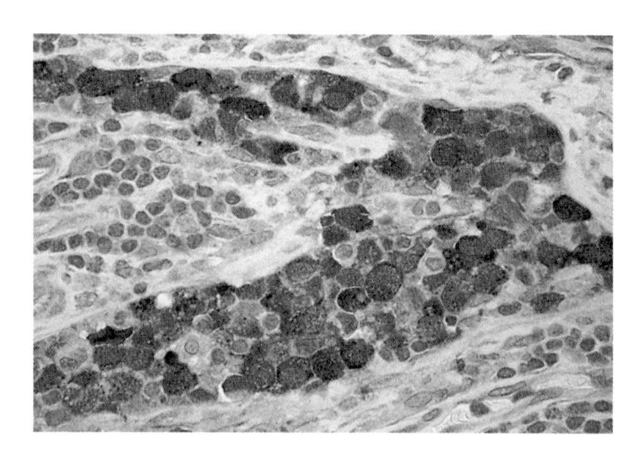

Figure 6.20 Anaplastic large cell lymphoma. Anaplastic lymphoma kinase (ALK) staining is both nuclear and cytoplasmic when the ALK/NPM (p80) translocation is present.

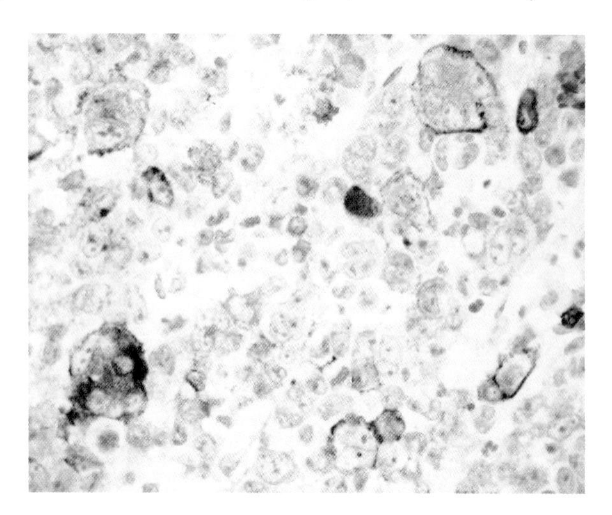

Figure 6.21 Expression of epithelial membrane antigen by anaplastic large cell lymphoma is variable and usually associated with anaplastic lymphoma kinase (ALK)-positive tumours. The staining pattern is similar to CD30 with membrane and Golgi positivity, but tends to be more granular.

GENETICS

The 2;5 translocation causes the NPM gene located at 5q35 to fuse with a gene at 2p23 encoding the tyrosine kinase receptor ALK. Transcription of the NPM-ALK hybrid gene results in production of NPM-ALK or p80. The presence of nuclear staining in addition to cytoplasmic staining is explained by the formation of dimers between NPM-ALK and wild-type nucleophosmin, which provides a nuclear signal and allows the dimer to enter the nucleus. The t(1;2) translocation occurs in about 10–20 per cent of ALK-positive ALCLs, leading to fusion of the ALK gene with the tropomyosin 3 (TPM3) gene, and results in cytoplasmic and cell membrane staining. Other translocations involving the ALK gene and producing different fusion proteins are rare. Those currently recognized and their staining patterns are shown in Table 6.3.

Secondary chromosomal abnormalities are frequent and variable, with ALK-positive and ALK-negative tumours tending to show different patterns. The difference between the two entities is also highlighted by gene expression profiling, which indicates different molecular signatures (Lamant *et al.* 2007). In addition, recent studies identified a recurrent translocation t(6;9)(p25.3;q32.3) in a subset of ALK-negative ALCL cases that results in downregulation of DUSP22 phosphatase gene expression located on chromosome 6p25.3, with some data suggesting that DUSP22 might function as a tumour suppressor gene (Feldman *et al.* 2011).

Despite the very variable expression of T-cell antigens, it is possible by polymerase chain reaction (PCR) to demonstrate that the majority (90 per cent) of ALCLs show clonal rearrangement of the TCR β or γ genes (Bonzheim *et al.* 2004).

Table 6.3 Translocation and anaplastic lymphoma kinase (ALK) staining patterns in (T/null) anaplastic large cell lymphoma (adapted from Swerdlow *et al.* 2008)

Chromosomal abnormality	ALK staining pattern	% of cases
t(2;5)(p23;q25)	Nuclear and diffuse cytoplasmic	84%
t(1;2)(q25;p23)	Diffuse cytoplasmic, most intense around periphery	13%
Inv (2)(p23q35)	Diffuse cytoplasmic	1%
t(2;3)(p23;q12)	Diffuse cytoplasmic	<1%
t(2;17)(p23;q23)	Granular cytoplasmic	<1%
t(X;2)(q11–12;p13.1)	Diffuse cytoplasmic	<1%
t(2;19)(p23;p13.1)	Diffuse cytoplasmic	<1%
t(2;22)(p23;q11.2)	Diffuse cytoplasmic	<1%
t(2;17)(p23;q25)	Diffuse cytoplasmic	<1%

DIFFERENTIAL DIAGNOSIS

When the pattern of infiltration of ALCL is partially or even predominantly within sinuses, metastatic carcinoma or malignant melanoma should be considered and can readily be excluded by staining for cytokeratins and melanoma markers such as S100, HMB45 and melan-A. In this situation EMA is not discriminatory because it is likely to be positive in both ALCL and malignant epithelial tumours, and occasionally in malignant melanoma.

Pleomorphic DLBCLs may express CD30, particularly on the more atypical cells, and occasionally show a sinusoidal pattern of spread. The cells may resemble H/RS cells or the hallmark cells of ALCL. Such tumours otherwise have the phenotype of DLBCL and do not express ALK. They are unrelated to ALCL and are regarded as anaplastic variants of DLBCL. A rare form of large B-cell lymphoma, which has immunoblastic or plasmablastic morphology and expresses ALK but not CD30, is discussed in Chapter 5. Other tumours in which ALK expression has been identified include inflammatory myofibroblastic tumours and some glioblastomas.

The distinction between ALCL and classical Hodgkin lymphoma has important implications for treatment and prognosis, and may be difficult. This applies particularly to tumours that lack expression of both ALK and CD15 and contain RS-like or multinucleate cells. By giving careful consideration to the expanded immunophenotype (Table 6.2), and, if necessary, to molecular techniques, it should be possible to make the distinction. The concept of a borderline group of 'ALCL–Hodgkin-like' lymphoma is no longer sustainable but there remain occasional tumours that defy the normal rules. These include tumours with Hodgkin-like cells expressing both CD30 and CD15, in which one or more of the T-cell markers or the cytotoxic granule markers are positive and B-cell markers are negative. Expression of CD45 and EBV-related antigens may be helpful in these cases as ALCL is usually negative for EBV and positive for CD45.

Cutaneous lesions of CD30+ lymphoma are seen in both primary cutaneous CD30+ ALCL of T-cell type (including lymphomatoid papulosis) and cutaneous manifestations of systemic ALCL. ALK expression is not usually a feature of the former and EMA is often negative. The presence of these markers suggests systemic disease although ALK-positive cutaneous ALCL has been described in children (Oschlies *et al.* 2013). If, on the other hand, a single lymph node shows ALK-negative ALCL or other CD30+ T-cell lymphoma, the possibility of a CD30+ primary cutaneous lymphoproliferative disease with nodal involvement should be considered, and a clinical history of skin lesions sought.

PERIPHERAL T-CELL LYMPHOMA NOT OTHERWISE SPECIFIED

In some respects, PTCL-NOS is a diagnosis of exclusion. The term covers the 50 per cent or so of peripheral T-cell lymphomas that do not fall into any of the other, better defined categories and therefore describes a heterogeneous group with very variable morphological features. Clinically, it is one of the more aggressive forms of non-Hodgkin lymphoma. It is primarily a disease of adults with an equal sex ratio. Most patients have systemic symptoms, and extranodal disease is frequent. Bone marrow, liver, spleen and skin are frequently involved sites and, rarely, there may be peripheral blood involvement. Less frequent manifestations are caused by abnormal production of cytokines and include eosinophilia and haemophagocytic syndrome (Ascani *et al.* 1997) (Box 6.9).

HISTOLOGY

In most cases, involved lymph nodes are diffusely infiltrated with complete effacement of normal nodal structure. The infiltrate may be monomorphic, consisting almost entirely of neoplastic cells, or polymorphic, with a mixed background of small lymphocytes, eosinophils, plasma cells and histiocytes. As in AITL, vessels may be prominent. Sometimes the pattern may be more clearly interfollicular with preservation of normal or even hyperplastic follicles (Suchi *et al.* 1987).

The malignant cells show a range of morphological features and sizes. In the majority of cases, nuclei are intermediate or large when compared with histiocyte nuclei. They frequently show some degree of irregularity or folding and tend to be hyperchromatic with indistinct nucleoli. In other examples, they may be vesicular with prominent nucleoli and resemble centroblasts or immunoblasts. The cytoplasm may be pale or clear with distinct cell borders. Markedly pleomorphic cells with polylobated nuclei may resemble those seen in ALCL and, if these are CD30+, the distinction from ALK-ALCL becomes blurred.

BOX 6.9: Peripheral T-cell lymphoma not otherwise specified: Clinical features

- Uncommon lymphomas with geographical and racial variations
- Usually adults with equal sex ratio
- Widespread disease and systemic symptoms frequent
- Poor response to treatment and low 5-year survival

The WHO classification resists the temptation to attempt further subclassification of PTCL-NOS on purely morphological characteristics but previous classifications have used cell size as the main criterion (Jaffe *et al.* 2001). The REAL classification provisionally divided lymphomas into those with medium-sized cells, mixed medium and large cells, and large cells (Harris *et al.* 1994); however, these subdivisions are subjective and do not at present appear to have clinical validity. Peripheral T-cell lymphomas in general have a poor 5-year survival, with age and International Prognostic Index being the main prognostic factors (Box 6.10 and Figures 6.22 to 6.26).

RECOGNIZED VARIANTS OF PERIPHERAL T-CELL LYMPHOMA NOT OTHERWISE SPECIFIED

Lymphoepithelioid (Lennert) lymphoma

In PTCL-NOS the background, non-neoplastic cell population may include clusters of epithelioid histiocytes. If these are prominent, such tumours are designated lymphoepithelioid T-cell lymphoma or Lennert lymphoma, a term first used in the Kiel classification (Suchi *et al.* 1987). The neoplastic T-cell population typically consists of a uniform population of intermediate-sized cells, but larger, more pleomorphic cells are often present. Some examples have a cytotoxic T-cell phenotype, expressing CD8 and one or more of the cytotoxic T-cell markers (Geissinger *et al.* 2004). Others express similar markers to AITL, suggesting an origin from TFH cells (Rodriguez-Pinilla *et al.* 2008). An aggressive course is reported in some patients with this variant, but the diagnosis has no established prognostic significance.

BOX 6.10: Peripheral T-cell lymphoma not otherwise specified: Morphology

- Infiltrate monomorphic or polymorphic
- Wide variation in cell size, typically medium to large
- Wide variation in nuclear morphology
- Epithelioid cell clusters often present

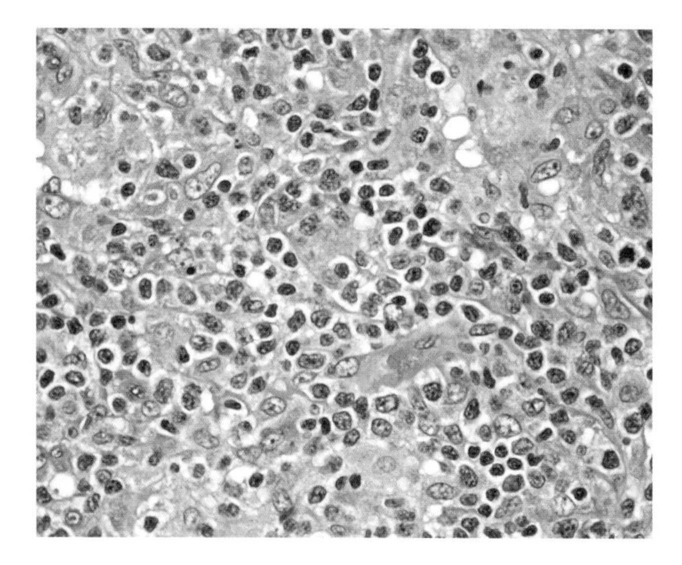

Figure 6.22 The infiltrate in peripheral T-cell lymphoma not otherwise specified is frequently polymorphous with neoplastic cells of varying sizes with irregular nuclei. In this example, some have clear cytoplasm similar to those seen in angioimmunoblastic lymphoma.

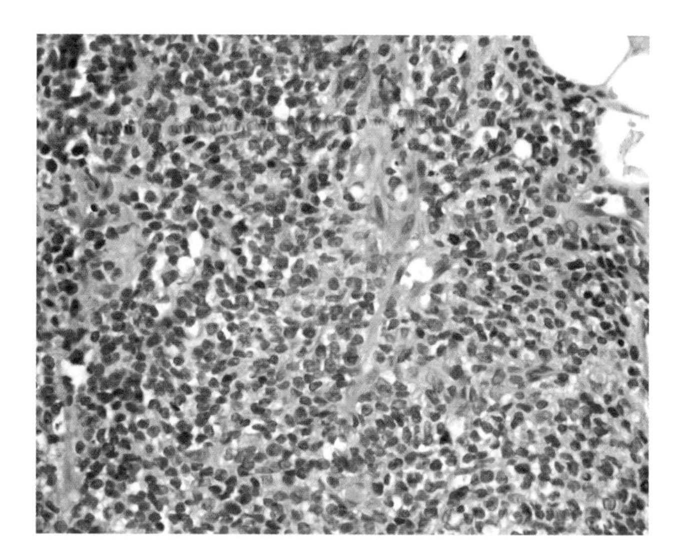

Figure 6.24 Peripheral T-cell lymphoma not otherwise specified: intermediate cell morphology. The tumour cells have irregular, small to medium-sized, uniformly hyperchromatic nuclei.

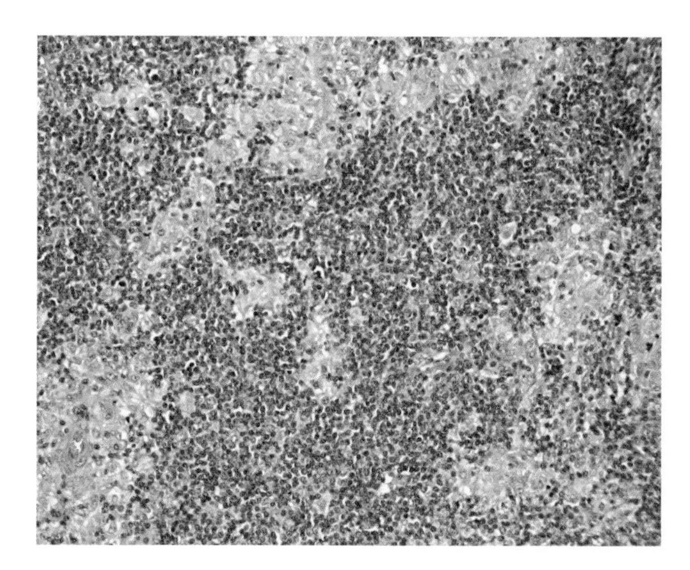

Figure 6.23 The so-called Lennert lymphoma is a peripheral T-cell lymphoma not otherwise specified in which cohesive clusters of epithelioid histiocytes are prominent.

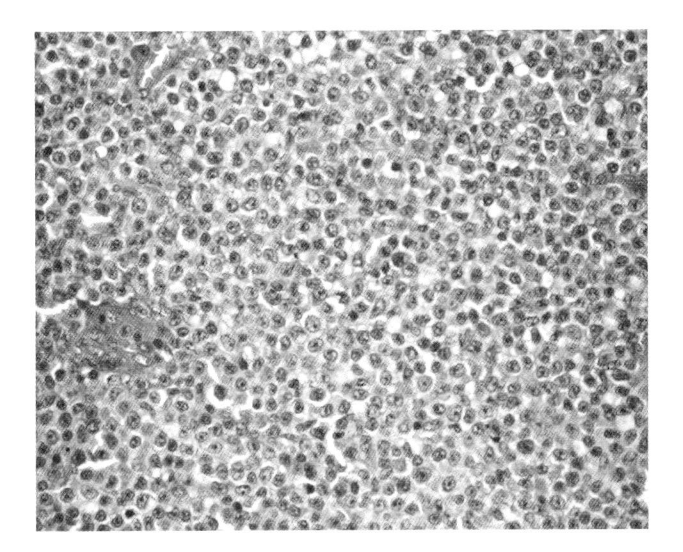

Figure 6.25 Peripheral T-cell lymphoma not otherwise specified: intermediate cell morphology. In this example, the cells have rounded, fairly uniform nuclei, most with a single prominent nucleolus.

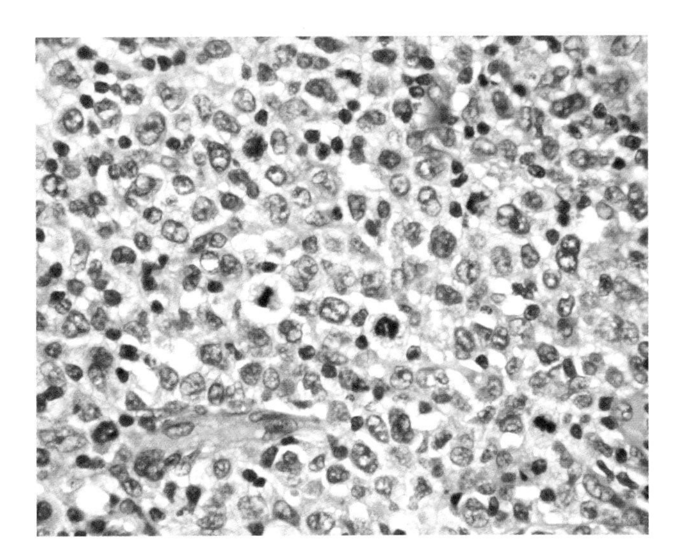

Figure 6.26 Peripheral T-cell lymphoma not otherwise specified: large cell morphology. In this example, cells have large, open, irregular, folded nuclei, one or more nucleoli and abundant pale or clear cytoplasm.

T-zone lymphoma

Residual follicles are frequently seen in T-cell lymphomas and do not usually constitute a diagnostic problem if other parts of the node are clearly infiltrated and effaced. When the follicles are largely preserved and the lymphoma has a uniformly paracortical growth pattern, the term T-zone lymphoma has been used. If the neoplastic cells are small or intermediate in size and atypia is not readily appreciated, the main differential diagnosis is reactive paracortical hyperplasia. The distinction depends on demonstrating an abnormal T-cell phenotype in the paracortical infiltrate. If the atypical cells have the phenotype of TFH cells, AITL with preservation of follicles (patterns I or II) is a consideration.

Follicular T-cell lymphoma

In contrast to T-zone lymphoma, in which the preserved follicles show normal reactive B-cell features, some T-cell lymphomas have a follicular or perifollicular growth pattern. Follicles consist wholly or partly of T-cells, which are typically medium-sized with clear cytoplasm, and the FDC network is preserved. The distinction from follicular lymphoma depends on recognizing the T-cell phenotype of the cells, which is often atypical. When the follicles are expanded and less well-defined, the pattern resembles progressive transformation of the germinal centres. In this situation, the presence of nodular aggregates of atypical cells may suggest nodular lymphocyte-predominant Hodgkin lymphoma. If the T-cells form an expanded perifollicular zone around residual reactive follicles, they resemble marginal zone lymphoma (Rudiger *et al.* 2000). The cells of follicular T-cell lymphoma have the immunophenotype of TFH cells (CD4+, CD10+, BCL-6+, PD-1+, CXCL13+), and there may be overlapping features with AITL. A minority of cases show a t(5;9) translocation (Huang *et al.* 2009). Similar to AITL, this entity occasionally has EBV-positive or EBV-negative Hodgkin-like cells, mimicking lymphocyte-rich classical Hodgkin lymphoma (Moroch *et al.* 2012). Clinically, the follicular variant of PTCL, NOS is distinct from AITL in that patients more often present with early stage disease with partial lymph node involvement and may lack the constitutional symptoms associated with AITL.

IMMUNOHISTOCHEMISTRY

In paraffin sections, using the commonly available T-cell markers CD3, CD4, CD5, CD7, CD8, CD43 and CD45RO, a variety of T-cell phenotypes can be identified and aberrant phenotypes are common. Most tumours express CD3, but CD5 and CD7 are frequently lost. CD4 and CD8 expression is also very variable and cells may express one, both or neither antigen. It is suggested that lymphoepithelioid T-cell lymphoma more often expresses CD8 and may be derived from cytotoxic T-cells (Geissinger *et al.* 2004). CD30 is frequently positive, particularly in large cell variants. It is unusual to find expression of the cytotoxic granule markers and CD56 in nodal PTCL-NOS.

Occasionally, scattered large cells resemble H/RS cells and may have an identical immunophenotype, expressing CD20, CD30 and even occasionally CD15. EBV-related antigens are also expressed and this may represent expansion of EBV-infected clones related to the immunodeficiency associated with the T-cell lymphoma (Quintanilla-Martinez *et al.* 1999).

BOX 6.11: Peripheral T-cell lymphoma not otherwise specified: Immunohistochemistry

- Aberrant T-cell phenotypes common
- CD3 usually positive
- CD5 and CD7 often lost
- CD4, CD8 very variable
- CD56 and cytotoxic granule markers unusual
- CD30 variably positive

Very rarely, neoplastic T-cells express CD20. This occurs mainly in PTCL-NOS but may also be seen in mycosis fungoides and enteropathy-associated T-cell lymphoma (EATL). CD20 is variably expressed in conjunction with T-cell markers, and molecular genetic studies may be required to clarify the apparently confusing immunophenotype (Rahemtullah *et al.* 2008) (Box 6.11).

GENETICS

Proof of clonality in T-cell proliferations depends on molecular techniques, and PCR can be used to demonstrate clonal rearrangement of the TCR genes. Peripheral T-cell lymphomas show a wide range of cytogenetic abnormalities but no consistent chromosomal abnormality has been identified. Gene expression profiling techniques have shown that the three main types of PTCL (AITL, ALCL and PTCL-NOS) share some abnormal gene patterns, but differences are seen in other groups of genes, in keeping with their differing morphological and clinical characteristics (Pittaluga *et al.* 2007b, Iqbal *et al.* 2010).

Recent studies have identified two molecular subgroups of PTCL-NOS characterized by high expression of either GATA-binding protein 3 (GATA-3) or T-box 21 (TBX21), with the GATA-3 subgroup being associated with distinctly worse prognosis (Iqbal *et al.* 2014). Both GATA-3 and TBX21 are transcription factors that are master regulators of gene expression profiles in T-helper (Th) cells, skewing Th polarization into Th2 and Th1 differentiation pathways, respectively. GATA-3–positive PTCLs are associated with eosinophilia that is typically mediated by Th2 cytokines. The TBX21 group might be more heterogeneous because it also includes a subset of cases with a cytotoxic profile, and there remain a proportion of cases in which the gene expression signature is indeterminate and cannot be assigned to either category.

DIFFERENTIAL DIAGNOSIS

In those tumours with a polymorphic background of non-neoplastic cells, particularly where follicles are preserved ('T-zone lymphoma'), it may be difficult to appreciate that the atypical cells are neoplastic T-cells rather than the activated blasts seen in a variety of reactive conditions. However, they tend to form a more clearly atypical population, and immunohistochemistry reveals their T-cell phenotype. If epithelioid histiocytes are prominent, other granulomatous conditions, such as toxoplasmosis, should be considered. The well-defined epithelioid and giant cell granulomas seen in sarcoidosis and granulomatous infections are a very occasional feature of T-cell lymphoma and they may even show some degree of central necrosis.

In some peripheral T-cell lymphomas, large pleomorphic cells may resemble Hodgkin or RS cells. CD30 is unreliable in making the distinction from classical Hodgkin lymphoma because many T-cell lymphomas express this antigen to a variable extent. It should also be borne in mind that some PTCL-NOSs express CD15. T-cell and B-cell markers are clearly useful in making this distinction: Hodgkin cells only rarely express T-cell antigens and, in about 30 per cent of cases, weak expression of CD20 is seen. Classical Hodgkin lymphoma is usually negative for CD45 and may be positive for EBV.

The distinction between PTCL-NOS and DLBCL should not present a problem once the basic immunohistochemical profile is completed. T-cell/histiocyte–rich, B-cell lymphoma may be more difficult to distinguish. The large, atypical B-cells are usually clearly seen against the background of small T-lymphocytes and histiocytes, but initial impressions, particularly on small and inadequate biopsies, may suggest a neoplastic T-cell population.

ALCL and angioimmunoblastic lymphoma both show some overlap with PTCL-NOS. ALCL should be considered if the tumour cells are large and pleomorphic with multilobated nuclei, and show strong and consistent expression of CD30. If ALK is negative, staining for EMA and the cytotoxic granule markers may help to decide whether the tumour should be categorized as ALK-negative ALCL. Recently it was found that 37 per cent of morphologically diagnosed PTCL-NOS cases were reclassified into other specific subtypes by molecular signatures. Subsequent re-examination of the morphology, immunohistochemistry, and IDH2 mutation analysis in these cases supported the validity of the reclassification (Iqbal *et al.* 2014).

ADULT T-CELL LEUKAEMIA/LYMPHOMA

ATLL is causally related to infection with the retrovirus HTLV-1. Sporadic cases do occur but the disease is concentrated in those areas where the virus is endemic. In parts of Japan, particularly on the island of Kyushu, where about 10 per cent of the population are seropositive, the virus is widespread and may be transmitted through blood and blood products, by breast milk from mother to child, or between sexual partners. Clusters of disease also occur in parts of West Africa, the Caribbean and the south-eastern United States. In New York and New Orleans, 10 to 25 per cent of intravenous drug abusers are seropositive for HTLV-1. In addition to ATLL, HTLV-1 is also linked to HTLV-1–associated myelopathy/tropical spastic paraparesis (HAM/TSP) syndrome. The disease has a long latency and it is estimated that 1 to 5 per cent of carriers of the virus develop ATLL, the incidence being greatest in those with the highest antibody titres. Neoplastic transformation does not depend only on the presence of the virus but is presumed to follow additional oncogenic events. The viral oncoprotein Tax is considered to be a major contributor to cell cycle deregulation by HTLV-1, transforming cells by directly disrupting cellular factors (protein–protein interactions) or altering their transcription profile. The virus infects the CD4+ (helper) T-cells and a definitive diagnosis can be made by demonstrating the HTLV-1 proviral DNA in the tumour cell DNA by PCR.

ATLL is a systemic disease with a variety of manifestations. Four clinical subtypes are recognized: acute, lymphomatous, chronic and smouldering (Box 6.12) (Shimoyama 1991, Matutes 2007). The acute form is the most common, usually presenting as stage IV disease, with widespread lymph node involvement and circulating atypical cells in the peripheral blood. About half of the patients also have skin involvement, hepatosplenomegaly is frequently present and the bone marrow is involved in about 60 per cent of cases. In Japanese patients, 'smouldering' or chronic forms of the disease are described in which circulating cells are uncommon. In the lymphomatous variant, lymphadenopathy is more prominent and circulating lymphoma cells are not seen. In the acute and lymphomatous types, therapy is largely ineffective and the prognosis is poor owing to high relapse rates (Shimoyama 1991).

One of the most characteristic features of ATLL is hypercalcaemia, often accompanied by lytic bone lesions. Bone biopsy shows increased osteoclastic activity and this may lead to a diagnosis of hyperparathyroidism. Blood eosinophilia is present in some cases.

BOX 6.12: Clinical subtypes of adult T-cell leukaemia/lymphoma

- Acute type: leukaemic with lymphocytosis, hypercalcaemia, lymphadenopathy and extranodal lesions
- Lymphomatous type: hypercalcaemia, lymphadenopathy and extranodal lesions; no lymphocytosis (<1 per cent abnormal cells)
- Smouldering type: skin or lung involvement. No hypercalcaemia or lymphadenopathy. No lymphocytosis (<5 per cent abnormal cells)
- Chronic type: lymphocytosis with $>3.5 \times 10^9$ T-cells, >5 per cent abnormal. No hypercalcaemia; lymphadenopathy may be present with skin, lung, liver and spleen lesions

The diagnosis of ATLL is usually made on the presence of the characteristic cells in the peripheral blood. These are pleomorphic and show considerable variation in size. Nuclear lobation, indicative of aneuploidy, is the most typical feature, giving rise to the term 'flower cells'.

HISTOLOGY

Lymph node biopsy from patients with acute and lymphomatous forms of the disease shows diffuse infiltration and apparent loss of architecture. However, reticulin stains may demonstrate retention of the underlying structure of the node and sinuses may be seen. Cell size is very variable. In some cases, small cells predominate or there is a mixture of small and large cells. In the majority of cases, they are intermediate to large with prominent nucleoli and clumped chromatin. Nuclear polylobation is characteristic and giant cells with very irregular nuclei may be present. In patients with early or smouldering chronic forms of the disease, the node may be only partially involved with infiltration of the paracortex by smaller cells and scattered giant cell forms resembling Hodgkin cells (Box 6.13 and Figures 6.27 to 6.29).

BOX 6.13: Adult T-cell leukaemia/lymphoma: Morphology

- Cell size very variable but distinctive polylobation of nuclei
- Often shows retention of underlying reticulin structure of node
- May mimic other forms of T-cell lymphoma or Hodgkin lymphoma

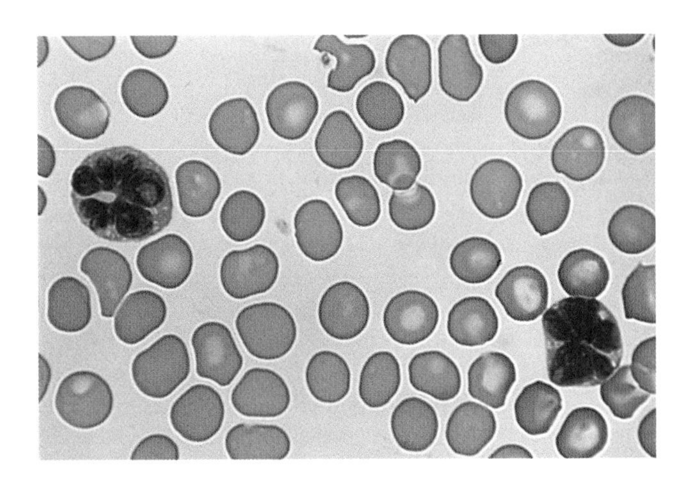

Figure 6.27 Adult T-cell leukaemia/lymphoma. Circulating neoplastic cells typically show complex lobation of the nuclei giving rise to the term 'flower cells'.

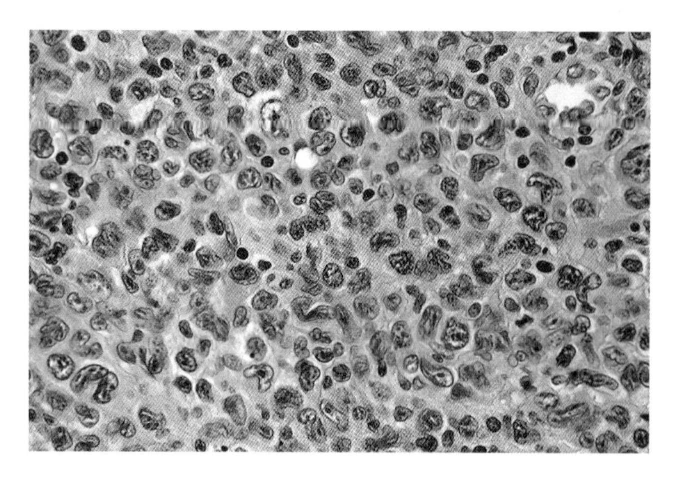

Figure 6.28 Adult T-cell leukaemia/lymphoma. In a typical case, the infiltrate is pleomorphic and many cells show nuclear folding and lobation.

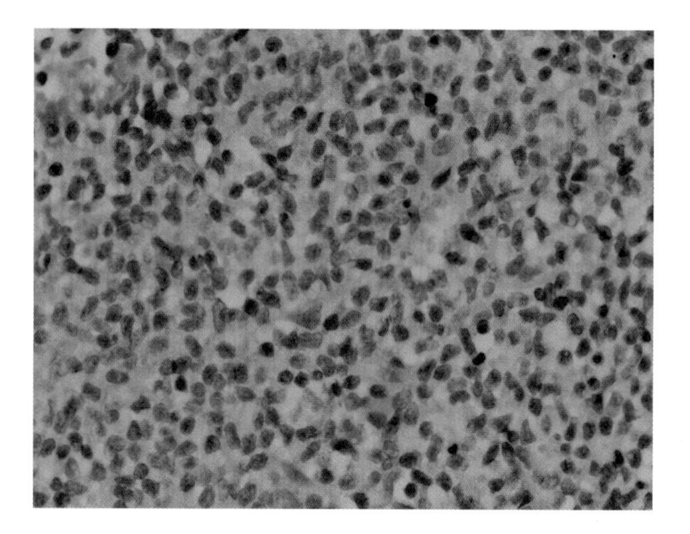

Figure 6.29 Adult T-cell leukaemia/lymphoma. In other examples, the infiltrate is more monomorphic and more closely resembles other forms of large cell lymphoma.

IMMUNOHISTOCHEMISTRY

The lymphoma cells express the T-cell antigens CD3 and CD5. They usually lack CD7 but stain strongly for the interleukin-2 (IL-2) receptor CD25. In most cases they are CD4+ and CD8– but exceptional cases are CD4– and CD8+ or express both antigens. CD52 is usually positive, which potentially has relevance for treatment with anti-CD52 humanised antibody (alemtuzumab). Some cases express FOXP3 in a proportion of the cells, which, in conjunction with CD25 and CD4, is a hallmark of regulatory T (Treg) cells. Large Hodgkin-like cells may express CD30 (Box 6.14).

BOX 6.14: Adult T-cell leukaemia/lymphoma: Immunohistochemistry

- CD3+, CD5+, CD7–
- CD25++
- Usually CD4+, CD8–
- Occasionally CD30+

GENETICS

Apart from clonal rearrangement of the TCR genes, no specific genetic abnormality has been identified.

DIFFERENTIAL DIAGNOSIS

Initial immunohistochemistry will indicate the T-cell origin of the cells. The distinction between ATLL and other types of unspecified peripheral T-cell lymphoma may be very difficult unless the diagnosis of ATLL is considered. In non-endemic areas, it is an unlikely diagnosis and immunohistochemistry is not sufficiently specific to suggest it as a possibility. The distinctive nuclear polylobation is usually present and should provide the clue to the diagnosis.

T-lymphoblastic leukaemia/lymphoma is a consideration, particularly in younger patients. In general, the cells in this condition are much more uniform. Although some may show nuclear irregularity and folding, polylobation is not a feature. Terminal deoxynucleotidyl transferase (TdT) is a very specific diagnostic marker of precursor B- and T-cell lymphomas, although not always positive (see Chapter 4).

T-cell prolymphocytic leukaemia (T-PLL) is discussed in the next section. This condition has similar clinical features and the circulating cells may have very irregular nuclei. The immunophenotype is not distinctive, apart from those cases that co-express CD4 and CD8, but the cells lack the strong CD25 expression seen in ATLL.

T-CELL PROLYMPHOCYTIC LEUKAEMIA

T-PLL, previously known as T-cell chronic lymphocytic leukaemia, is a more aggressive disease than B-cell chronic lymphocytic leukaemia (Matutes *et al.* 1991). Although long survival is occasionally recorded, most patients survive for only about a year after diagnosis. Like ATLL, it is a systemic disease in which leukaemia is associated with hepatosplenomegaly and generalized lymphadenopathy. The lymphocyte count is usually high, over 100×10^6, and there is associated anaemia and thrombocytopenia. The bone marrow is diffusely infiltrated and about 20 per cent of patients have skin infiltrates.

The diagnosis is usually made on the peripheral blood. The cells may resemble small lymphocytes or may be larger and 'prolymphocytic' with more prominent nucleoli.

The nuclear outline may be round or irregular. A characteristic feature is the presence of cytoplasmic protrusions or blebs, seen in blood or marrow films.

HISTOLOGY

In lymph node biopsies, the appearance resembles B-cell small lymphocytic lymphoma (B-SLL/CLL) but, in contrast to this disease, there are usually more residual follicles. Proliferation centres are absent and vessels, specifically high endothelial venules, are more conspicuous. The cells are slightly larger than normal lymphocytes and have more open chromatin with small nucleoli. They show a variable degree of nuclear irregularity.

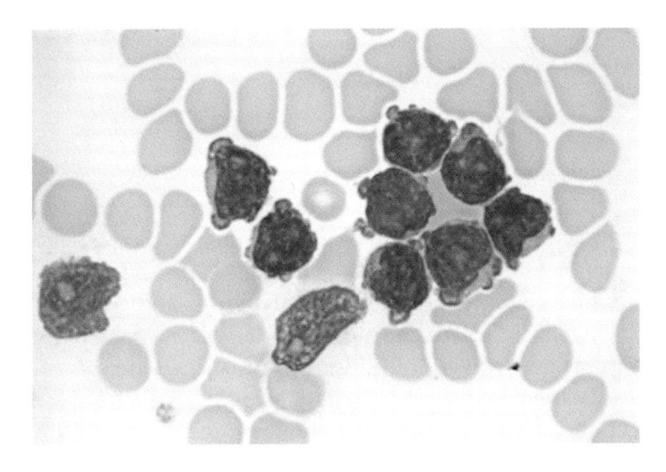

Figure 6.30 T-cell prolymphocytic leukaemia. Leukaemic cells show nuclear irregularity and cytoplasmic blebs.

IMMUNOHISTOCHEMISTRY

The cells express the T-cell antigens CD3, CD5 and CD7. CD3 membrane staining may be weak but there is usually cytoplasmic staining. The majority of cases are CD4+ CD8–, a small number are CD8+ and CD4– and some express both antigens. Antibodies are now available to the T-cell leukaemia/lymphoma 1 (TCL1) oncoprotein, which is overexpressed in the majority of cases of T-PLL (Herling *et al.* 2004).

Figure 6.31 T-cell prolymphocytic leukaemia. A low-power view of the node shows occasional spared follicles and high endothelial venules within the infiltrate.

GENETICS

T-PLL has several characteristic and recurring molecular abnormalities. In a majority of patients there is an abnormality involving the TCRαβ locus at 14q11, usually with an inversion of chromosome 14, which causes juxtaposition of this locus with the TCL1 and TCL1b oncogenes and resulting deregulation. Abnormalities of chromosome 8 are also present in about 80 per cent of cases. Other frequent abnormalities include deletions on chromosome 11 involving the ataxia telangiectasia mutated (ATM) gene, and gains or losses on chromosome 8.

DIFFERENTIAL DIAGNOSIS

In the context of a high lymphocyte count and generalized lymphadenopathy, the main differential diagnosis of T-PLL is the much more common SLL/CLL. The presence of unusual features, such as follicular sparing and lack of proliferation centres, should throw doubt on this diagnosis and immunohistochemistry will confirm the T-cell phenotype. In practice, the distinction from other types of T-cell lymphoma is rarely a problem once the presence of leukaemia is recognized. High lymphocyte counts with lymph node involvement are seen in Sézary syndrome, and in some examples of T-PLL marked nuclear irregularity is present with cerebriform nuclear outlines giving an appearance similar to Sézary cells. Secondary erythroderma is a rare feature of T-PLL with skin involvement, again suggesting the possibility of Sézary syndrome (Herling *et al.* 2004). In this situation TCL1 expression serves to confirm the diagnosis of T-PLL (Figures 6.30 to 6.32).

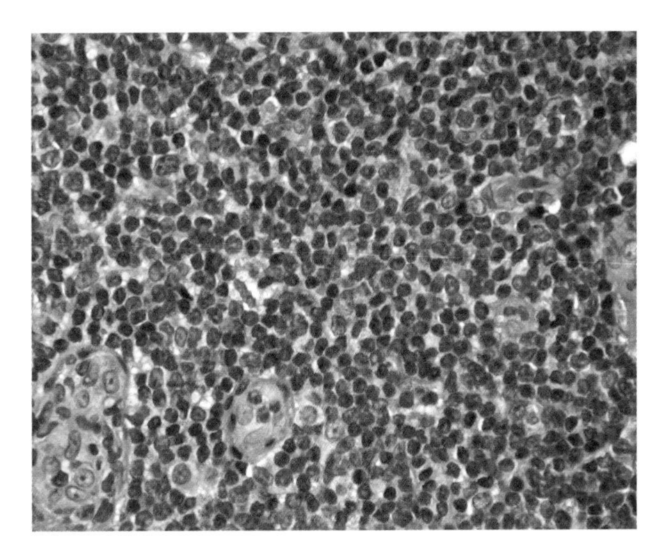

Figure 6.32 T-cell prolymphocytic leukaemia. Cells are slightly larger than lymphocytes with slightly irregular nuclei and a single nucleolus.

MYCOSIS FUNGOIDES AND SÉZARY SYNDROME

Mycosis fungoides and Sézary syndrome are now regarded as separate diseases that are often considered together for historical reasons and some overlap in the clinical presentation. Both are primary cutaneous T-cell lymphomas in which there is infiltration of the dermis and epidermis by neoplastic cells showing characteristic morphological features and epidermotropism. Mycosis fungoides is usually an indolent disease, progressing over a period of years from erythematous patches resembling dermatitis or psoriasis, through a plaque phase to a tumour phase. Sézary syndrome is a rare, more aggressive variant in which there is generalized reddening of the skin (erythroderma), a leukaemic phase, with circulating atypical cells and bone marrow involvement, and lymphadenopathy. Involvement of extracutaneous organs is a late manifestation of mycosis fungoides and indicates advanced disease with a poor prognosis, and an average survival time of about 12 months from that point. In addition to the lymph nodes, the organs most often involved are the lungs, spleen and liver.

The characteristic cells in mycosis fungoides and Sézary syndrome are mature T-lymphocytes of helper cell type and are usually CD4+. Large and small variants are recognized. The small cells, sometimes known as Lutzner cells, are slightly larger than normal lymphocytes and have deeply indented or convoluted nuclei, usually described as cerebriform. These are best seen in thin, plastic-embedded sections or by electron microscopy, but they can be appreciated in good-quality paraffin sections. Larger cells, or classical Sézary cells, have nuclei comparable in size to those of histiocytes. They are hyperchromatic and show marked nuclear irregularity and folding. Large and small cells are seen in both diseases, but small cells tend to predominate in the skin lesions of mycosis fungoides, at least in the early stages. In the late stages of the disease, blastic transformation may occur with the appearance of large CD30+ cells with less irregular nuclei and more prominent nucleoli. In Sézary syndrome, the circulating cells are usually of mixed small and large cell types.

HISTOLOGY

Assessment of lymph node involvement in mycosis fungoides is recognized to be a difficult and controversial area. It is important because it has been shown that lymph node involvement is associated with a worsening prognosis. Most patients with mycosis fungoides have some degree of lymphadenopathy, usually involving axillary and inguinal nodes draining the areas of involved skin. Histologically, the nodes show features of dermatopathic lymphadenopathy (see Chapter 3) and it is against this background that involvement by neoplastic cells has to be assessed, as the two processes often co-exist. In nodes showing clear evidence of involvement, there is partial or complete infiltration by mycosis fungoides cells of both small and large cell types. The paracortical area is selectively involved initially, but then there is progression to total effacement with spread beyond the capsule and foci of necrosis. A confident diagnosis of early involvement is more difficult as small clusters of neoplastic T-cells may be difficult to recognize in a background of dermatopathic lymphadenopathy. The degree of atypia and nuclear irregularity may be insufficient to distinguish the neoplastic cells from benign T-cells and the immunophenotype is not specific. Good-quality thin sections from well-fixed material are essential in assessing the cell morphology. WHO recommends a modified grading system for lymph node involvement in mycosis fungoides in which the Dutch system is combined with the classification recommended by the National Cancer Institute (NCI) (Box 6.15) (Ralfkiaer *et al.* 2008).

Involved lymph nodes show similar features in Sézary syndrome. The assessment of early involvement may be less of a problem as there is usually diffuse infiltration by the lymphoma cells. Transformation to a large cell lymphoma may occur (Figures 6.33 and 6.34).

BOX 6.15: Lymph node involvement in mycosis fungoides and Sézary syndrome (adapted from Olsen *et al.* 2007)

- N1: Dermatopathic lymphadenopathy with no atypical cells or small clusters of atypical lymphocytes (includes Dutch system[a] grade 1—dermatopathic lymphadenopathy [DL], and NCI-VA[b] classification LN1 and LN2)
- N2: Early involvement by mycosis fungoides with DL, preservation of nodal architecture, cerebriform nuclei and aggregates of atypical lymphocytes in paracortex (includes Dutch system grade 2 and NCI-VA LN3)
- N3: Partial or complete effacement of node by atypical or frankly malignant cells, many with cerebriform nuclei (includes Dutch system grades 3 and 4 and NCI-VA LN4)

[a] Dutch system: Scheffer *et al.* 1980.
[b] NCI-VA (National Cancer Institute-Veterans Administration classification): Clendenning and Rappaport 1979, Colby *et al.* 1981, Sausville *et al.* 1985.

IMMUNOHISTOCHEMISTRY

Neoplastic T-cells in mycosis fungoides and Sézary syndrome express CD3, CD5 and CD4. They are usually but not invariably negative for CD8. The absence of CD7 is a frequent finding but may also be seen in reactive conditions and therefore is of limited diagnostic value. Large, CD30+ blasts may be present. Cytotoxic granule antigens, TIA-1 and granzyme, are usually negative but are sometimes expressed in more atypical cells in advanced lesions (Figure 6.35).

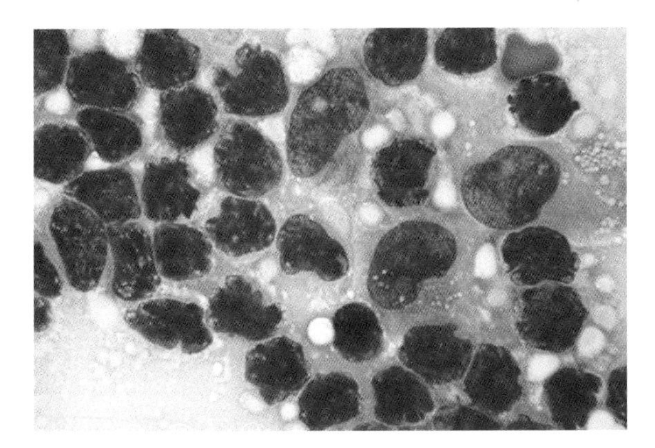

Figure 6.33 Mycosis fungoides/Sézary syndrome. In touch preparations, the cells show marked complex nuclear folding or convolutions.

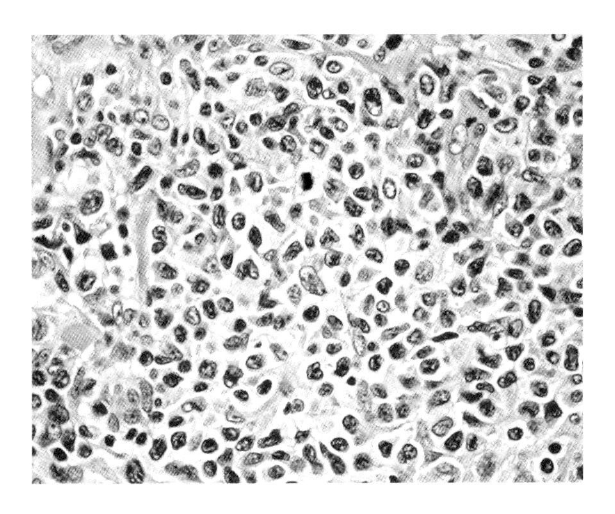

Figure 6.34 Mycosis fungoides/Sézary syndrome. In grade 3 involvement, the infiltrating cells include large blasts and show considerable variation in size. Many have irregular folded nuclei.

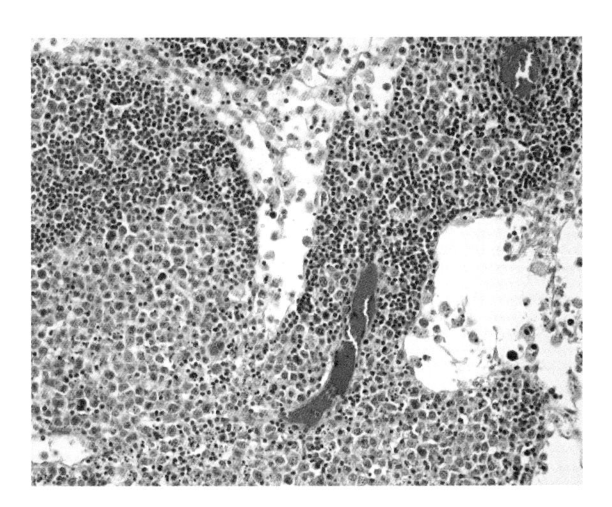

Figure 6.35 Enteropathy-associated T-cell lymphoma (EATL). In mesenteric nodes involved by EATL, the infiltrate is similar to that seen in the bowel wall. In this example of early involvement, the sinuses contain an infiltrate of pleomorphic cells.

GENETICS

There is clonal rearrangement of TCR genes in most cases and this provides a useful way of assessing early nodal involvement. The molecular alterations in Sézary syndrome differ from those of mycosis fungoides.

ENTEROPATHY-TYPE T-CELL LYMPHOMA (ENTEROPATHY-ASSOCIATED T-CELL LYMPHOMA)

First described as malignant histiocytosis of the intestine, EATL is now recognized as a form of T-cell lymphoma and is one of the more common forms seen in Europe. Its name derives from its close link to adult-onset coeliac disease, which may be clinically occult before presentation of the lymphoma. A minority of patients show only minimal evidence of gluten enteropathy with an increase in intraepithelial lymphocytes. Patients with or without a history of coeliac disease usually present with abdominal pain or small bowel perforation and are found to have ulceration of the small bowel, often with one or more tumour masses. Typically, the diagnosis is made on resected bowel but occasionally abdominal lymph nodes may be the only material available. The prognosis is very poor.

Two distinct variants of EATL are now recognized: type I and type II. In both types neoplastic cells diffusely infiltrate the wall of the small bowel. In type I EATL these are intermediate to large with rounded or irregular nuclei and prominent nucleoli. The cell population may appear polymorphic owing to the presence of inflammatory cells, including eosinophils and histiocytes. Some lymphomas show more marked pleomorphism with multinucleate cells. The adjacent mucosa often shows the changes of coeliac disease and large numbers of intraepithelial lymphocytes are typically present (Wright 1997). Type II EATL is not associated with coeliac disease and lacks the inflammatory background of type I EATL. The tumour is composed of a monomorphic population of medium-sized lymphoid cells with clear cytoplasm and inconspicuous nucleoli; there may be prominent epitheliotropism. It has been proposed that it should be regarded as a distinct entity and given the alternative name *monomorphic intestinal T-cell lymphoma* or *epitheliotropic intestinal T-cell lymphoma* (Tan *et al.* 2013).

HISTOLOGY

Involved mesenteric lymph nodes are often completely effaced by an infiltrate similar to that seen in the bowel wall but the degree of infiltration is variable and, in nodes showing early involvement, the lymphoma cells may show a sinus pattern of spread.

IMMUNOHISTOCHEMISTRY

The lymphoma cells typically express CD3 and CD7, and larger, more anaplastic cells express CD30 to a variable degree (Wright 1997). CD5 and CD4 are negative and CD8 is negative in type 1 EATL. Type II EATL usually expresses CD8 and CD56 (Chott *et al.* 1998, Tan *et al.* 2013). The cytotoxic granule markers TIA-1, granzyme and perforin are positive in both types. The cells may also express the homing receptor CD103.

DIFFERENTIAL DIAGNOSIS

In the unusual event of there being no definite evidence of small bowel pathology, or if only a needle core biopsy is available, EATL has to be distinguished from other forms of T-cell lymphoma. Depending on the morphological features, these include ALCL and PTCL-NOS. ALCL is a consideration if the lymphoma shows considerable pleomorphism and strong, consistent CD30 expression. The immunophenotype is similar, with the important exception of CD3 and CD7, which are usually negative in ALCL. In those tumours showing more uniform cell morphology with weak or absent CD30 expression, the distinction from PTCL-NOS is difficult but the presence of cytotoxic granule-associated proteins points to EATL.

T-CELL LEUKAEMIAS AND RELATED SYSTEMIC DISORDERS

T-cell leukaemias and those disorders involving multiple organ systems frequently involve lymph nodes. Lymphadenopathy is not usually the presenting symptom or a major manifestation, and the biopsy appearances may be misleading or inconclusive if they are not taken in the context of other findings. T-cell lymphoblastic leukaemia/lymphoma is discussed in Chapter 4. Two other rare diseases, both associated with EBV, merit brief mention here.

AGGRESSIVE NK-CELL LEUKAEMIA

Aggressive NK-cell leukaemia (ANKL) is a rare neoplasm of NK cells that is seen almost exclusively in Asian races, although occasional cases are described in the West. It is closely related to nasal-type NK-cell lymphoma and gene expression profiling studies show similarities among all of the EBV-positive NK-cell malignancies. However, there are distinct clinical differences between the two diseases. Patients with ANKL are generally younger, with a median age of 42 years, and the pattern of organ involvement differs. Hepatosplenic involvement is frequent and skin lesions are unusual. Whereas lymphadenopathy is not a feature of nasal-type NK-cell lymphoma, it is present in about 40 per cent of patients with ANKL. The disease has an aggressive course and its complications include haemophagocytic syndrome (Suzuki *et al.* 2004, Oshimi 2007)

ANKL usually presents with B-symptoms, particularly fever. The diagnostic cells are seen in variable numbers in the peripheral blood and bone marrow, and patients may be overtly leukaemic, anaemic or pancytopenic. In peripheral blood preparations or bone marrow smears the cells resemble large granular lymphocytes with lightly staining cytoplasm and azurophilic granules. Nuclei are large, atypical and irregularly folded. In tissue sections, including involved lymph nodes, the infiltrating cells have round or irregular, hyperchromatic nuclei with small nucleoli (Suzuki *et al.* 2004, Oshimi 2007).

Immunohistochemistry

The neoplastic cells are positive for CD2 and CD56. Both CD7 and the NK-cell marker CD16 are positive in about 75 per cent of cases and CD57 is occasionally expressed. The majority of T-cell markers, including CD3 and CD5, are negative. However, cytoplasmic CD3 (CD3ε) is usually positive. EBV can be detected by *in-situ* hybridization in most cases.

EPSTEIN–BARR VIRUS–POSITIVE T-CELL LYMPHOPROLIFERATIVE DISORDERS OF CHILDHOOD

The response to infection with EBV is complex and to a large extent dependent on individual immune response. In young children, primary infection is typically asymptomatic, whereas young adults may develop infectious mononucleosis. Fatal infectious mononucleosis is very rare in immunocompetent individuals, whereas immunocompromised patients, particularly those with X-linked lymphoproliferative disease (Duncan's disease), are prone to develop a fulminant form of the disease. EBV-related lymphoproliferative disorders can involve B-, T- or NK cells: their nomenclature remains confusing and a detailed description is beyond the scope of this book (Cohen *et al.* 2009).

Chronic active Epstein–Barr virus (CAEBV) disease occurs after primary infection in patients without known immunodeficiency. It is defined as a systemic, EBV-positive lymphoproliferative disease characterized by fever, lymphadenopathy and splenomegaly. Patients have high blood levels of EBV DNA and elevated levels of EBV RNA or viral proteins in the tissues. It occurs mainly in children and has a strong racial predisposition, with most cases occurring in Japan, Korea, Mexico or Central America. The term covers a broad spectrum of disease, both polyclonal and clonal (Cohen *et al.* 2009). The current WHO classification recognizes two neoplastic variants: hydroa vacciniforme–like lymphoma and systemic EBV-positive T-cell lymphoproliferative disorder of childhood (Quintanilla-Martinez *et al.* 2008). The latter is a fulminating illness of children or young adults that shows some overlap with aggressive NK-cell lymphoma. Patients present with fever and later develop hepatosplenomegaly with liver failure, sometimes accompanied by lymphadenopathy. Median survival is less than a year. The liver and spleen are infiltrated by small T-cells with no significant atypia. The lymph nodes are largely unremarkable with preserved architecture and open sinuses. Hyperplasia of the follicular or paracortical compartments is lacking. A variable degree of sinus histiocytosis with erythrophagocytosis is present and small granulomas are described.

Immunohistochemistry

The infiltrating T-cells are typically positive for CD2, CD3 and TIA-1, but negative for CD56. In the majority of cases they are CD8+, but some, typically those associated with acute primary EBV infection, are CD4+. Some show both CD4 and CD8 subsets. The small lymphoid cells are positive for EBV by *in-situ* hybridization.

REFERENCES

Aladily TN, Medeiros J, Amin MB, *et al.* 2012 Anaplastic large cell lymphoma associated with breast implants: A report of 13 cases. *American Journal of Surgical Pathology* **36:** 1000–1008.

Alobeid B, Sevilla DW, El-Tamer, *et al.* 2009 Aggressive presentation of breast implant–associated ALK-1 negative anaplastic large cell lymphoma with bilateral axillary lymph node involvement. *Leukemia and Lymphoma* **50:** 831–833.

Ascani S, Zinzani PL, Gherlinzoni F, *et al.* 1997 Peripheral T-cell lymphomas. Clinicopathologic study of 168 cases diagnosed according to the REAL Classification. *Annals of Oncology* **8:** 583–592.

Attygalle AD, Al-Jehani R, Diss TC, *et al.* 2002 Neoplastic T-cells in angioimmunoblastic T-cell lymphoma express CD10. *Blood* **99:** 627–633.

Attygalle AD, Cabeçadas J, Gaulard P, *et al.* 2014 Peripheral T-cell and NK-cell lymphomas and their mimics; taking a step forward—report on the lymphoma workshop of the XVIth meeting of the European Association for Haematopathology and the Society for Hematopathology. *Histopathology* **64:** 171–199.

Attygalle AD, Diss TC, Munson P, *et al.* 2004 CD10 expression in extranodal dissemination of angioimmunoblastic T-cell lymphoma. *American Journal of Surgical Pathology* **28:** 54–61.

Bajor-Dattilo EB, Pittaluga S, Jaffe ES 2013 Pathobiology of T-cell and NK-cell lymphomas. *Best Practice and Research Clinical Haematology* **26:** 75–87.

Benharroch D, Meguerian-Bedoyan Z, Lamant L, *et al.* 1998 ALK-positive lymphoma a single disease with a broad spectrum of morphology. *Blood* **91:** 2076–2084.

Bonzheim I, Geissinger E, Roth S, *et al.* 2004 Anaplastic large cell lymphomas lack the expression of T-cell receptor molecules or molecules of proximal T-cell receptor signalling. *Blood* **104:** 3358–3360.

Cheuk W, Hill RW, Bacchi C, *et al.* 2000 Hypocellular anaplastic large cell lymphoma mimicking inflammatory lesions of lymph nodes. *American Journal of Surgical Pathology* **24:** 1537–1543.

Chott A, Haedicke W, Mosberger I, *et al.* 1998 Most CD56+ intestinal lymphomas are CD8+ CD5 negative T-cell lymphomas of monomorphic small to medium size histology. *American Journal of Pathology* **153:** 1483–1490.

Clendenning WE, Rappaport HW 1979 Report of the committee on pathology of cutaneous T cell lymphomas. *Cancer Treatment Reports* **63:** 719–724.

Cohen JI, Kimura H, Nakamura S, *et al.* 2009 Epstein–Barr virus-associated lymphoproliferative disease in non-immunocompromised hosts: A status report and summary of an international meeting, 8–9 September 2008. *Annals of Oncology* **20:** 1472–1482.

Colby TV, Burke JS, Hoppe RT 1981 Lymph node biopsy in mycosis fungoides. *Cancer* **47:** 351–359.

de Leval L, Savilo E, Longtine J, *et al.* 2001 Peripheral T-cell lymphoma with follicular involvement and a CD4+/bcl-6+ phenotype. *American Journal of Surgical Pathology* **25:** 395–400.

Dogan A 2003 Angioimmunoblastic T-cell lymphoma. *British Journal of Haematology* **121:** 681–691.

Feldman AL, Dogan A, Smith DI, *et al.* 2011 Discovery of recurrent t(6;7)(p25.3;q32.3) translocations in ALK-negative anaplastic large cell lymphomas by massively parallel genomic sequencing. *Blood* **117:** 915–919.

Geissinger E, Odenwald T, Lee SS, *et al.* 2004 Nodal peripheral T-cell lymphomas and, in particular, their lymphoepithelioid (Lennert's) variant are often derived from CD8(+) cytotoxic T-cells. *Virchows Archiv* **445:** 334–343.

Grogg KL, Attygalle AD, Macon WR, *et al.* 2006 Expression of CXCL13, a chemokine highly upregulated in germinal centre T-helper cells, distinguishes angioimmunoblastic T-cell lymphoma from peripheral T-cell lymphoma unspecified. *Modern Pathology* **19:** 1101–1107.

Harris NL, Jaffe ES, Stein H, *et al.* 1994 A revised European-American classification of lymphoid neoplasms: A proposal from the International Lymphoma Study Group. *Blood* **84:** 1361–1392.

Herling M, Khoury JD, Washington LT, *et al.* 2004 A systematic approach to diagnosis of mature T-cell lymphomas reveals heterogeneity among WHO categories. *Blood* **104:** 328–335.

Huang Y, Moreau A, Dupuis J, *et al.* 2009 Peripheral T-cell lymphomas with a follicular growth pattern are derived from follicular helper T-cells (TFH) and may show overlapping features with angioimmunoblastic T-cell lymphoma. *American Journal of Surgical Pathology* **33:** 682–690.

Iqbal J, Weisenburger DD, Greiner TC, *et al.* 2010 Molecular signatures to improve diagnosis in peripheral T-cell lymphoma and prognostication in angioimmunoblastic T-cell lymphoma. *Blood* **115:** 1026–1036.

Iqbal J, Wright G, Wang C, *et al.* 2014 Gene expression signatures delineate biological and prognostic subgroups in peripheral T-cell lymphoma. *Blood* **123:** 2915–2923.

Jaffe ES 2001 Anaplastic large cell lymphoma: The shifting sands of diagnostic hematopathology. *Modern Pathology* **14:** 219–228.

Jaffe ES, Harris NL, Stein H, Vardiman JW (Eds) 2001 *Tumours of the Haematopoietic and Lymphoid Tissues.* Lyon: IARC Press.

Jones D, Jorgensen JL, Shahsafaei A, *et al.* 1998 Characteristic proliferations of reticular and dendritic cells in angioimmunoblastic lymphoma. *American Journal of Surgical Pathology* **22:** 956–964.

Lamant L, de Reynies A, Duplantier MM, *et al.* 2007 Gene-expression profiling of systemic anaplastic large-cell lymphoma reveals differences based on ALK status and two distinct morphologic ALK+ subtypes. *Blood* **109:** 2156–2164.

Matutes E 2007 Adult T-cell leukaemia/lymphoma. *Journal of Clinical Pathology* **60:** 1373–1377.

Matutes E, Brito-Babapulle V, Swansbury J, *et al.* 1991 Clinical and laboratory features of 78 cases of T-prolymphocytic leukaemia. *Blood* **78:** 3269–3274.

Miranda RN, Aladily TN, Prince HM, *et al.* 2014. Breast implant-associated anaplastic large-cell lymphoma: Long-term follow-up of 60 patients. *Journal of Clinical Oncology* **32:** 114–120.

Moroch J, Copie-Bergman C, de Leval L, *et al.* 2012 Follicular peripheral T-cell lymphoma expands the spectrum of classical Hodgkin lymphoma mimics. *American Journal of Surgical Pathology* **36:** 1636–46.

Non-Hodgkin's Lymphoma Classification Project 1997 A clinical evaluation of the International Lymphoma Study Group classification of Non-Hodgkin's lymphoma. The Non-Hodgkin's Lymphoma Classification Project. *Blood* **89:** 3908–3918.

Olsen E, Vonderheit E, Pimpinelli N, *et al.* 2007 Revisions to the staging and classification of mycosis fungoides and Sézary syndrome: A proposal of the International Society for Cutaneous Lymphomas (ISCL) and the cutaneous lymphoma task force of the European Organization of Research and Treatment of Cancer (EORTC). *Blood* **110:** 1713–1722.

Onciu M, Behm FG, Raimondi SC, *et al.* 2003 ALK-positive anaplastic large cell lymphoma with leukemic peripheral blood involvement is a clinicopathological entity with an unfavourable prognosis. Report of three cases and review of the literature. *American Journal of Clinical Pathology* **120:** 617–625.

Oschlies I, Lisfeld J, Lamant L, *et al.* 2013 ALK-positive anaplastic large cell lymphoma limited to the skin: Clinical, histopathological and molecular analysis of 6 pediatric cases. A report from the ALCL99 study. *Haematologica* **98:** 50–56.

Oshimi K 2007 Progress in understanding and managing natural killer-cell malignancies. *British Journal of Haematology* **139:** 532–544.

Ottaviani G, Buesco-Ramos CE, Seilstad K, *et al.* 2004 The role of the perifollicular sinus in determining the complex immunoarchitecture of angioimmunoblastic T-cell lymphoma. *American Journal of Surgical Pathology* **28:** 1632–1640.

Pittaluga PP, Agostinelli C, Califano A, *et al.* 2007a Gene expression analysis of angioimmunoblastic lymphoma indicates derivation from T follicular helper cells and vascular endothelial growth factor deregulation. *Cancer Research* **67:** 10703–10710.

Pittaluga PP, Agostinelli C, Califano A, *et al.* 2007b Gene expression analysis of peripheral T-cell lymphoma, unspecified, reveals distinct profiles and new therapeutic agents. *Journal of Clinical Investigation* **117:** 823–834.

Quintanilla-Martinez L, Fend F, Rodriquez-Martinez L, *et al.* 1999 Peripheral T-cell lymphoma with Reed–Sternberg–like cells of B-cell phenotype and genotype associated with Epstein–Barr virus infection. *American Journal of Surgical Pathology* **23:** 1233–1240.

Quintanilla-Martinez L, Kimura H, Jaffe ES 2008 EBV-positive T-cell lymphoproliferative disorders of childhood. In Swerdlow SH, Campo E, Harris NL, *et al.* (Eds) *WHO Classification of Tumours of Haematopoietic and Lymphoid Tissues.* Lyon: IARC Press.

Rahemtullah A, Longtine JA, Harris NL, *et al.* 2008 CD20+ T-cell lymphoma: Clinicopathologic analysis of 9 cases. *American Journal of Surgical Pathology* **32:** 1593–1607.

Ralfkiaer E, Cerroni L, Sander CA, *et al.* 2008 Mycosis fungoides. In Swerdlow SH, Campo E, Harris NL, *et al.* (Eds) *WHO Classification of Tumours of Haematopoietic and Lymphoid Tissues.* Lyon: IARC Press.

Ree HJ, Kadin ME, Kikuchi M, *et al.* 1998 Angioimmunoblastic lymphoma (AILD-type T-cell lymphoma) with hyperplastic germinal centers. *American Journal of Surgical Pathology* **22:** 643–655.

Rodriguez-Justo M, Attygalle AD, Munson, *et al.* 2009 Angioimmunoblastic T-cell lymphoma with hyperplastic germinal centres: A neoplasia with origin in the outer zone of the germinal centres. *Modern Pathology* **22:** 753–761.

Rodriguez-Pinilla SM, Atienza L, Murillo C, *et al.* 2008 Peripheral T-cell lymphoma with follicular T-cell markers. *American Journal of Surgical Pathology* **32:** 1789–1799.

Rudiger T, Ichinohasama R, Ott MM, *et al.* 2000 Peripheral T-cell lymphoma with distinct perifollicular growth pattern. A distinct subtype of T-cell lymphoma? *American Journal of Surgical Pathology* **24:** 117–122.

Sausville EA, Worsham GF, Matthews MJ, *et al.* 1985 Histologic assessment of lymph nodes in mycosis fungoides/Sézary syndrome (cutaneous T-cell lymphoma): Clinical correlations and prognostic import of a new classification system. *Human Pathology* **16:** 1098–1109.

Scheffer E, Meijer CJ, vanVloten WA 1980 Dermatopathic lymphadenopathy and lymph node involvement in mycosis fungoides. *Cancer* **45:** 137–148.

Shimoyama M 1991 Diagnostic criteria and classification of clinical subtypes of adult T-cell leukaemia lymphoma. A report from the Lymphoma Study Group (1984–87). *British Journal of Haematology* **79:** 428–437.

Stein H, Foss HD, Durkop H, *et al.* 2000 CD30+ anaplastic large cell lymphoma: A review of its histopathologic, genetic, and clinical features. *Blood* **96:** 3681–3695.

Stein H, Mason DY, Gerdes J, *et al.* 1985 The expression of the Hodgkin disease associated antigen Ki-1 in reactive and neoplastic lymphoid tissue: Evidence that Reed–Sternberg cells and histiocytic malignancies are derived from activated lymphoid cells. *Blood* **66:** 848–858.

Suchi T, Lennert K, Tu LY, *et al.* 1987 Histopathology and immunohistochemistry of peripheral T-cell lymphomas: A proposal for their classification. *Journal of Clinical Pathology* **40:** 995–1015.

Summers TA, Moncor JT 2010 The small cell variant of anaplastic large cell lymphoma. *Archives of Pathology and Laboratory Medicine* **134:** 1706–1710.

Suzuki R, Suzumiya J, Nakamura S, *et al.* 2004 Aggressive natural killer cell leukemia revisited: Large granular lymphocyte leukemia of cytotoxic NK cells. *Leukemia* **18:** 763–770.

Swerdlow SH, Campo E, Harris NL, *et al.* (Eds) 2008 *WHO Classification of Tumours of Haematopoietic and Lymphoid Tissues*. Lyon: IARC Press.

Tan SY, Chuang SS, Tang T, *et al.* 2013 Type II EATL (epitheliotropic intestinal T-cell lymphoma): a neoplasm of intra-epithelial T-cells with predominant CD8$\alpha\alpha$ phenotype. *Leukemia* **27:** 1688–1696.

Taylor CR, Siddiqi IN, Brody GS 2013 Anaplastic large cell lymphoma occurring in association with breast implants: Review of pathologic and immunohistochemical features in 103 cases. *Applied Immunohistochemistry and Molecular Morphology* **21:** 13–20.

Wright DH 1997 Enteropathy-associated T-cell lymphoma. *Cancer Surveys* **30:** 249–261.

Ye X, Shokrollahi K, Rozen WM, *et al.* 2014 Anaplastic large cell lymphoma (ALCL) and breast implants: Breaking down the evidence. *Mutation Research/Reviews in Mutation Research* **762:** 123–132.

Youd E, Boyde AM, Attanoos RI, *et al.* 2009 Small cell variant of anaplastic large cell lymphoma: A 10-year review of the All Wales Lymphoma Panel Database. *Histopathology* **55:** 355–357.

7

Immunodeficiency-associated lymphoproliferative disorders

Any form of defect of the immune system increases the risk of lymphoproliferative disease. This is mainly because of impairment of T-cell surveillance, allowing unrestricted proliferation of B-cells in response to an antigenic stimulus or viral infection of which the Epstein–Barr virus (EBV) is most common. Hyperplasia ultimately leads to clonal proliferation and the development of lymphoma, usually of B-cell type. The risk of neoplasia and the patterns of lymphoid proliferation or lymphoma vary according to the type of immune deficiency. Some types of disease more often affect extranodal sites but others show frequent lymph node involvement.

LYMPHOPROLIFERATIVE DISEASES ASSOCIATED WITH PRIMARY IMMUNODEFICIENCY DISEASES

The primary immune disorders are rare and have diverse underlying pathologies. They occur predominantly in childhood, although common variable immunodeficiency (CVID) is seen in adults.

NON-NEOPLASTIC LESIONS

Fatal infectious mononucleosis is seen in patients with X-linked lymphoproliferative disorder (XLP or Duncan syndrome) and with severe combined immunodeficiency (SCID). Infection with EBV is associated with unrestrained proliferation of EBV-positive plasmacytoid and immunoblastic cells at nodal and extranodal sites. Death may be caused by haemophagocytic syndrome, pancytopenia and infection.

CVID includes a heterogeneous group of primary immunodeficiency disorders that may present in adulthood and are characterised by hypogammaglobulinaemia. Striking lymphadenopathy, usually cervical, mediastinal or abdominal, is a common finding in CVID, and in the majority of cases lymph node biopsies are benign. Histological patterns recognised include reactive lymphoid hyperplasia, atypical lymphoid hyperplasia and granulomatous inflammation (Sander *et al.* 1992). The alteration of lymphoid tissues seen in patients with immune disorders may make the distinction between benign and malignant lymphoid proliferations ambiguous (da Silva *et al.* 2011) and clonal lymphocyte populations may occur in CVID patients without lymphoma; therefore, careful attention to the immunohistochemical findings is required.

Children with the X-linked hyper-immunoglobulin M (IgM) syndrome have normal or elevated serum IgM, with reduced levels of IgG and IgA, and defective T-cell function. Lymph node biopsies show absence of germinal centres from follicles and plasma cells producing only IgM and IgD. The incidence of lymphoma is low.

Lymph node changes associated with the autoimmune lymphoproliferative syndrome (ALPS) are discussed in Chapter 3.

LYMPHOMAS

Patients with most forms of primary immune disorder have an increased incidence of malignant lymphoma, ranging from less than 10% in CVID to nearly 100% in XLP.

In the majority of cases the lymphomas are associated with EBV infection and are of B-cell type, with diffuse large B-cell lymphoma (DLBCL) and Burkitt lymphoma being most frequent. Their morphological and immunohistochemical features do not differ significantly from those of lymphomas arising in immunocompetent individuals.

In both CVID and ALPS, the range of lymphomas encountered is wider than with other types of primary immunodeficiency. They include peripheral T-cell lymphoma, lymphocyte-predominant Hodgkin lymphoma, classical Hodgkin lymphoma and low-grade neoplasms such as extranodal marginal zone lymphoma, small lymphocytic lymphoma and lymphoplasmacytic lymphoma. Classical Hodgkin lymphoma has also been reported in patients with Wiskott–Aldrich syndrome and ataxia telangiectasia. Ataxia telangiectasia is unusual in that T-cell neoplasms, usually either T-acute lymphoblastic leukaemia or T-prolymphocytic leukaemia, are more commonly seen than B-cell neoplasms (Gompels *et al.* 2003, Cunningham-Rundles 2012).

Lymphomatoid granulomatosis (LYG) is a rare complication of primary immunodeficiency described occasionally in patients with Wiskott–Aldrich syndrome and CVID.

LYMPHOMAS ASSOCIATED WITH INFECTION BY THE HUMAN IMMUNODEFICIENCY VIRUS

In addition to a spectrum of benign changes in lymph nodes, including those associated with opportunistic infections (see Chapter 3), patients with human immunodeficiency virus (HIV) infection have an increased incidence of lymphomas, although this has fallen since the introduction of effective antiretroviral therapy in 1996 (Kirk *et al.* 2001, Carbone *et al.* 2009). Lymphomas constitute an acquired immune deficiency syndrome (AIDS)–defining feature in HIV-positive patients.

The most common lymphomas seen in patients with AIDS are DLBCL and Burkitt lymphoma (Gabarre *et al.* 2001). These most frequently present with extranodal tumours, most commonly in the gastrointestinal tract. Between 30 and 60 per cent of AIDS-related Burkitt lymphomas are EBV positive. Burkitt lymphoma usually develops early in the course of the disease, before the drop in CD4 counts, and may be related to the lymphoproliferation that occurs at this stage.

There is an increased incidence of Hodgkin lymphoma in patients with HIV infection, with a greater prevalence of the poor prognosis subtypes (mixed cellularity and lymphocyte depleted) than is seen in immunocompetent patients. HIV-related Hodgkin lymphoma is frequently associated with EBV, and the incidence of Hodgkin lymphoma has increased since introduction of combined antiretroviral therapy (cART). A study of 31,056 HIV-infected male veterans showed that the Hodgkin lymphoma risk was highest in the first 12 months after cART initiation, and in the moderately, rather than severely, immunosuppressed individuals as determined by CD4 counts, supporting the hypothesis that immune reconstitution provides a more suitable environment for Hodgkin lymphoma development (Kowalkowski *et al.* 2013).

Two rare lymphomas are specifically associated with AIDS: primary effusion lymphoma (PEL), which remains invariably extranodal even when it extends beyond serous cavities, and plasmablastic lymphoma, which is discussed in Chapter 5.

POST-TRANSPLANT LYMPHOPROLIFERATIVE DISORDERS

The incidence of post-transplant lymphoproliferative disorders (PTLDs) varies with the type of allograft and the immunosuppressive regimen used. Historically, renal transplants have a risk of less than 1 per cent, whereas after cardiac transplantation the risk is 7 per cent. The risk of PTLD after bone marrow transplantation is less than 1 per cent except in patients with human leukocyte antigen (HLA) mismatch with high levels of immunosuppression. Multivisceral transplantation has been associated with a PTLD incidence of 38% (Madariaga *et al.* 2000). PTLD is of host origin in 90 per cent of cases and donor origin in 10 per cent, except after bone marrow transplantation, wherein almost all cases are of donor origin.

As with other immunodeficiency-associated lymphoproliferations, impaired T-cell immunity to EBV appears to be a major causal factor in PTLD. Patients who acquire EBV infection after transplantation are at greater risk of developing PTLD than those in whom the latent virus is reactivated and who had previous immunity. Approximately 20 per cent of PTLDs are EBV negative. These occur later (4–5 years post-transplant) than EBV positive cases (6–10 months post-transplant). Both may regress with reduction in immunosuppression. PTLD may be nodal or extranodal, often developing in sites such as the gastrointestinal tract or central nervous system. The grafted organ may be involved, particularly in lung transplant recipients, suggesting that additional local factors may be involved in the development of PTLD.

A recent study of 62 cases of PTLD after solid organ transplantation (Vase *et al.* 2015) showed that CD30 is commonly expressed in all PTLD subgroups, including early/polymorphic PTLDs, and that its expression is associated with superior outcome. With the development of anti-CD30 therapeutic agents, CD30 has become a readily targetable biomarker, which may offer a potential treatment option, particularly for relapsed PTLD (Box 7.1).

BOX 7.1: Post-transplant lymphoproliferative disease (PTLD)

- Early lesions:
  - Non-specific lymphoid hyperplasia: 'post-transplant florid follicular hyperplasia'—architecture preserved with residual reactive follicles
  - Plasmacytic hyperplasia: polyclonal (nodes, tonsils, adenoids), numerous plasma cells, immunoblasts, lymphocytes
  - Infectious mononucleosis–like changes
- Polymorphic PTLD—polyclonal or monoclonal; often occurs in children after primary Epstein–Barr virus infection; nodal or extranodal; nodal architecture effaced by plasma cells, immunoblasts, lymphocytes and intermediate forms
- Monomorphic PTLD—diffuse large B-cell lymphoma or Burkitt lymphoma
- T-cell lymphoma
- Other lymphomas—Hodgkin lymphoma, plasmacytoma-like lesions and myeloma

EARLY LESIONS

Non-specific lymphoid hyperplasia

Some transplant recipients develop a non-specific form of lymphoid hyperplasia in which the nodal architecture is preserved. Follicles show varying degrees of hyperplasia, which may be florid.

Plasmacytic hyperplasia and infectious mononucleosis-like post-transplant lymphoproliferative disorder

These proliferations typically involve lymph nodes and the tissues of Waldeyer's ring. They occur most commonly among young patients who were EBV negative at the time of transplantation.

Histology usually shows partial preservation of the normal nodal structure with a proliferation of polytypic plasma cells or of blast cells. The features may be similar to those seen in primary immunodeficiency-associated fatal infectious mononucleosis, when large numbers of EBV-positive blasts are present, together with plasma cells and plasmacytoid cells.

POLYMORPHIC POST-TRANSPLANT LYMPHOPROLIFERATIVE DISORDER

Polymorphic PTLD is the most common type of PTLD seen in children and follows primary EBV infection post-transplant. These proliferations usually show loss of the normal nodal architecture with a mixed infiltrate of small lymphocytes with angulated nuclei, plasma cells and immunoblasts. Many cells show intermediate morphology with some plasma cell features, indicating a full range of B-cell maturation (Figure 7.1). Bizarre cells may be seen, including some that are CD30+ and resemble Hodgkin/Reed–Sternberg cells. Mitoses are numerous and there may be areas of necrosis. More monomorphic areas may indicate progression to frank lymphoma (monomorphic PTLD). Immunohistochemistry may demonstrate light chain restriction, often confined to parts of the node, and many cells are EBER positive. Immunoglobulin gene analysis will usually show the proliferation to be clonal, the majority having mutated immunoglobulin genes. Some cases may regress after reduction of immunosuppression, and others progress despite treatment for lymphoma.

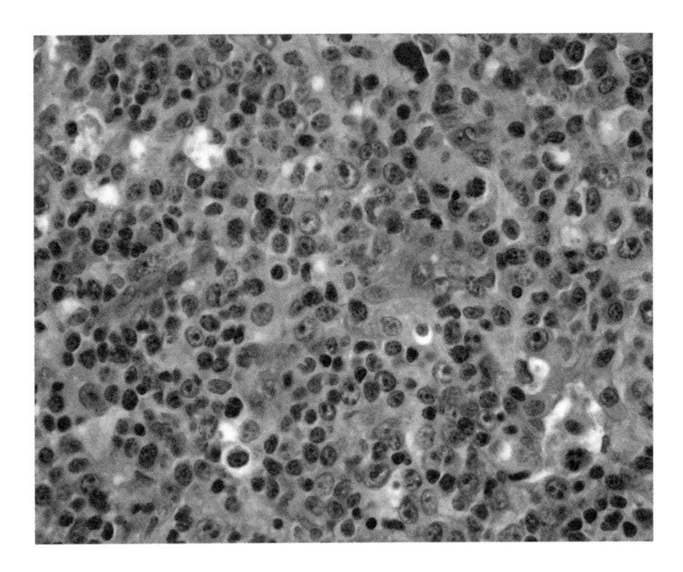

Figure 7.1 Polymorphic post-transplant lymphoproliferative disorder. The B-cell proliferation includes immunoblasts, plasma cells and many showing intermediate differentiation.

MONOMORPHIC POST-TRANSPLANT LYMPHOPROLIFERATIVE DISORDER

The development and expansion of malignant clones in PTLD leads to proliferations that closely resemble lymphomas and should be reported as such, although PTLD should appear in the diagnosis since they may regress on reduction of immunosuppression. The majority of cases are of B-cell origin, usually DLBCL, less commonly Burkitt lymphoma. More rarely, the tumours resemble plasma cell neoplasms, sometimes indistinguishable from plasmacytoma. The majority of lymphomas are EBV positive, but EBV negative cases are seen, particularly in late-onset lymphomas (Nelson *et al.* 2000). Follicular lymphomas and marginal zone/mucosa associated lymphoid tissue (MALT) lymphomas have also been reported in transplant patients but are currently not considered PTLD (Figures 7.2 to 7.4).

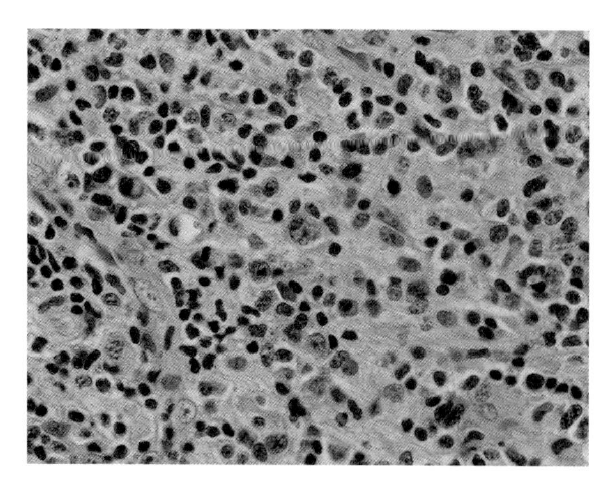

Figure 7.2 Methotrexate-associated malignant lymphoma. Although some cells closely resemble Hodgkin/Reed–Sternberg cells, the background cell population is not typical of classical Hodgkin lymphoma.

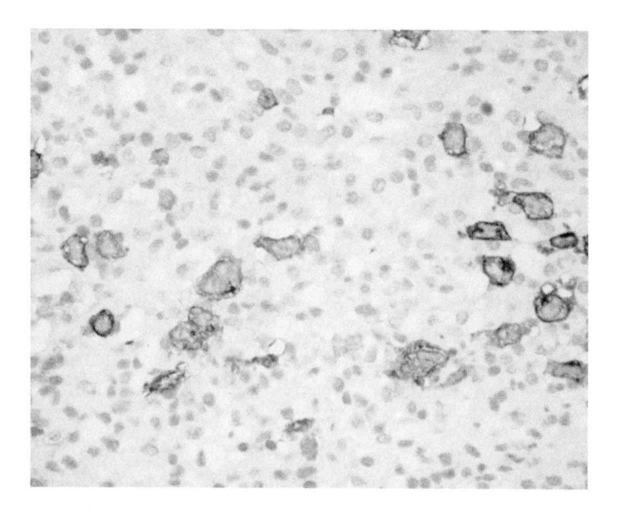

Figure 7.3 Methotrexate-associated malignant lymphoma stained for CD30. The larger cells are positive for CD30 with a typical pattern of membrane and Golgi staining.

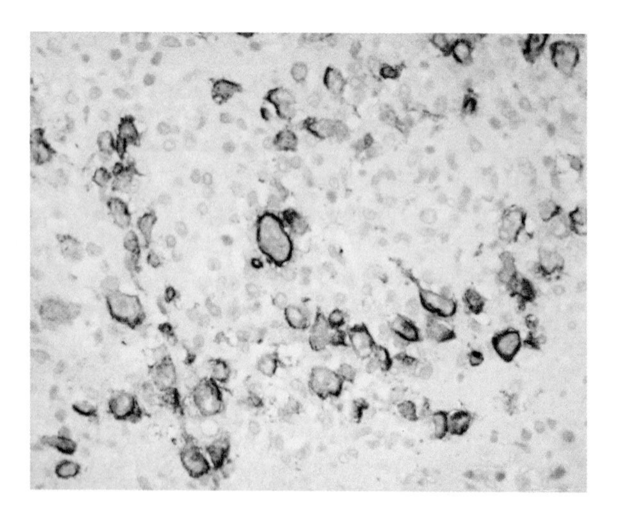

Figure 7.4 Methotrexate-associated malignant lymphoma stained for LMP1. Many of the cells express LMP1, indicating latent Epstein–Barr virus infection.

T-cell PTLD is much less common than B-cell PTLD and is more often EBV negative. A wide spectrum of morphology has been reported, T-cell lymphoma not otherwise specified being most frequent. Many are extranodal lymphomas, including subcutaneous panniculitis-like T-cell lymphoma, anaplastic large cell lymphoma and hepatosplenic T-cell lymphoma. Concurrent or composite T-cell and B-cell clonal proliferations are also described. Post-transplant Hodgkin lymphoma has been reported. This should not be confused with other forms of PTLD containing occasional CD30+ multinucleated cells, and the presence of CD15 is helpful in making this distinction.

In the majority of cases, monomorphic B-cell PTLD is indistinguishable from B-cell lymphoma (usually DLBCL) in non-transplant patients. EBV-positive tumours show a non-germinal centre pattern of immunostaining (CD10–, BCL-6+/–, MUM1+), and EBV-negative tumours are usually of germinal centre type (CD10+, BCL-6+, MUM1–). The presence of larger bizarre cells, sometimes resembling Reed–Sternberg cells, or cells showing plasmacytic features suggests a residual polymorphic element. PTLD can be regarded as a spectrum of disease with no sharp division between polymorphic and monomorphic variants (Swerdlow *et al.* 2008b).

LYMPHOPROLIFERATIVE DISORDERS ASSOCIATED WITH IATROGENIC IMMUNODEFICIENCY

In addition to those patients immunosuppressed for transplantation, a relatively small number of patients receiving long-term immunosuppressive therapy for autoimmune disease (rheumatoid arthritis [RA], psoriasis, dermatomyositis) have been reported to develop lymphomas or lymphoma-like proliferations. This is most often seen with methotrexate therapy, but is also described with azathioprine, used in the treatment of inflammatory bowel disease, and a group of drugs that act as antagonists to tumour necrosis factor alpha (TNFα)—infliximab, adalimumab and etanercept—often used in conjunction with methotrexate. A large study of patients with ulcerative colitis showed that they were four times more likely to develop lymphoma while treated with thiopurines compared with patients who had not been treated with thiopurines, and that the risk of lymphoma increased gradually for successive years of therapy (Khan *et al.* 2013).

Patients with RA are known to be at increased risk of lymphoma independent of therapy, and the risk increases with age and disease severity. The increased incidence directly resulting from therapy, which is usually reserved for patients with severe disease, may be small or even negligible (Wolfe and Michaud 2007). Nevertheless, the response of many of the reported cases to the withdrawal of methotrexate therapy suggests that the drug does have a role in lymphoma development in some cases.

A meta-analysis of all randomized controlled clinical trials published (2000–2009) in RA patients receiving anti-TNFα therapy showed a slight excess of lymphomas in the treated group, although the number of lymphomas was very small and this difference did not reach statistical significance (Wong *et al.* 2012).

The most common type of tumour seen with methotrexate therapy is DLBCL, but almost a third of patients have Hodgkin lymphoma or Hodgkin-like lesions (Kamel *et al.* 1996, Gaulard *et al.* 2008). In some patients the response closely resembles polymorphic PTLD. T-cell proliferations, some resembling angioimmunoblastic T-cell lymphoma, are less frequently seen.

LYMPHOPROLIFERATIVE DISEASE CLOSELY ASSOCIATED WITH IMMUNE DISORDERS

LYMPHOMATOID GRANULOMATOSIS

Lymphomatoid granulomatosis (LYG) is included in the World Health Organization (WHO) classification as a neoplasm of mature B-cells (Swerdlow *et al.* 2008a). It is an uncommon disease that almost invariably involves the lungs, frequently involves other extranodal sites (skin, brain, kidney, liver) but rarely involves lymph nodes. Patients with LYG usually show an underlying inherited or acquired immunodeficiency state, and associations include the Wiskott–Aldrich syndrome, AIDS and organ transplantation. It is also described in patients treated for acute myeloid or lymphoblastic leukaemia and in those undergoing therapy with methotrexate or imatinib. Those who do not have a specific overt immunodeficiency disorder usually show evidence of impaired immune function. The disease course may fluctuate spontaneously or in response to immunomodulation.

The infiltrate is composed of a polymorphous background, consisting mainly of small T-cells, and variable numbers of EBV-positive B-cells. An angiocentric and angiodestructive growth pattern is characteristic, with infiltration of vessel walls by a polymorphous infiltrate, and this may result in areas of infarction and necrosis. Vascular damage may also be induced by EBV-mediated chemokine release, resulting in a true vasculitis. If lymph nodes are involved, their architecture is effaced and the infiltrate with associated necrosis extends into the perinodal soft tissues. The diagnosis and grading of LYG depends on the numbers of B-cells present and their degree of atypia. In grade 1 lesions, necrosis is not seen and CD20+ B-cells are often inconspicuous. They show little atypia, and EBV-positive cells are infrequent (fewer than 5 per high power field). In grade 2 lesions, necrosis is usually a feature and B-cells are more numerous, larger and more atypical. EBV-positive B-cells are more readily identified, numbering between 5 and 50 per high power field. If large, EBV-positive B-cells are more numerous with more extensive areas of necrosis, the lesion should be regarded as grade 3. The B-cells have blast-like features and show variable degrees of atypia, often including some that may resemble Hodgkin/Reed–Sternberg cells.

Differential diagnosis

Although the distribution of the disease may suggest LYG, grade 3 disease is regarded as a form of DLBCL: extensive areas of necrosis with preservation of tumour cells in and around vessel walls are seen in both entities.

Grade 2 LYG is more easily mistaken for classical Hodgkin lymphoma, particularly as many of the B-cells are CD30+. However, CD15 is negative in LYG and the distribution of the disease is unusual for Hodgkin lymphoma.

LYG differs from polymorphic PTLD in both its clinical manifestations and its histology. Polymorphic PTLD lacks the typical vascular changes and shows a range of B-cells from plasma cells to immunoblasts, whereas LYG has a predominant background of small T-cells.

REFERENCES

Carbone A, Cesarman E, Spina M, *et al.* 2009 HIV-associated lymphomas and gamma-herpes viruses. *Blood* **113:** 1213–1224.

Cunningham-Rundles C 2012 The many faces of common variable immunodeficiency. *Hematology* **1:** 301–305.

da Silva SP, Resnick E, Lucas M, *et al.* 2011 Lymphoid proliferations of indeterminate malignant potential arising in adults with common variable immunodeficiency disorders: Unusual case studies and immunohistological review in the light of possible causative events. *Journal of Clinical Immunology* **31:** 784–791.

Gabarre J, Raphael M, Lepage E, *et al.* 2001 Human immunodeficiency virus–related lymphoma: Relation between clinical features and histologic subtypes. *American Journal of Medicine* **111:** 704–711.

Gaulard P, Swerdlow SH, Harris NL, *et al.* 2008 Other iatrogenic immunodeficiency-associated lymphoproliferative disorders. In Swerdlow SH, Campo E, Harris NL, *et al.* (Eds) *WHO Classification of Tumours of Haematopoietic and Lymphoid Tissues.* Lyon: IARC Press.

Gompels MM, Hodges E, Lock RJ, *et al.* 2003 Lymphoproliferative disease in antibody deficiency: A multi-centre study. *Clinical and Experimental Immunology* **134:** 314–320.

Kamel OW, Weiss LM, van de Rijn M, *et al.* 1996 Hodgkin's disease and lymphoproliferations resembling Hodgkin's disease in patients receiving long-term methotrexate therapy. *American Journal of Surgical Pathology* **20:** 1279–1287.

Khan N, Abbas AM, Lichtenstein GR, *et al.* 2013 Risk of lymphoma in patients with ulcerative colitis treated with thiopurines: A nationwide retrospective cohort study. *Gastroenterology* **145:** 1007–1015.

Kirk O, Pedersen C, Cozzi-Lepri A, *et al.* 2001 EuroSIDA Study Group. Non-Hodgkin lymphoma in HIV-infected patients in the era of highly active antiretroviral therapy. *Blood* **98:** 3406–3412.

Kowalkowski MA, Mims MP, Amiran ES, *et al.* 2013 Effect of immune reconstitution on the incidence of HIV-related Hodgkin lymphoma. *PLoS One* **8:** e77409.

Madariaga JR, Reyes J, Mazariegos G, *et al.* 2000 The long-term efficacy of multivisceral transplantation. *Transplant Proceedings* **32:** 1219–1220.

Nelson BP, Nalesnik MA, Bahler DW, *et al.* 2000 Epstein-Barr virus–negative post-transplant lymphoproliferative disorders: A distinct entity? *American Journal of Surgical Pathology* **24:** 375–385.

Sander CA, Medeiros LJ, Weiss LM, *et al.* 1992 Lymphoproliferative lesions in patients with common variable immunodeficiency syndrome. *American Journal of Surgical Pathology* **16:** 1170–1182.

Swerdlow SH, Campo E, Harris NL, *et al.* (Eds) 2008a *WHO Classification of Tumours of Haematopoietic and Lymphoid Tissues.* Lyon: IARC Press.

Swerdlow SH, Webber SA, Chadburn A, *et al.* 2008b Post-transplant lymphoproliferative disorders. In Swerdlow SH, Campo E, Harris NL, *et al.* (Eds) *WHO Classification of Tumours of Haematopoietic and Lymphoid Tissues.* Lyon: IARC Press.

Vase MØ, Maksten EF, Knud B, *et al.* 2015 Occurrence and prognostic relevance of CD30 expression in post-transplant lymphoproliferative disorders. *Leukemia and Lymphoma* **56:** 1677–1685.

Wolfe F, Michaud K 2007 The effect of methotrexate and anti-tumour necrosis factor therapy on the risk of lymphoma in rheumatoid arthritis in 19,562 patients during 89,710 person-years of observation. *Arthritis and Rheumatism* **56:** 1433–1439.

Wong AK, Kerkoutian S, Said J, *et al.* 2012 Risk of lymphoma in patients receiving antitumor necrosis factor therapy: A meta-analysis of published randomized controlled studies. *Clinical Rheumatology* **31:** 631–636.

8

Hodgkin lymphoma

INTRODUCTION

Hodgkin lymphoma occupies an enigmatic place among malignant lymphomas. Thomas Hodgkin was appointed to the post of Inspector of the Dead and Curator of the Museum at Guy's Hospital in 1826. In his role as a morbid anatomist, he noted a peculiar appearance of the lymph nodes and spleen in six patients and was shown paintings of a similar case from France by his friend Robert Carswell, the first professor of pathology at University College Hospital. Hodgkin presented details of these cases in two lectures given to the Medical-Chirurgical Society in 1832 that were subsequently published in the transactions of that society under the title 'On some morbid appearances of the absorbent glands and spleen'. This paper would have probably faded into obscurity had not Sir Samuel Wilks, a distinguished physician at Guy's Hospital, described similar cases between 1856 and 1877 and, noting Dr. Hodgkin's precedence, named them as *Hodgkin's disease.*

In 1926 an American pathologist, Herbert Fox, traced tissue from three of the six cases described by Hodgkin. Histology of these tissues showed two to be what he considered to be Hodgkin disease and one to be a non-Hodgkin lymphoma (NHL). It is amazing that, from such modest beginnings, all malignant lymphomas are categorized as either Hodgkin lymphoma or NHL at the beginning of the twenty-first century.

In the twentieth century, the Reed–Sternberg (RS) cell was recognized as the hallmark cell of Hodgkin disease. After a spate of reports of RS cells in a diverse range of reactive and neoplastic conditions, it became accepted that the RS cell was diagnostic of Hodgkin disease only if it was in a background of reactive cells characteristic of one of the subtypes of this disease. Immunohistochemistry further refined the diagnosis of Hodgkin disease with the identification of antigens characteristic of RS cells. There remain, however, grey areas in which the distinction between Hodgkin disease and NHL is difficult (see Chapter 5).

The feature that distinguishes most cases of Hodgkin lymphoma from NHL is that the neoplastic cells (both mononuclear Hodgkin and binucleate RS (H/RS) cells) usually form a minority population within the tumour, sometimes accounting for less than 1 per cent of the cells. The reactive cells (lymphocytes, plasma cells, histiocytes, neutrophils, eosinophils, fibroblasts) appear to be attracted into the proliferation by cytokines and chemokines secreted by or induced by the neoplastic cells (Teruya-Feldstein *et al.* 1999, Mani and Jaffe 2009). The interaction of H/RS cells with their microenvironment plays an important role in the pathogenesis of Hodgkin lymphoma through the action of these cytokines and chemokines, which provide the H/RS cells with an anti-apoptotic and proliferative phenotype as well as providing immune escape mechanisms (Liu *et al.* 2014, Matsuki and Younes 2015).

The nature of the H/RS cell remained a mystery for most of the twentieth century, with many candidates being proposed as the cell of origin. Immunohistochemistry and molecular genetics indicate that the majority of RS cells are derived from germinal centre B-cells that have lost their ability to synthesize immunoglobulin and to express many B-cell antigens. In rare cases a clonal T-cell receptor rearrangement has been detected in the H/RS cells (Seitz *et al.* 2000).

Hodgkin lymphoma accounts for approximately 10 per cent of all lymphomas. It shows a bimodal age

BOX 8.1: World Health Organization classification of Hodgkin lymphoma

Nodular lymphocyte-predominant Hodgkin lymphoma

Classical Hodgkin lymphoma

- Nodular sclerosis classical Hodgkin lymphoma
- Lymphocyte-rich classical Hodgkin lymphoma
- Mixed cellularity classical Hodgkin lymphoma
- Lymphocyte-depleted classical Hodgkin lymphoma

distribution with the majority of cases occurring in young adult life. Cervical lymph nodes are the most common site of presentation and of biopsy. Mediastinal and other axial lymph node groups may be involved but involvement of the mesenteric nodes is rare. The spleen is involved in approximately 20 per cent of cases; however, involvement of other extranodal sites is uncommon except in advanced disease. The mode of spread of Hodgkin lymphoma, by contiguity, is a feature that differentiates it from most NHLs, which tend to be more widely disseminated. This contiguous mode of spread provides the rationale for the Ann Arbor staging system on which the treatment of Hodgkin lymphoma is based.

In the World Health Organization (WHO) classification of Hodgkin lymphoma there are two entities: nodular lymphocyte-predominant Hodgkin lymphoma (NLPHL) and classical Hodgkin lymphoma, both of which derive from germinal centre B-cells, but which differ with respect to their clinical features, immunohistochemistry and molecular genetics. There is an argument for placing NLPHL among the B-cell NHLs. However, the WHO working group felt that, at the present time, NLPHL should remain as a subtype of Hodgkin lymphoma pending further investigation (Box 8.1).

NODULAR LYMPHOCYTE-PREDOMINANT HODGKIN LYMPHOMA

NLPHL is an uncommon neoplasm accounting for 3–8 per cent of all cases of Hodgkin lymphoma. It was almost certainly not represented among the cases described by Thomas Hodgkin, and it differs from classical Hodgkin lymphoma in its clinical behaviour, histology, immunohistochemistry and molecular genetics. Nevertheless, it has in the past been confused with classical Hodgkin lymphoma, particularly with lymphocyte-rich classical Hodgkin lymphoma (LRCHL), and, at least for the time being, it has been categorized as Hodgkin lymphoma in the WHO classification.

NLPHL shows a marked male predominance. It can occur from childhood to late adult life but shows a peak incidence in the fourth decade of life. The majority of patients present with lymphadenopathy in the cervical, axillary or, less commonly, inguinal regions. Involvement of other sites, including bone marrow, is uncommon. Lymphadenopathy may be of long duration. Over 80 per cent of patients have stage I or II disease with a 10-year survival of 80 per cent. In some patients, the excision biopsy alone may be curative. In most patients, the disease runs an indolent course in which recurrence is common but disease-related death is rare. Advanced disease is seen in 6–23 per cent of cases (Anagnostopoulos *et al.* 2000, Hartmann *et al.* 2013) and is more often progressive and fatal.

Rarely, NLPHL may show synchronous or metachronous association with progressive transformation of germinal centres (PTGC). The majority of cases of PTGC, however, run an entirely benign course, although recurrence is not uncommon.

High-grade transformation of NLPHL to diffuse large B-cell lymphoma (DLBCL) or T-cell/histiocyte-rich B-cell lymphoma (THRLBCL) may occur in up to 17 per cent of cases and is associated with extranodal disease involving liver, spleen and bone marrow (Eyre *et al.* 2015). Rare patients in whom NLPHL occurs subsequent to a DLBCL have been recorded. The DLBCLs associated with NLPHL have been reported to have a similar long-term outcome to primary DLBCL.

Classical cases of NLPHL have a nodular growth pattern with large, relatively uniform nodules displacing the normal nodal structures, which may be seen as a compressed rim at one end of the biopsy. The nodularity may be highlighted by a reticulin stain or by immunohistochemistry for B-cells or dendritic reticulum cells. The current WHO classification of lymphomas demands that at least one nodule be recognized for the diagnosis of NLPHL.

The tumour cells of NLPHL are frequently referred to as L&H cells after the Lukes and Butler classification (Lukes *et al.* 1966), which used the designation lymphocytic and/or histiocytic (L&H) predominance Hodgkin lymphoma for what is now known as NLPHL. The WHO has now designated these cells as LP cells. LP cells are usually found within nodules but may be extranodular. They may be scanty or abundant. The nuclei are open with fine chromatin and are typically multilobated (popcorn cells). Nucleoli are clearly visible but are rarely as large as those seen in classic H/RS cells. For many years, it was the rule that classic RS cells should be seen before a diagnosis of NLPHL was made. It is now recognized that classic H/RS cells may occur in NLPHL but are relatively infrequent (Anagnostopoulos *et al.* 2000).

The LP cells are seen within the nodules, which are composed predominantly of small B-lymphocytes. Histiocytes, singly or in clusters, may be seen, but plasma cells are uncommon and neutrophils and eosinophils are not a feature of NLPHL. Varying degrees of banded sclerosis may be present and, rarely, residual follicle centres are recognizable.

Fan *et al.* (2003) described six immuno-architectural patterns in 137 biopsy samples of NLPHL. The classic nodular, predominantly small B-cell pattern involving all or part of the node was present in 136 of the biopsies. A mixture of patterns in a single biopsy was more common than a single pure pattern. A diffuse (T-cell–rich large B-cell–like) pattern was present in 35 cases either as a major (16) or minor (19) component. The German Hodgkin Study Group (Hartmann *et al.* 2015) showed that the variant histological patterns were associated with a higher stage at presentation and a higher relapse rate at 5 years (Boxes 8.2 to 8.4 and Figures 8.1 to 8.5).

BOX 8.2: Nodular lymphocyte-predominant Hodgkin lymphoma: Clinical features

- Peak age incidence in fourth decade of life
- Male predominance
- Most common presentation with enlarged cervical, axillary or, less commonly, inguinal lymph nodes
- Lymphadenopathy may be of long duration
- Most patients have stage I or stage II disease
- Recurrence common but rarely fatal
- Three to 5 per cent develop diffuse large B-cell lymphoma

BOX 8.3: Nodular lymphocyte-predominant Hodgkin lymphoma: Morphology

- Nodular or nodular and diffuse growth pattern
- Neoplastic cells (LP cells) often have multilobated nuclei (popcorn cells)
- LP cells may be scanty or abundant
- Background cells mainly small lymphocytes and histiocytes
- Histiocytes may have epithelioid morphology and form clusters or ring nodules
- Neutrophils and eosinophils absent
- Plasma cells usually scanty
- Sclerosis may be present

BOX 8.4: Nodular lymphocyte-predominant Hodgkin lymphoma: Immunohistochemistry

LP CELLS

- CD45+
- CD20+ and CD79a+
- B-cell transcription factors – BOB.1+, Oct-2+ and BSAP+
- J chain positive; immunoglobulin heavy and light chains may be detected
- EMA+ in 50 per cent of cases CD15−, CD30 weak or negative

BACKGROUND CELLS

- Predominantly polyclonal small B-cells
- Small T-cells rosette the L&H cells and form more of the background cells in the diffuse areas
- Many of the T-cells express CD57 and PD-1
- CD21 identifies expanded dendritic cell network within nodules

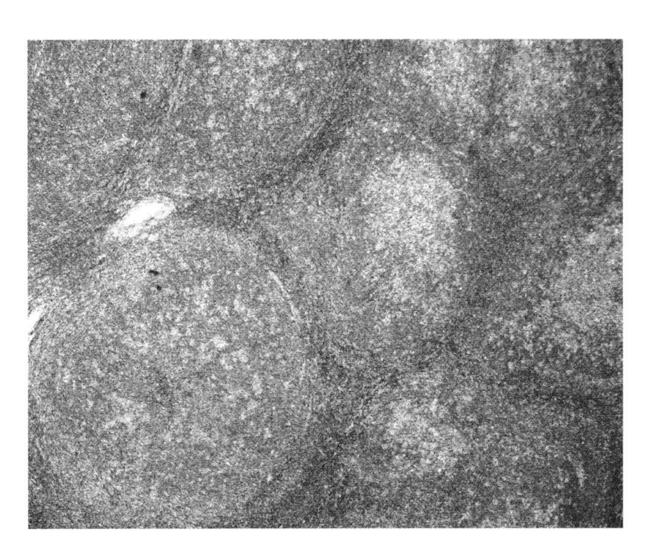

Figure 8.1 Nodular lymphocyte-predominant Hodgkin lymphoma. Low-power view, showing nodular growth pattern.

Figure 8.2 Section of nodular lymphocyte-predominant Hodgkin lymphoma, stained for reticulin, showing the characteristic nodular growth pattern.

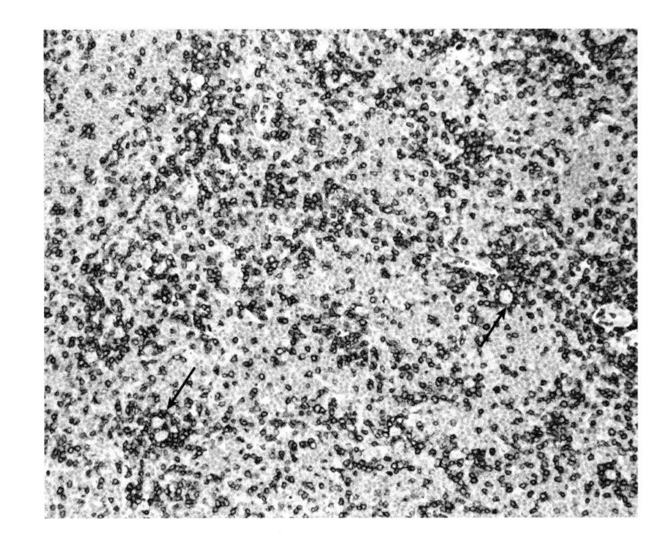

Figure 8.3 Nodular lymphocyte-predominant Hodgkin lymphoma showing the centre of a nodule stained for CD3. Note the T-cell rosetting of the LP cells (arrows).

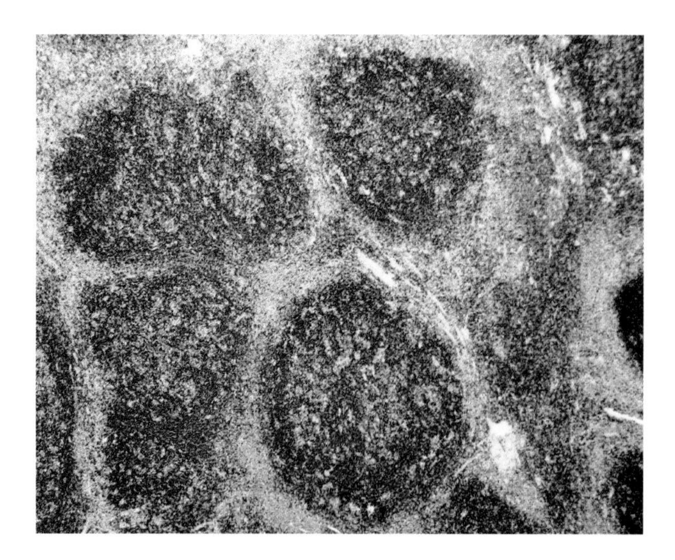

Figure 8.4 Nodular lymphocyte-predominant Hodgkin lymphoma, stained for CD20, showing nodular proliferation composed predominantly of B-cells.

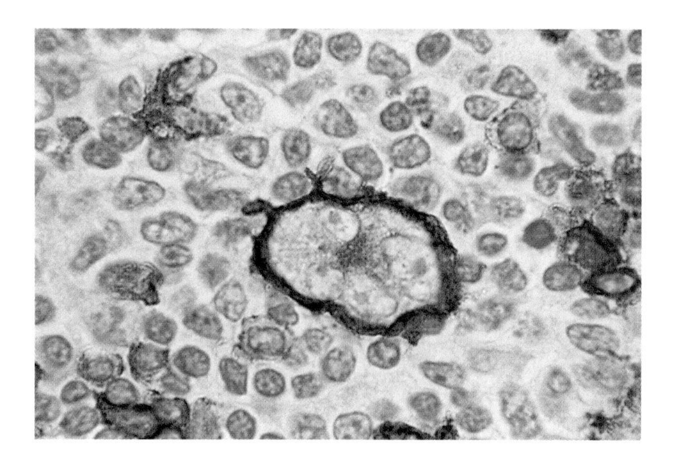

Figure 8.5 Section of nodular lymphocyte-predominant Hodgkin lymphoma, stained for CD20, showing a strongly labelled multinucleated LP cell surrounded by unstained T-lymphocytes. Cytoplasmic spikes from the LP cell extend between the T-cells.

IMMUNOPHENOTYPE

The neoplastic cells of NLPHL are B-cells. They express CD45, CD20, CD79a and BCL-6. They show striking Oct-2 positivity and are also positive for the other B-cell transcription factors, BSAP (PAX-5) and BOB.1. MUM1 is inconsistently positive and, in well-fixed tissues, J chains as well as immunoglobulin heavy and light chains may be detected in the LP cells (Goel *et al.* 2014).

Staining for CD15 is almost invariably negative and staining for CD30 is usually negative or weak. Careful evaluation will usually show that any CD30+ cells are background mononuclear cells similar to those seen around reactive follicles. Most LP cells show strong nuclear staining for Ki67, indicating that they are in cycle. Approximately 50 per cent express epithelial membrane antigen (EMA). LP cells do not contain Epstein–Barr virus–encoded RNAs (EBERs); neither do they express Epstein–Barr virus (EBV)–associated antigens.

The background cells are predominantly polyclonal small B-lymphocytes with a mantle zone phenotype (IgM+, IgD+). However, within the sea of small B-cells, the LP cells are surrounded in a rosette by small T-cells. The number of small T-cells appears to increase as the disease progresses and they are often the predominant population within the diffuse areas. A variable proportion of the T-cells, including those in a rosette around the LP cells, express CD57 and PD-1. Each nodule is encompassed by an expanded network of follicular dendritic cells (FDCs), identifiable with CD21 and CD23.

GENETICS

Analysis of the immunogenetics of LP cells has only been possible since the introduction of micromanipulation to isolate single cells. The LP cells in any one case show monoclonal rearrangements of the immunoglobulin genes. The variable region genes show a high degree of mutation and show ongoing mutations characteristic of follicle centre B-cells. The immunoglobulin rearrangements appear to be functional. IgH/BCL-6 translocations and other BCL-6 rearrangements are common (Wlodarska *et al.* 2004, Renne *et al.* 2005).

DIFFERENTIAL DIAGNOSIS

The nodular form of NLPHL is a distinctive proliferation. The differential diagnosis includes PTGC in which the germinal centres are infiltrated and expanded by small mantle B-cells and may resemble the individual nodules of NLPHL. Also, small clusters or individual centroblasts are surrounded by small lymphocytes, but popcorn cells are not seen. The formation of rosettes around LP cells by T-lymphocytes, a feature of NLPHL, is rarely seen in progressive transformation of germinal centres (Nguyen *et al.* 1999). These two conditions are usually easily distinguished on low-power microscopy, since PTGC involves one or more follicles within a reactive lymph node, whereas NLPHL displaces and compresses the residual nodal tissue.

The differentiation of NLPHL from classical Hodgkin lymphoma is important from a patient management point of view and can be difficult, particularly in the differential diagnosis from LRCHL, which also has a nodular background of small IgM+, IgD+ B cells. In well-fixed tissue, LP cells can usually be differentiated from classic H/RS cells or lacunar cells on the basis of their morphology. They can also be differentiated on the basis of their immunohistochemistry since LP cells usually show expression of CD45 and B-cell antigens CD20, CD79a and PAX5, and usually lack CD15 and CD30. The identification of small regressed germinal centres at the periphery of the nodules in LRCHL is also helpful in distinguishing it from the nodular structures of NLPHL, which do not usually contain germinal centres.

It is recognised that NLPHL may show overlapping features with T-cell/histiocyte-rich B-cell lymphoma (THRLBCL), with some cases of NLPHL showing a variant diffuse T-cell–rich large B-cell–like pattern (Fan *et al.* 2003), suggesting a biological relationship between them (Pittaluga and Jaffe 2010). It would appear to be an unlikely relationship between a relatively benign lymph node–based proliferation and an aggressive lymphoma that frequently involves the bone marrow and spleen; however, array comparative genomic hybridization has shown that there are similarities between NLPHL and THRLBCL at the genomic level, confirming that these entities are part of a pathobiological spectrum with common molecular features but varying clinical presentations (Hartmann *et al.* 2015).

In the differential diagnosis of NLPHL and THRLBCL the presence of small IgD+ B-cells and FDC networks favours the former. Even a small focus of nodularity in association with FDC networks precludes a diagnosis of THRLBCL in an otherwise diffuse growth pattern. The presence of CD4+, CD57+, PD-1+ T-cells favours NLPHL whereas CD8+, TIA-1+ T-cells favour THRLBCL (Boudova *et al.* 2003).

CLASSICAL HODGKIN LYMPHOMA

Histologically, classical Hodgkin lymphoma is composed of variable proportions of RS cells, mononuclear Hodgkin cells and their variants (H/RS cells) set in a background population of small lymphocytes (predominantly T-cells), eosinophils, neutrophils, histiocytes, plasma cells, fibroblasts and collagen. The four subtypes of classical Hodgkin lymphoma, largely defined by this background population, show broad differences in their clinical characteristics and such features as EBV association. Although classical Hodgkin lymphoma appears to be a distinct lymphoma entity overall, H/RS cells of its histologic subtypes diverge in their similarity to other related lymphomas. It has been proposed (Levy *et al.* 2002, Mani and Jaffe 2009) that nodular sclerosis classical Hodgkin lymphoma (NSCHL) is possibly derived from thymic B-cells and is related to primary mediastinal large B-cell lymphoma (PMBL). Genome-wide expression studies on micro-dissected H/RS cells have shown that both NSCHL and lymphocyte-depleted classical Hodgkin lymphoma (LDCHL) map more closely to PMLBCL, whereas those of mixed cellularity classical Hodgkin lymphoma (MCCHL) and LRCHL map with NLPHL (Tiacci *et al.* 2012). Surprisingly, there were not significant differences between expression profiles of H/RS cells with and without EBV, suggesting that similar mechanisms of oncogenesis are operating regardless of the EBV status.

Classical Hodgkin lymphoma may occur at any age from early childhood onwards. It shows a bimodal age distribution with a peak incidence between 15 and 35 years, and a later less well-defined peak in older age. There is an increased incidence of classical Hodgkin lymphoma in patients with a previous history of infectious mononucleosis and in patients with human immunodeficiency virus (HIV) infection.

Classical Hodgkin lymphoma occasionally occurs in patients with pre-existing B-cell chronic lymphocytic leukaemia/small lymphocytic lymphoma (B-CLL/SLL) or follicular lymphoma. Immunoglobulin gene rearrangements in such cases may show clonal identity between the NHL cells and the H/RS cells.

Classical Hodgkin lymphoma typically involves the axial lymph node groups (cervical, mediastinal, axillary, inguinal and para-aortic) and occurs rarely in the mesenteric lymph nodes. Mediastinal tumour, which may involve lymph nodes or thymus, occurs in 80 per cent of patients with NSCHL. Splenic tumour is found in 20 per cent of patients and appears to be necessary for further haematogenous spread to the liver and bone marrow. Primary involvement of Waldeyer's ring (tonsils and adenoids) and the gastrointestinal tract is rare.

Classical Hodgkin lymphoma is staged according to its spread, broadly as follows:

- Stage I: involvement of a single lymph node group or lymphoid structure.
- Stage II: involvement of two or more lymph node groups, or lymphoid structures, on the same side of the diaphragm.
- Stage III: involvement of lymph node groups, or lymphoid structures, on both sides of the diaphragm.
- Stage IV: involvement of extranodal sites, such as liver and bone marrow.

Patients with classical Hodgkin lymphoma may experience a number of systemic symptoms probably related to the production of chemokines and cytokines. Fever, drenching night sweats and weight loss are categorized as 'B-symptoms' and have adverse prognostic significance.

The classic RS cell is binucleate or has a bilobed nucleus. Each nucleus or nuclear lobe contains a large eosinophilic nucleolus giving the cell an 'owl-eye' appearance. Delicate strands of chromatin often radiate from the nucleolus to the well-defined nuclear membrane. In Giemsa-stained preparations, the cytoplasm is basophilic. Mononuclear Hodgkin cells have similar nuclear and cytoplasmic features and some of them probably represent RS cells cut in a plane that shows only one nucleus or nuclear lobe. Shrunken, pyknotic H/RS cells are frequently encountered in classical Hodgkin lymphoma; they are referred to as mummified cells. Variant forms of H/RS cells known as lacunar cells are found in NSCHL and are described in the section on that subtype (Figures 8.6 to 8.10).

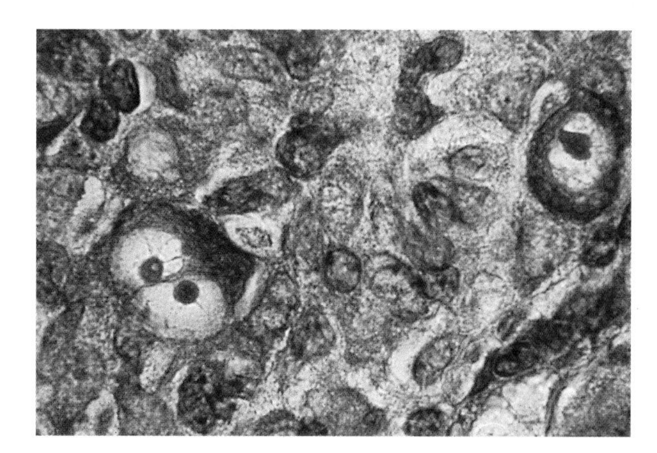

Figure 8.6 Giemsa-stained section of classical Hodgkin lymphoma showing a Reed–Sternberg cell and a mononuclear Hodgkin cell. Both cells show deep cytoplasmic basophilia.

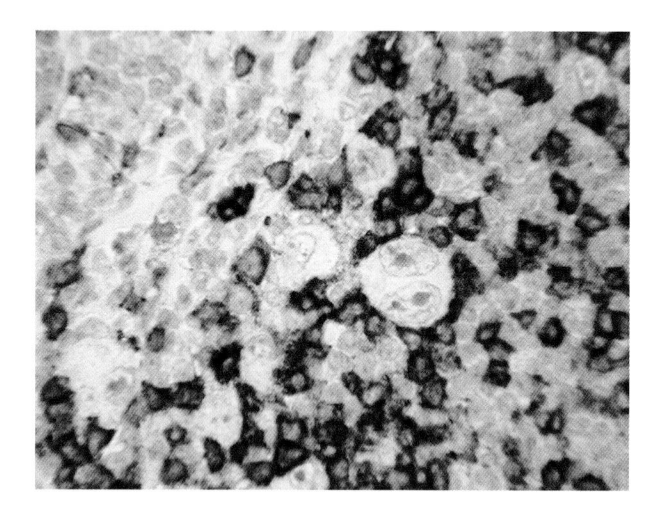

Figure 8.7 Classical Hodgkin lymphoma stained for CD3. T-cells are the predominant cell type and show tight rosetting of a Reed–Sternberg cell.

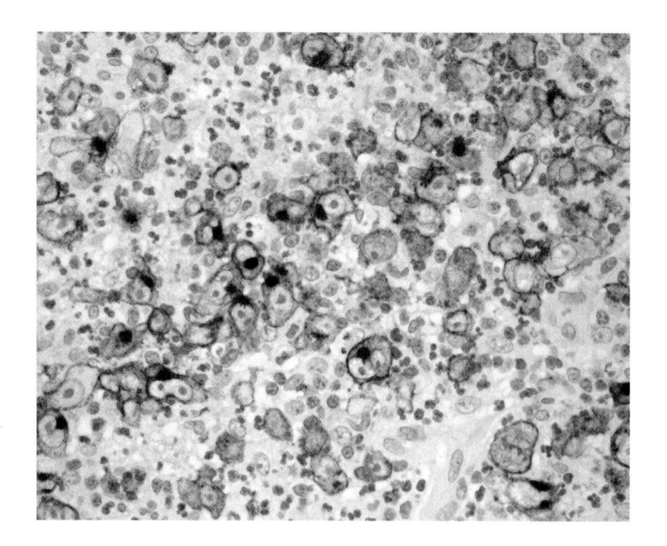

Figure 8.8 Classical Hodgkin lymphoma stained for CD30. Note the strong positive staining of the cell membrane and Golgi region of Hodgkin/Reed–Sternberg cells.

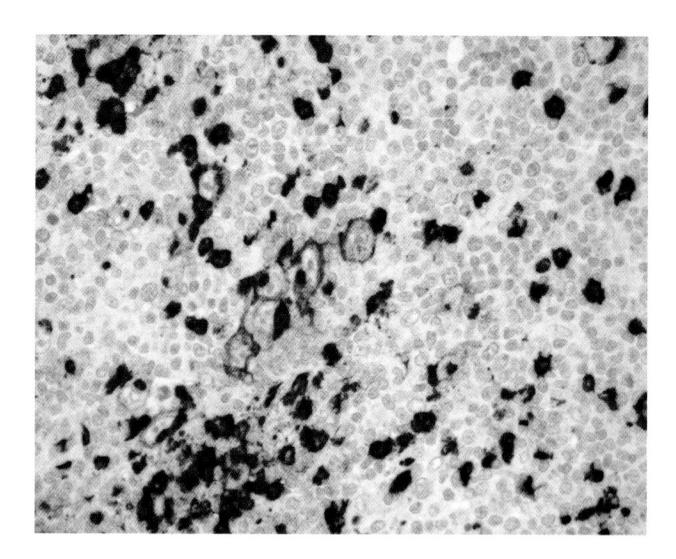

Figure 8.9 Classical Hodgkin lymphoma stained for CD15. Note the membrane and cytoplasmic staining with Golgi zone accentuation of Hodgkin/Reed–Sternberg (H/RS) cells. Large numbers of granulocytes show strong staining. These can obscure the positive staining of H/RS cells.

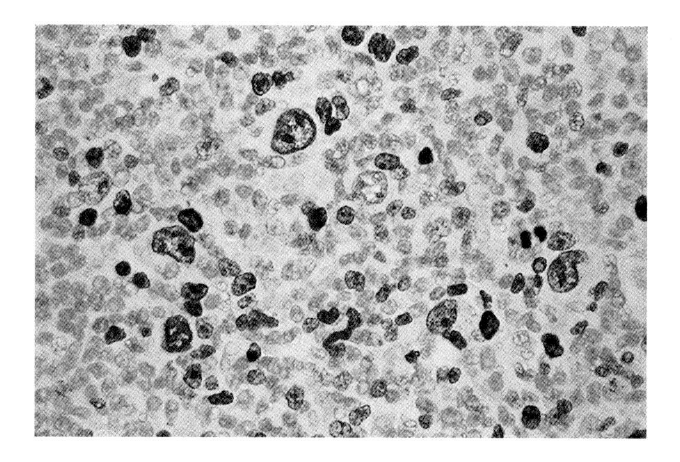

Figure 8.10 Classical Hodgkin lymphoma stained for Ki67. All Hodgkin/Reed–Sternberg cells are in cycle and show positive nuclear staining. The nuclei of these cells are recognizable by their large size.

IMMUNOHISTOCHEMISTRY

In almost all cases of Hodgkin lymphoma, at least a proportion of the H/RS cells express CD30. This takes the form of membrane staining with additional positivity in the Golgi region. CD15 shows a similar staining pattern in 75–85 per cent of cases. Weak expression of the PAX5 gene product B-cell–specific activation protein (BSAP) is a characteristic feature of H/RS cells and MUM1/IRF4 is usually positive. In more than 20 per cent of cases of Hodgkin lymphoma, the H/RS cells may show variable CD20 expression. CD79a is less frequently expressed. H/RS cells are CD45–, although this is often difficult to determine because of the close apposition of CD45+ rosette-forming lymphocytes. H/RS cells are J chain negative. They may show strong cytoplasmic staining for immunoglobulin heavy and light chains, but this is a result of passive diffusion into the cell, rather than synthesis. The inability of H/RS cells to synthesize immunoglobulin has been attributed to defective gene transcription. The transcription factors BOB.1 and Oct-2, found in normal B-cells, may be expressed individually in H/RS cells, but it is claimed that they do not occur together. Gene expression profiles of Hodgkin lymphoma cell lines have shown downregulation of genes affecting multiple components of signalling pathways active in B-cells (Schwering *et al.* 2003).

Routine testing for EBV is recommended, particularly when Hodgkin-like morphology is seen at an extranodal site where the differential includes post-transplant lymphoproliferative disease (PTLD) or EBV+ DLBCL of the elderly. Since EBER is uniformly present in all types of latency, its assessment by *in-situ* hybridization is the gold standard.

Rare examples of H/RS cells that express T-cell antigens have been reported. Despite their T-cell antigens, these cells have been shown to have clonal immunoglobulin gene rearrangements. A study from Japan reported that in almost 10 per cent of cases of Hodgkin lymphoma the H/RS cells expressed FDC markers and showed germline T-cell receptor and immunoglobulin genes (Nakamura *et al.* 1999). Thus, it appears that whereas the majority of cases of classical Hodgkin lymphoma have H/RS cells of B-cell origin, a very small minority of cases may have a different histogenesis. Almost all H/RS cells show positive staining for Ki67 and thus appear to be in cycle.

GENETICS

The large majority of classical Hodgkin lymphoma cases studied have shown monoclonal immunoglobulin gene rearrangements. The variable region genes show a high degree of intraclonal mutations characteristic of germinal centre B-cells. Very few cases of classical Hodgkin lymphoma in which the H/RS cells show clonal T-cell gene rearrangements have been observed. Conventional cytogenetics, fluorescence *in-situ* hybridization (FISH) and comparative genomic hybridization have shown aneuploidy in H/RS cells, and recurrent gains and losses of specific chromosome regions. Molecular profiling shows considerable heterogeneity among classical Hodgkin lymphoma cases; however, constitutive activation of the NF-κB pathway appears to play a central role, and the loss of a B-cell phenotype is directly reflected in its genetic and epigenetic profile, which shows downregulation of almost all B-cell–specific genes by various mechanisms (King *et al.* 2014, Matsuki and Younes, 2015).

Oudejans *et al.* (1996) first documented a lack of expression of major histocompatibility complex class I (MHC-I) and beta-2-microglobulin (B2M) in the H/RS cells of a significant proportion of classical Hodgkin lymphoma cases and reported that EBV-positive cases expressed significantly higher levels of MHC-I and B2M molecules than cases lacking EBV. Reichel *et al.* (2015) performed exome sequencing on flow-sorted H/RS cells from ten classical Hodgkin lymphoma specimens and described consistent alterations in genes responsible for interactions with the immune system, preservation of genomic stability, and transcriptional regulation. They showed that B2M is the most commonly altered gene in H/RS cells, with seven of ten cases having inactivating mutations that led to loss of MHC-I expression. B2M-deficient cases encompassed most of the NSCHL cases and only a minority of MCCHL cases, suggesting that B2M deficiency may determine the tumour micro-environment.

EPSTEIN–BARR VIRUS

Latent EBV infection is found in a proportion of patients with classical Hodgkin lymphoma. The H/RS cells in these cases contain EBERs and express latent membrane protein 1 (LMP1) and Epstein–Barr nuclear antigen 1 (EBNA1; latency pattern II). LMP1 aggregates on the cell membrane, mimicking an active CD40 receptor, leading to the constitutive activation of NF-κB pathway. Since almost all adults have been infected with EBV, scattered EBER-positive small lymphocytes may be present in cases in which the H/RS cells are negative. The incidence of EBV latency in H/RS cells is highest in patients with acquired immune deficiency syndrome (AIDS)–related classical Hodgkin lymphoma (90 per cent) and in MCCHL (70 per cent), and lowest in NSCHL (10–40 per cent). Children and older adults are more likely to have EBV+ H/RS cells, as are males.

NODULAR SCLEROSIS CLASSICAL HODGKIN LYMPHOMA

NSCHL is the most common subtype of classical Hodgkin lymphoma, accounting for 70 per cent of cases in developed countries. It is more common in females, in contrast with other subtypes of classical Hodgkin lymphoma, which show a male predominance. The peak age incidence is between 15 and 34 years. Patients with AIDS who develop NSCHL do so before showing a marked fall in CD4 counts, suggesting that an intact immune system is necessary for its development (Biggar *et al.* 2006).

Gene expression studies have shown that primary mediastinal large B-cell lymphomas (PMBL) share features with NSCHL (Rosenwald *et al.* 2003, Savage *et al.* 2003). The relationship between NSCHL and PMBL is also shown by the sporadic occurrence of composite tumours, the metachronous occurrence of both lymphomas in the same patient and the existence of mediastinal grey zone lymphomas that have intermediate features between PMBL and NSCHL (Traverse-Glehen *et al.* 2005). This has led to the proposal that NSCHL should form a separate entity within the category of classical Hodgkin lymphoma (Levy *et al.* 2002, Mani and Jaffe 2009) (Box 8.5).

BOX 8.5: Nodular sclerosis classical Hodgkin lymphoma: Clinical features

- Median age 28 years
- More common in females
- Mediastinal involvement common
- Most patients present with stage I or II disease

The WHO classification defines NSCHL as a subtype of classical Hodgkin lymphoma with lacunar-type H/RS cells and with collagen bands that surround at least one nodule. Lacunar H/RS cells, as seen in imprint cytology and in electron micrographs, have abundant, clear, organelle-poor cytoplasm. In histological sections of formalin-fixed tissues, this cytoplasm shrinks and retracts, leaving the H/RS cell in a clear space, a feature that gives rise to the term *lacunar cell.* The nuclei of these cells may show considerable morphological variation. They often have numerous nuclear lobes with finer chromatin and smaller nucleoli than those of classic H/RS cells. They have the same immunohistochemical profile as classic H/RS cells. For many years it was dogma that classical binucleate H/RS cells as well as lacunar-type H/RS cells must be seen before making a diagnosis of NSCHL. It is now accepted that cells with the morphology and immunophenotype of lacunar H/RS cells in an appropriate cellular background are in themselves diagnostic of NSCHL. In some cases of NSCHL, the lacunar H/RS cells may form syncytial sheets, sometimes with central areas of necrosis.

Occasional biopsies will be seen in which there are lacunar cells, but in which there is little or no sclerosis. It has been proposed that these should be designated as cellular-phase NSCHL. Support for this designation comes from the observation that some patients with cellular-phase NSCHL are found to have banded sclerosis in subsequent biopsies. The WHO demands that at least one nodule surrounded by collagen bands is necessary for the diagnosis of NSCHL and would place cellular phase NSCHL into the category of MCCHL (Box 8.6 and Figures 8.11 to 8.16).

BOX 8.6: Nodular sclerosis classical Hodgkin lymphoma: Morphology

- Classic and/or lacunar Hodgkin/Reed–Sternberg cells
- Nodular growth pattern
- Collagen bands (surround at least one nodule)
- Lacunar cells may form syncytial aggregates, which may show central necrosis
- Association with primary mediastinal large B-cell lymphoma (grey zone lymphomas)

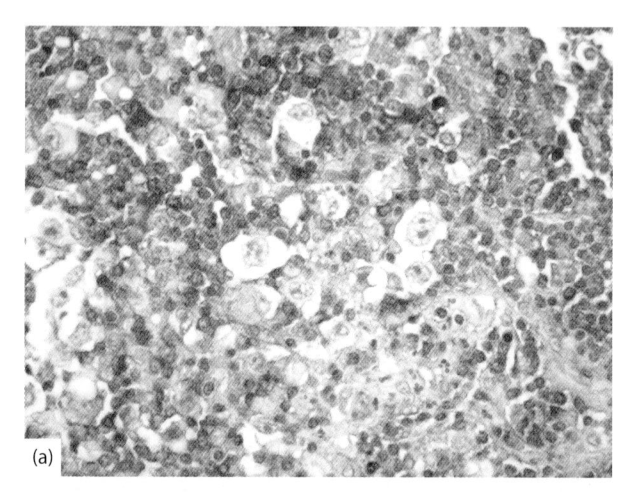

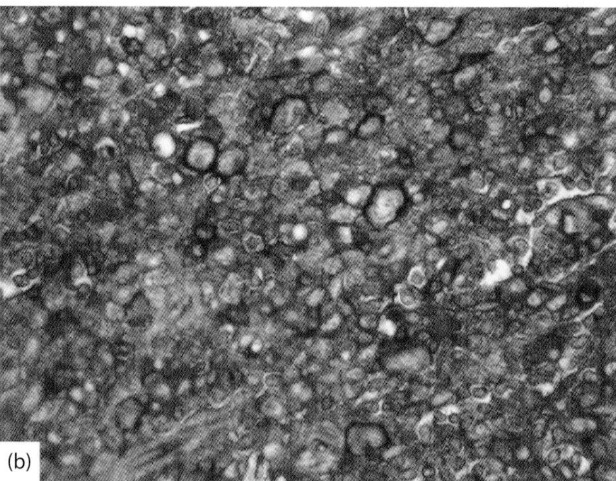

Figure 8.11 Immunohistochemical staining for beta-2-microglobulin (B2M) in two representative cases of classical Hodgkin lymphoma. (a) B2M staining in a case with mutated B2M and a corresponding absence of B2M expression in Hodgkin/Reed–Sternberg (H/RS) cells and (b) B2M staining in a case with wild-type B2M showing Golgi and membrane staining in H/RS cells.

GRADING

The British National Lymphoma Investigation (MacLennan *et al.* 1992) introduced a grading system that divided NSCHL into grade 1, in which 75 per cent or more of the nodules show scattered H/RS cells in a lymphocyte-rich, mixed cellular or fibrohistiocytic background, and grade 2, in which 25 per cent or more of the nodules showed H/RS cells in sheets (syncytial variant), showing pleomorphism and associated with lymphocyte depletion. These grades appeared to predict substantial survival differences. However, subsequent studies showed that these survival differences were eliminated for all except those with advanced stage disease if all patients were given optimal treatment. A study of 965 patients with NSCHL, all of whom had been staged and treated using rigorous protocols, was published by the German Hodgkin Lymphoma Study Group (von Wasielewski *et al.* 2003). These researchers found that grading based on eosinophilia, lymphocyte depletion and atypia of the H/RS cells gave a significant indication of prognosis in intermediate- and advanced-stage disease (Box 8.7 and Figures 8.17 and 8.18).

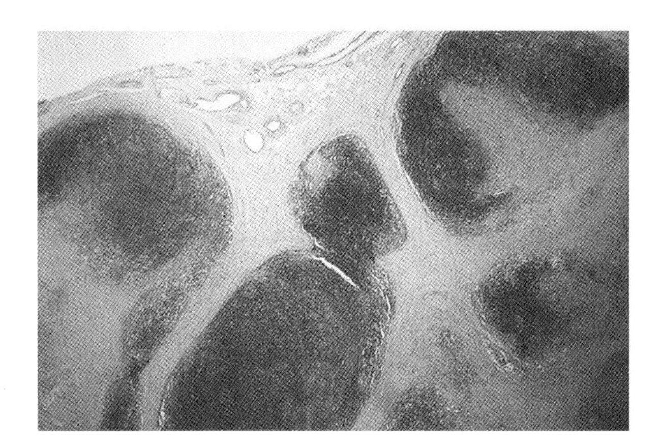

Figure 8.12 Low-power view of a section of nodular sclerosis classical Hodgkin lymphoma showing nodular growth pattern and banded sclerosis.

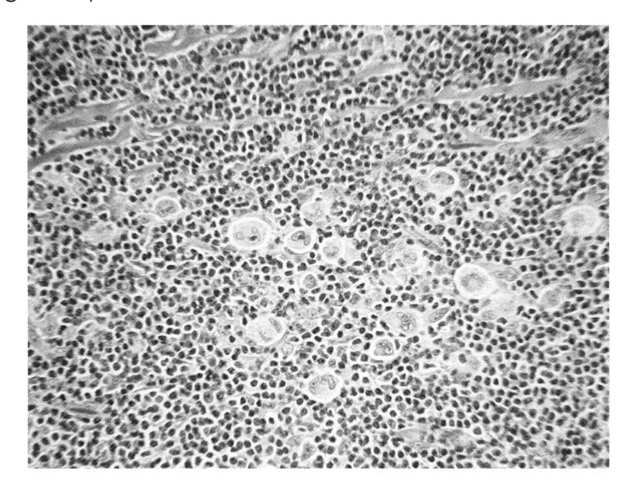

Figure 8.13 Nodular sclerosis classical Hodgkin lymphoma showing characteristic lacunar cells. Note collagen fibres at top of picture and the absence of classic Reed–Sternberg cells.

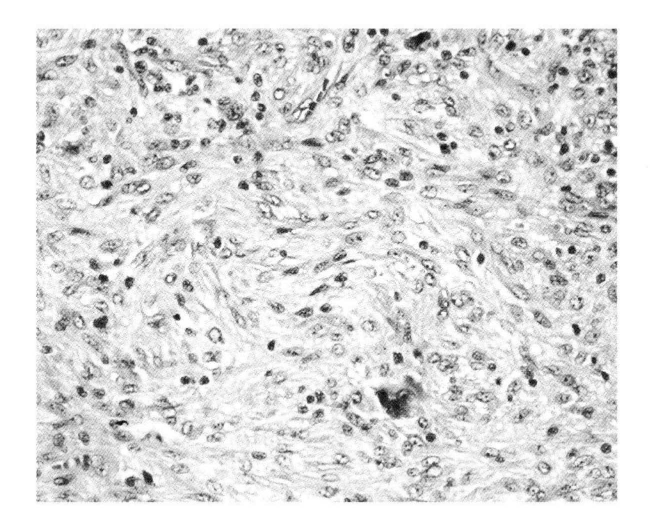

Figure 8.14 Nodular sclerosis classical Hodgkin lymphoma, fibrohistiocytic variant. Degenerate Hodgkin/Reed–Sternberg cells are seen in a background of spindle-shaped histiocytes and scattered eosinophils with few small lymphocytes.

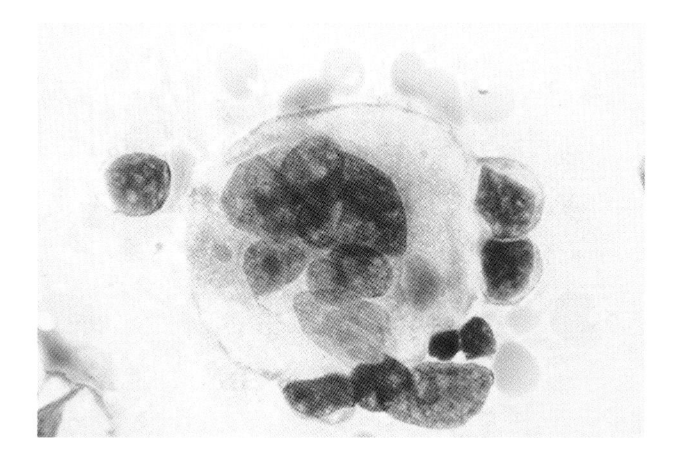

Figure 8.15 Imprint preparation of nodular sclerosis classical Hodgkin lymphoma showing a lacunar Hodgkin/Reed–Sternberg cell. Note the abundant pale-staining cytoplasm, multilobed nucleus and small nucleoli. Rosetting lymphocytes can be seen adhering to the cell membrane.

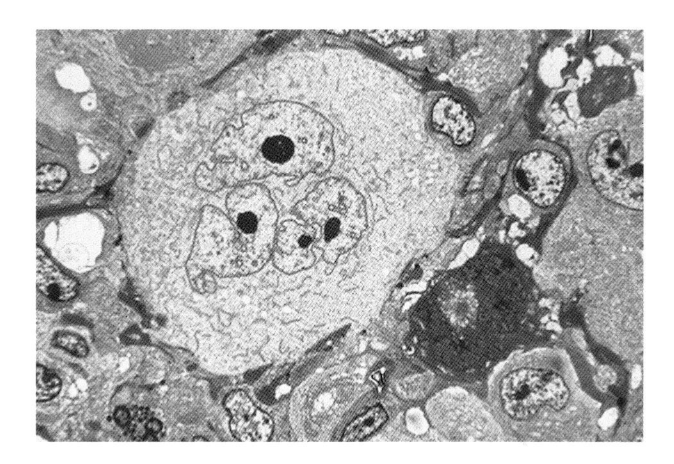

Figure 8.16 Plastic-embedded section of nodular sclerosis classical Hodgkin lymphoma showing a lacunar Hodgkin/Reed–Sternberg cell. The cytoplasm of this cell is relatively abundant and contains few organelles.

BOX 8.7: Prognostically adverse histological features in nodular sclerosis classical Hodgkin lymphoma

- Morphological atypia of the Hodgkin/Reed–Sternberg (H/RS) cells: >25 per cent bizarre and highly anaplastic-appearing H/RS cells with pleomorphic nuclear features, hyperchromatism and highly irregular nuclear outlines
- Relative number of lymphocytes: lymphocytes <33 per cent of all cells in the whole section
- Tissue infiltration by eosinophils: eosinophils constitute more than 5 per cent of all cells in at least five high-power fields

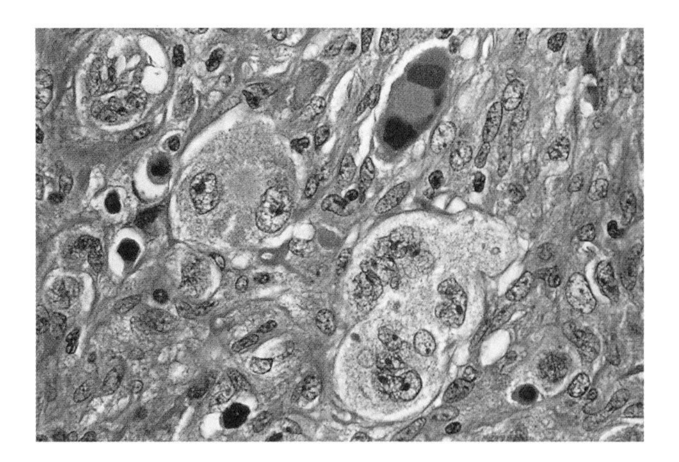

Figure 8.17 Nodular sclerosis classical Hodgkin lymphoma showing atypical Hodgkin/Reed–Sternberg cells.

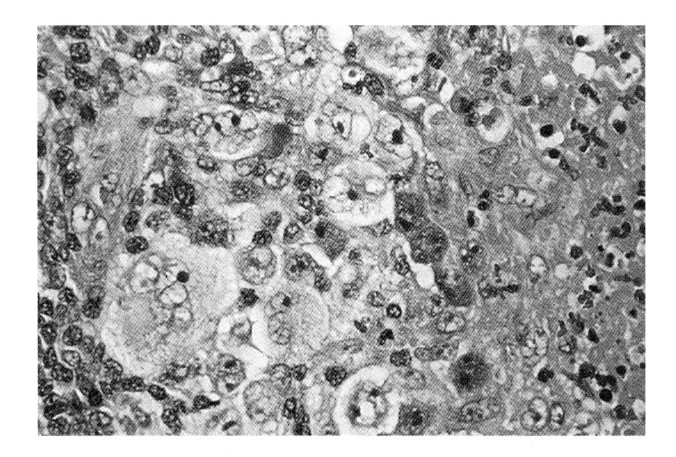

Figure 8.18 Nodular sclerosis classical Hodgkin lymphoma showing sheets of Hodgkin/Reed–Sternberg cells with an area of necrosis (so-called 'syncytial Hodgkin lymphoma').

LYMPHOCYTE-RICH CLASSICAL HODGKIN LYMPHOMA

The LRCHL subtype was introduced by Ashton-Key *et al.* (1995). It usually has a nodular growth pattern and in the past was often misdiagnosed as NLPHL. It accounts for approximately 5 per cent of all cases of classical Hodgkin lymphoma and shows a male predominance of 70 per cent. Peripheral lymphadenopathy is the most common presentation, with mediastinal involvement in 15 per cent. Most patients have stage I or II disease.

The tumour almost always has a nodular growth pattern with small, inactive germinal centres in many of the nodules. The nodules abut one another with little internodular tissue. H/RS cells are scattered in the expanded mantle zone and appear smaller than H/RS cells in other subtypes of classical Hodgkin lymphoma. Eosinophils and neutrophils are absent or scanty (Box 8.8 and Figures 8.19 and 8.20).

BOX 8.8: Lymphocyte-rich classical Hodgkin lymphoma

- Five per cent of all classical Hodgkin lymphomas
- Seventy per cent male predominance
- Usually present in stage I or II with peripheral lymphadenopathy
- Commonly nodular, rarely diffuse
- Hodgkin/Reed–Sternberg cells small, CD30+ and CD15+, occur in expanded mantle/marginal zones consisting predominantly of IgM+, IgD+ small B-lymphocytes
- Eosinophils and neutrophils usually absent
- In the diffuse form the surrounding small lymphocytes show T-cell predominance

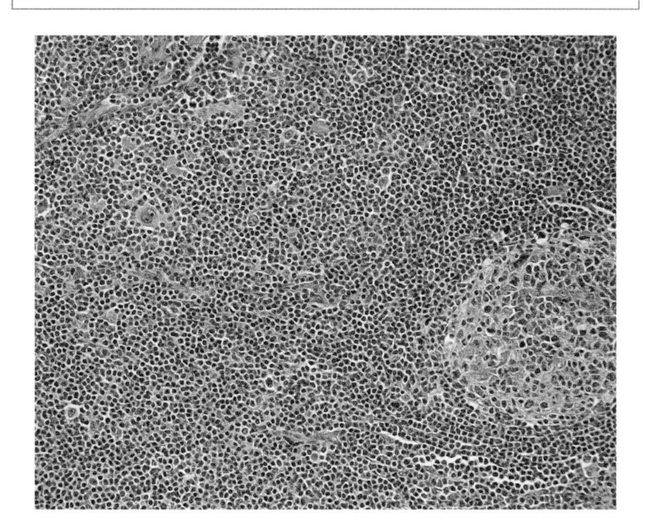

Figure 8.19 Lymphocyte-rich classical Hodgkin lymphoma showing scattered Hodgkin/Reed–Sternberg cells in the mantle/marginal zone of a follicle. Note the absence of inflammatory cells apart from small lymphocytes.

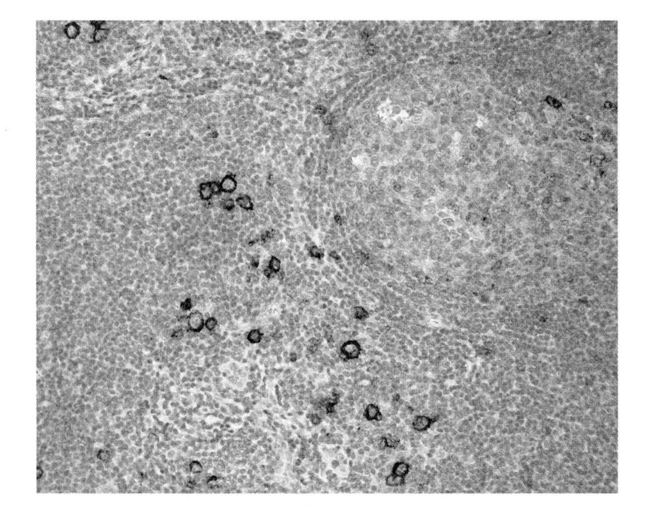

Figure 8.20 Lymphocyte-rich classical Hodgkin lymphoma showing numerous CD30+ Hodgkin/Reed–Sternberg cells in the mantle/marginal zone of a follicle with occasional positive cells at the periphery of the follicle centre.

IMMUNOHISTOCHEMISTRY

The H/RS cells in LRCHL have the same immunophenotype as those of other classical subtypes of Hodgkin lymphoma. The germinal centres in the nodules are BCL-2–, and contain scattered T-cells and a dendritic cell network demonstrable with CD21 and CD23. The majority of the small lymphocytes are B-cells and have the immunophenotype of mantle cells, expressing both surface IgM and IgD. As in other types of classical Hodgkin lymphoma, the small lymphocytes forming a rosette around the H/RS cell are mostly T-cells. In the rare diffuse forms of LRCHL, the number of background T-cells is increased.

DIFFERENTIAL DIAGNOSIS

The European Task Force on Lymphomas (ETFL) (Anagnostopoulos *et al.* 2000) showed that before the routine use of immunohistochemical stains, LRCHL and THRLBCL were frequently misdiagnosed as NLPHL on morphological grounds. In this study, 30 per cent of cases originally diagnosed as NLPHL were found to be examples of LRCHL on critical immunohistochemical evaluation. This almost certainly accounts for some of the cases reported as NLPHL that ran an unfavourable course, since relapsed LRCHL has a less favourable prognosis than NLPHL. The finding of the classical H/RS phenotype (CD30+, CD15+, CD20–\+, J chain–) differentiates LRCHL from NLPHL, in which the LP cells are CD15–, CD20+ and J chain+ and are usually CD30– or show only weak staining.

MIXED CELLULARITY CLASSICAL HODGKIN LYMPHOMA

MCCHL is defined in the WHO classification as a subtype of classical Hodgkin lymphoma with scattered classical H/RS cells in a diffuse or vaguely nodular background of mixed inflammatory cells without nodular sclerosing fibrosis. In this definition, cellular-phase NSCHL is categorized as MCCHL. The border between MCCHL and LDCHL is usually, but not always, easily determined. Mixed cellularity Hodgkin lymphoma comprises approximately 20 per cent of cases of classical Hodgkin lymphoma in the developed world. It shows a male predominance and is the subtype of classical Hodgkin lymphoma most commonly associated with HIV infection (Bellas *et al.* 1996).

Within the basic parameters set out in the definition, the morphology of MCCHL varies considerably. The background infiltrate to the H/RS cells includes lymphocytes (mainly T-cells), plasma cells, eosinophils, neutrophils and histiocytes, but the relative proportions of these cells vary. Histiocytes may form loose epithelioid cell granulomas and may fuse to form Langhans-type giant cells. This granulomatous reaction is more prominent in EBV-positive cases. Interfollicular Hodgkin lymphoma is included in the category of MCCHL. In this variant, the lymph node biopsy shows prominent reactive follicles with small islands of Hodgkin lymphoma, including classic RS cells, in the interfollicular zone. It may be associated with more typical MCCHL in part of the node or in other biopsies. The importance of recognizing this variant is that it should be distinguished from a purely reactive proliferation (Boxes 8.9 and 8.10 and Figures 8.21 and 8.22).

BOX 8.9: Mixed cellularity Hodgkin lymphoma: Clinical features

- Seventy per cent male predominance
- Associated with human immunodeficiency virus (HIV) infection
- Peripheral lymphadenopathy most common presentation
- Spleen involved in 30 per cent, bone marrow in 10 per cent
- Often presents in advanced stage with B-symptoms

BOX 8.10: Mixed cellularity Hodgkin lymphoma: Morphology

- May have interfollicular growth pattern
- Hodgkin/Reed–Sternberg cells usually easy to find
- Capsule not thickened
- No banded sclerosis
- Variable background of T-lymphocytes, plasma cells, eosinophils, neutrophils and histiocytes
- Histiocytes may form loose granulomas

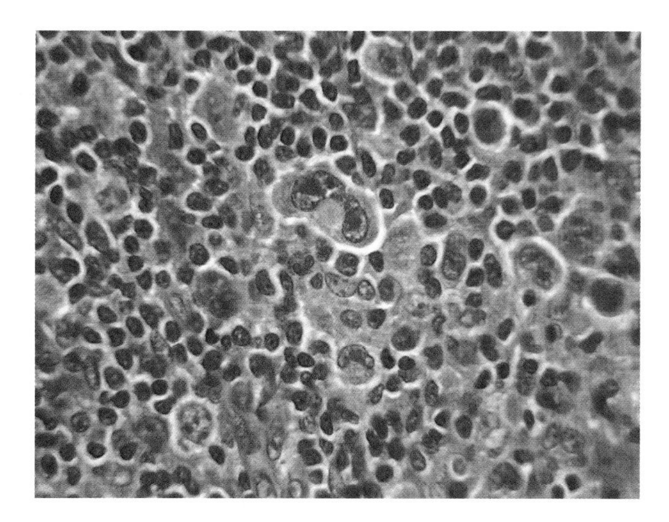

Figure 8.21 Mixed cellularity classical Hodgkin lymphoma showing a classical Reed–Sternberg cell together with H/RS cells in a background of small lymphocytes, eosinophils and histiocytes.

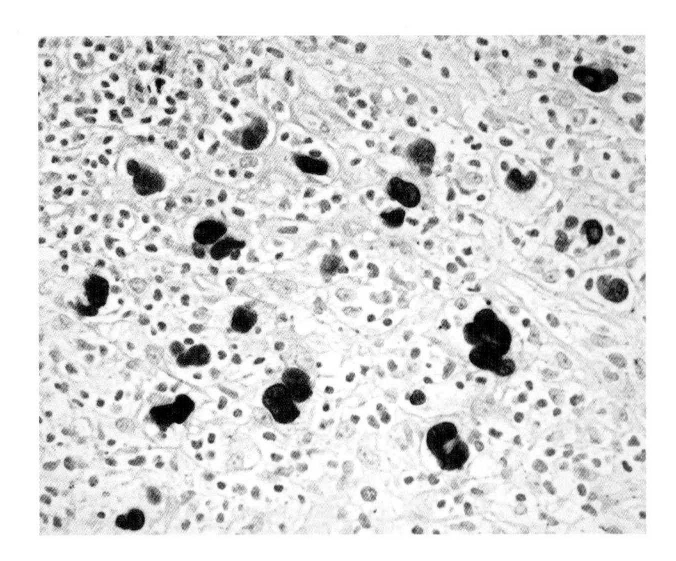

Figure 8.22 Mixed cellularity classical Hodgkin lymphoma stained to show Epstein–Barr virus–encoded RNAs (EBERs). All Hodgkin/Reed–Sternberg cells present show nuclear positivity.

LYMPHOCYTE-DEPLETED CLASSICAL HODGKIN LYMPHOMA

The Lukes and Butler classification of Hodgkin lymphoma (1966) recognized two forms of lymphocyte-depleted Hodgkin lymphoma (LDCHL), both of which have a diffuse growth pattern (Lukes *et al.* 1966). The reticular type is composed of sheets of H/RS cells with sparse lymphocytes and other reactive cells. This disease was also known in the past as Hodgkin sarcoma. The second type was designated as diffuse fibrosis in which H/RS cells are found in a paucicellular background composed mainly of weakly eosinophilic, periodic acid–Schiff (PAS)–positive ground substance rather than mature collagen.

The advent of immunohistochemistry revealed that many cases diagnosed as reticular Hodgkin lymphoma were in fact anaplastic carcinomas or NHLs that contained multinucleate cells bearing some resemblance to H/RS cells. Some cases were lymphocyte-depleted variants of NSCHL.

LDCHL is the least common subtype of classical Hodgkin lymphoma (less than 1 per cent), although some of the AIDS-associated cases of classical Hodgkin lymphoma fall into this category.

LDCHL is said to occur more frequently in abdominal organs (liver, spleen and retroperitoneal lymph nodes) than other subtypes of classical Hodgkin lymphoma. It frequently affects the bone marrow. Patients often present at an advanced stage, therefore, and frequently have B-symptoms.

The diagnosis of LDCHL is dependent on finding H/RS cells, often pleomorphic, with the classic immunophenotype in a lymphocyte-depleted background and without evidence of banded fibrosis (Boxes 8.11 and 8.12 and Figures 8.23 and 8.24).

BOX 8.11: Lymphocyte-depleted Hodgkin lymphoma: Clinical features

- Seventy-five per cent male predominance
- Association with human immunodeficiency virus (HIV) infection
- Relative sparing of peripheral lymph nodes
- Frequent involvement of abdominal organs and bone marrow
- Usually presents in advanced stage with B-symptoms

BOX 8.12: Lymphocyte-depleted Hodgkin lymphoma: Morphology

- Reticular form has sarcomatous appearance with many Hodgkin/Reed–Sternberg (H/RS) cells and pleomorphic forms
- Diffuse fibrosis form shows scattered H/RS cells in a paucicellular eosinophilic, periodic acid–Schiff (PAS)–positive background

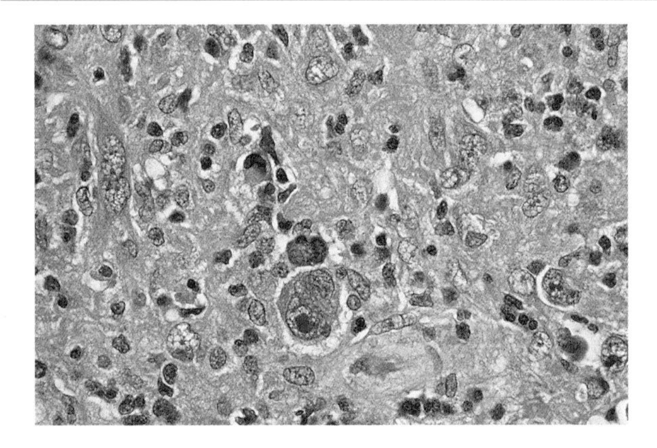

Figure 8.23 Lymphocyte-depleted classical Hodgkin lymphoma showing Hodgkin/Reed–Sternberg cells in a paucicellular background of amorphous eosinophilic material (previously called diffuse fibrosis).

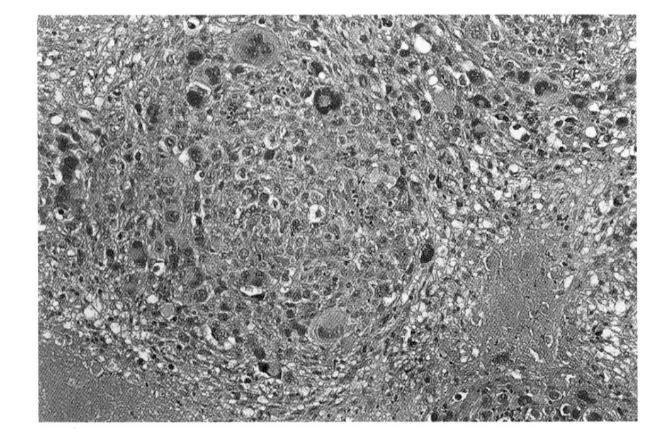

Figure 8.24 Lymphocyte-depleted classical Hodgkin lymphoma showing sheets of atypical Hodgkin/Reed–Sternberg cells and areas of necrosis without banded sclerosis (previously called Hodgkin sarcoma).

ASSOCIATION OF CLASSICAL HODGKIN LYMPHOMA WITH NON-HODGKIN LYMPHOMA

Classical Hodgkin lymphoma may uncommonly show synchronous or metachronous association with NHL. This association is seen most frequently with follicular lymphoma and less commonly with B-CLL/SLL and DLBCL. The H/RS cells have the morphology and immunophenotype of H/RS cells seen in classical Hodgkin lymphoma. They are usually surrounded by a rosette of small T-lymphocytes and associated with a mixed background infiltrate characteristic of classical Hodgkin lymphoma. In B-CLL/SLL, isolated H/RS cells are rarely encountered dispersed in a background of neoplastic small B-lymphocytes. Such cases should be categorized as B-CLL/SLL containing H/RS like cells. The diagnosis of classical Hodgkin lymphoma in association with NHL should not be equated with finding large, multinucleated CD30+ cells, which might be indicative of high-grade transformation.

Analysis of the immunoglobulin gene rearrangements in the NHL cells and in micro-dissected H/RS cells has shown clonal identity in many cases studied. The H/RS cells frequently show evidence of latent EBV infection. A report described a patient with composite mantle cell lymphoma (MCL) and classical Hodgkin lymphoma (Tinguely *et al.* 2003). Immunoglobulin gene analysis showed clonal identity between the MCL cells and the H/RS cells. Interestingly, the immunoglobulin genes in the MCL cells were unmutated, whereas those in the H/RS cells were mutated, suggesting that the progression from MCL cells (or their common precursors) to H/RS cells had occurred in a germinal centre. A further point of interest in this case was the finding that one clone of the H/RS cells was EBV positive whereas a clone showing fewer IgV gene mutations was EBV negative, suggesting that EBV infection was a late event in the pathogenesis of the classical Hodgkin lymphoma.

Analysis of the rearranged IgV genes in some cases of composite classical Hodgkin lymphoma and B-cell NHL has also shown both shared mutations and mutations present in only one of the two lymphomas, indicating that one lymphoma clone is not the direct descendent of the other. This mutation pattern suggests that both lymphomas shared a common origin from a mutated germinal centre B cell, from which the two lymphomas developed independently in the germinal centre micro-environment (Küppers *et al.* 2014).

A grey zone exists between PMBL and NSCHL and both may occur as synchronous or metachronous tumours (see Chapter 5).

DIFFERENTIAL DIAGNOSIS OF CLASSICAL HODGKIN LYMPHOMA

Cells with the morphology and immunophenotype of classic H/RS in a background setting of one of the recognized subtypes of classical Hodgkin lymphoma provide a firm basis for the diagnosis of the disease. Difficulties may arise owing to the small size of the biopsy; however, even with adequate biopsies, grey areas exist between classical Hodgkin lymphoma and other lymphoid proliferations. Cells with the morphology of classic H/RS cells may be found in biopsies from patients with infectious mononucleosis (more commonly in the tonsil than in the lymph node). These cells may be CD30+ but do not express CD15, and are usually found in a background of blast cells inconsistent with classical Hodgkin lymphoma.

The Revised European-American Lymphoma (REAL) classification included a provisional category of 'Hodgkin-related anaplastic large cell lymphoma' in which differentiation between these two lymphomas could not be reliably made. This category was not included in the WHO classification. Most cases that were included in this category were examples of syncytial NSCHL. Separation between classical Hodgkin lymphoma and anaplastic large cell lymphoma can usually be made with little difficulty on the basis of their morphology and immunophenotype (Table 8.1).

DLBCLs may contain pleomorphic cells resembling H/RS cells and these are usually CD30+. Such tumours have a background population of blast cells that usually strongly express B-cell antigens and are thus inconsistent with any subtype of classical Hodgkin lymphoma. THRLBCL may, however, have a greater resemblance to classical Hodgkin lymphoma. The large B-cells in a minority of these tumours may resemble H/RS cells and express CD30. They are, however, negative for CD15, and almost always express CD20 and other B-cell antigens much more strongly than is usually seen in H/RS cells.

The differentiation of NLPHL from LRCHL has already been discussed. These two entities are easily confused on morphology but can usually be clearly separated with immunohistochemistry.

The increased incidence of lymphomas in primary and acquired immunodeficiency states as well as in iatrogenic immunosuppression and long-term methotrexate therapy for autoimmune disease includes Hodgkin lymphoma. In such cases, the disease is usually associated with EBV infection. A proportion of these proliferations regress when immunosuppression is reversed, suggesting that they are EBV (and possibly other virus)–driven lymphoproliferations. When appropriate, immunosuppressive drugs may be reduced or stopped for a trial period to determine whether regression will occur.

EBV+ DLBCL of the elderly is regarded as a lymphoproliferative disorder associated with immunosenescence and it may be misdiagnosed as classical Hodgkin lymphoma. Although the majority of cases with systemic involvement are aggressive, it is important to identify variants presenting with localised mucosal and cutaneous disease, which have a good prognosis and require little treatment (Dojcinov SD *et al.* 2011).

Table 8.1 Differentiation between classical Hodgkin lymphoma (CHL) and T/null anaplastic large cell lymphoma (ALCL)

	ALCL	CHL
Histological features		
Sinusoidal infiltration	Frequent	Rare
Perivascular infiltration	Sometimes seen	Not seen
Hallmark ALCL cells	Present	Not seen
Classic H/RS cells	Very rare	Present
Cells form cohesive sheets	Usual	Rare
Cells surrounded by rosette of T-lymphocytes	Unusual	Usual
Sclerosis	Unusual	Present in NSCHL
Immunophenotype		
CD45	Usually positive	Negative
CD30	Strongly positive	Positive
CD15	Rarely positive	75 per cent positive
EMA	Usually positive	Negative
CD20	Negative	Variable positivity in 20 per cent
ALK-1	60–85 per cent positive	Negative
T-cell antigens	Often positive	Negative
TIA-1, perforin, granzyme	Often positive	Rarely positive
EBV antigens	Negative	Often positive

ALK, anaplastic lymphoma kinase; EBV, Epstein–Barr virus; EMA, epithelial membrane antigen; H/RS, Hodgkin/Reed–Sternberg; NSCHL, nodular sclerosis classical Hodgkin lymphoma; TIA, T-cell intracellular antigen.

REFERENCES

Anagnostopoulos I, Hansmann ML, Franssila K, *et al.* 2000 European Task Force on Lymphoma project on lymphocyte predominance Hodgkin disease: Histologic and immunohistologic analysis of submitted cases reveals 2 types of Hodgkin disease with a nodular pattern and abundant lymphocytes. *Blood* **96:** 1889–1899.

Ashton-Key M, Thorpe PA, Allen JP, Isaacson PG 1995 Follicular Hodgkin's disease. *American Journal of Surgical Pathology* **19:** 1294–1299.

Bellas C, Santon A, Manzanal A, *et al.* 1996 Pathological, immunological and molecular features of Hodgkin's disease associated with HIV infection. Comparison with ordinary Hodgkin's disease. *American Journal of Surgical Pathology* **20:** 1520–1524.

Biggar RJ, Jaffe ES, Goedert JJ, *et al.* 2006 Hodgkin lymphoma and immunodeficiency in persons with HIV/AIDS. *Blood* **108:** 3786–3791.

Boudova L, Torlakovic E, Delabie J, *et al.* 2003 Nodular lymphocyte-predominant Hodgkin lymphoma with nodules resembling T-cell/histiocyte rich B-cell lymphoma: Differential diagnosis between nodular lymphocyte-predominant Hodgkin lymphoma and T-cell/histiocyte rich B-cell lymphoma. *Blood* **102:** 3753–3758.

Dojcinov SD, Venkataraman G, Pittaluga S *et al.* 2011 Age-related EBV-associated lymphoproliferative disorders in the Western population: A spectrum of reactive lymphoid hyperplasia and lymphoma. *Blood* **117:** 4726–4735.

Eyre TA, Gatter K, Collins GP *et al.* 2015 Incidence, management and outcome of high grade transformation of nodular lymphocyte predominant Hodgkin lymphoma: Long-term outcomes from a 30-year experience. *American Journal of Hematology* **90:** E103–E110.

Fan Z, Natkunam Y, Bair E, *et al.* 2003 Characterization of variant patterns of nodular lymphocyte predominant Hodgkin lymphoma with immunohistologic and clinical correlation. *American Journal of Surgical Pathology* **27:** 1346–1356.

Goel A, Fan W, Patel AA, *et al.* 2014 Nodular lymphocyte predominant Hodgkin lymphoma: Biology, diagnosis and treatment. *Clinical Lymphoma, Myeloma and Leukemia* **14:** 261–270.

Hartmann S, Döring C, Vucic E *et al.* 2015 Array comparative genomic hybridization reveals similarities between nodular lymphocyte predominant Hodgkin lymphoma and T cell/histiocyte rich large B cell lymphoma. *British Journal of Haematology* **169:** 415–422.

Hartmann S, Eichenauer DA, Plutschow A, *et al.* 2013 The prognostic impact of variant histology in nodular lymphocyte predominant Hodgkin lymphoma: A report from the German Hodgkin Study Group (GHSG). *Blood* **122:** 4246–4252.

King RL, Howard MT, Bagg A 2014 Hodgkin lymphoma: Pathology, pathogenesis, and a plethora of potential prognostic predictors. *Advances in Anatomic Pathology* **21:** 12–25.

Küppers R, Dührsen U, Hansmann ML 2014 Pathogenesis, diagnosis, and treatment of composite lymphomas. *Lancet Oncology* **15:** e435–e446.

Levy A, Armon Y, Gopas J, *et al.* 2002 Is classical Hodgkin's disease indeed a single entity? *Leukemia and Lymphoma* **43:** 1813–1818.

Liu Y, Sattarzadeh A, Diepstra A, *et al.* 2014 The microenvironment in classical Hodgkin lymphoma: An actively shaped and essential tumor component. *Seminars in Cancer Biology* **24:** 15–22.

Lukes R, Butler J, Hicks E 1966 Natural history of Hodgkin's disease as related to its pathological picture. *Cancer* **19:** 317–344.

MacLennan KA, Bennett MH, Vaughan HB, *et al.* 1992 Diagnosis and grading of nodular sclerosing Hodgkin's disease: A study of 2190 patients. *International Review of Experimental Pathology* **33:** 27–51.

Mani H, Jaffe ES 2009 Hodgkin lymphoma: An update on its biology with newer insights into classification. *Clinical Lymphoma and Myeloma* **9:** 206–216.

Matsuki E, Younes A. 2015 Lymphomagenesis in Hodgkin lymphoma. *Seminars in Cancer Biology* **34:** 14–21.

Nakamura S, Nagahama M, Kagami Y, *et al.* 1999 Hodgkin's disease expressing dendritic cell marker CD21 without any other B-cell marker. *American Journal of Surgical Pathology* **23:** 363–376.

Nguyen PL, Ferry JA, Harris NL 1999 Progressive transformation of germinal centers and nodular lymphocyte predominance Hodgkin's disease. A comparative immunohistochemical study. *American Journal of Surgical Pathology* **23:** 27–33.

Oudejans JJ, Jiwa NM, Kummer JA, *et al.* 1996 Analysis of major histocompatibility complex class I expression on Reed–Sternberg cells in relation to the cytotoxic T-cell response in Epstein-Barr virus–positive and –negative Hodgkin's disease. *Blood* **87:** 3844–3851.

Pittaluga S, Jaffe ES 2010 T-cell/histiocyte rich large B-cell lymphoma. *Haematologica* **95:** 352–356.

Reichel J, Chadburn A, Rubinstein PG, *et al.* 2015 Flow sorting and exome sequencing reveal the oncogenome of primary Hodgkin and Reed–Sternberg cells. *Blood* **125:** 1061–1072.

Renne C, Martin-Subero JI, Hansmann ML, Siebert R 2005 Molecular cytogenetic analysis of immunoglobulin loci in lymphocyte predominant Hodgkin's lymphoma reveals a recurrent IGH-BCL6 juxtaposition. *Journal of Molecular Diagnosis* **7:** 352–356.

Rosenwald A, Wright G, Leroy K, *et al.* 2003 Molecular diagnosis of primary mediastinal B-cell lymphoma identifies a clinically favorable subgroup of diffuse large B cell lymphoma related to Hodgkin lymphoma. *Journal of Experimental Medicine* **198:** 851–862.

Savage KJ, Monti S, Kutok JL, *et al.* 2003 The molecular signature of mediastinal large B-cell lymphoma differs from that of other diffuse large B-cell lymphomas and shares features with classical Hodgkin lymphoma. *Blood* **102:** 3871–3879.

Schwering I, Brauninger A, Klein U, *et al.* 2003 Loss of the B-lineage-specific gene expression program on Hodgkin and Reed–Sternberg cells of Hodgkin lymphoma. *Blood* **101:** 1505–1512.

Seitz V, Hummel M, Marafioti T, *et al.* 2000 Detection of clonal T-cell receptor gamma-chain gene rearrangements in Reed–Sternberg cells of classic Hodgkin disease. *Blood* **95:** 3020–3024.

Teruya-Feldstein J, Jaffe E, Burd PR, *et al.* 1999 Differential chemokine expression in tissues involved by Hodgkin's disease: Direct correlation of eotaxin expression and tissue eosinophilia. *Blood* **93:** 2463–2470.

Tiacci E, Döring C, Brune V *et al.* 2012 Analyzing primary Hodgkin and Reed-Sternberg cells to capture the molecular and cellular pathogenesis of classical Hodgkin lymphoma. *Blood* **120:** 4609–4620.

Tinguely M, Rosenquist R, Sundstrom C, *et al.* 2003 Analysis of a clonally related mantle cell and Hodgkin lymphoma indicates Epstein–Barr virus infection of a Hodgkin/Reed–Sternberg precursor in a germinal center. *American Journal of Surgical Pathology* **27:** 1483–1488.

Traverse-Glehen A, Pittaluga S, Gaulard P, *et al.* 2005 Mediastinal gray zone lymphoma. The missing link between classic Hodgkin's lymphoma and mediastinal large B-cell lymphoma. *American Journal of Surgical Pathology* **29:** 1411–1421.

von Wasielewski S, Franklin J, Fischer R, *et al.* 2003 Nodular sclerosing Hodgkin disease: New grading predicts prognosis in intermediate and advanced stages. *Blood* **101:** 4063–4069.

Wlodarska I, Stul M, Wolf-Peeters C, Hagemeijer A 2004 Heterogeneity of BCL6 rearrangements in nodular lymphocyte predominant Hodgkin's lymphoma. *Haematologica* **89:** 965–972.

9

Histiocytic and dendritic cell neoplasms, mastocytosis and myeloid sarcoma

HISTIOCYTIC AND DENDRITIC CELL NEOPLASMS

The World Health Organization (WHO) classification of histiocytic and dendritic cell neoplasms is summarized in Box 9.1 (Swerdlow *et al.* 2008).

These are uncommon neoplasms, often represented in the literature as small series or single-case reports. They include neoplasms of macrophages and interdigitating dendritic cells (IDCs), both derived from haemopoietic precursors; and neoplasms of follicular dendritic cells, which are of mesenchymal origin and arise from ubiquitous perivascular precursor cells (Krautler *et al.* 2012). Neoplasms of IDCs and follicular dendritic cells have in the past been designated as 'sarcoma/tumour' to encompass the variable cytological grade and behaviour of these tumours: not all are clearly sarcomatous, but the majority are prone to multiple recurrences and metastasis.

> **BOX 9.1: World Health Organization classification of histiocytic and dendritic cell neoplasms (Swerdlow *et al.* 2008)**
>
> - Histiocytic sarcoma
> - Langerhans cell tumours
>   - Langerhans cell histiocytosis
>   - Langerhans cell sarcoma
> - Interdigitating dendritic cell sarcoma
> - Follicular dendritic cell sarcoma
> - Other rare dendritic cell tumours
>   - Fibroblastic reticular cell tumour
>   - Indeterminate dendritic cell tumour
> - Disseminated juvenile xanthogranuloma

The International Lymphoma Study Group (ILSG) collected 61 cases of histiocytic and dendritic cell neoplasms and subjected them to detailed morphological and immunohistochemical analysis (Pileri *et al.* 2002). They were able to categorize 57 of these cases into four groups, using a panel of six antibodies (Table 9.1). The remaining four cases were allocated to groups on the basis of light microscopic and ultrastructural features. The cases of Langerhans cell histiocytosis (LCH) included in this report are not representative of paediatric disease because a high proportion of adult and unusual cases were selected for the study.

There are increasing numbers of reports of histiocytic and IDC lesions occurring synchronously or subsequent to a B- or T-cell neoplasm. Feldman and colleagues (2008) reported seven cases of histiocytic and dendritic cell tumours, either metachronous or synchronous, with follicular lymphoma (FL), and all members of the histiocytic/dendritic group showed t(14;18) or other evidence of a clonal relationship. Shao *et al.* (2011) used laser-capture microdissection and polymerase chain reaction (PCR) to show identical clonal immunoglobulin gene rearrangements in seven cases of chronic lymphocytic leukaemia/small lymphocytic lymphoma (CLL/SLL) associated with histiocytic and IDC sarcomas. Castro *et al.* (2010) described histiocytic lesions following acute lymphoblastic leukaemia (ALL) of T- or B-cell type in 15 patients and were able to demonstrate clonal identity between the primary leukaemia and the histiocytic lesion in all seven cases in which material was available.

Table 9.1 Results from a study of 61 cases of histiocytic and dendritic cell neoplasms

Neoplasm	*n*	CD68 (%)	LYS (%)	CD1a (%)	S100 (%)	CD21/35	Median age (years)	DOD (%)
Histiocytic sarcoma	18	100	94	0	33	0	46	58
Langerhans cell tumours	26	96	42	100	100	0		
LCH	17			33	31			
LCS	9			41	44			
Follicular dendritic cell tumour/sarcoma	13	54	8	0	16	100	65	9
Interdigitating dendritic cell tumour/sarcoma	4	50	25	0	100	0	71	0

Source: Pileri *et al.* 2002.
DOD, died of disease; LCH, Langerhans cell histiocytosis; LCS, Langerhans cell sarcoma; Lys, lysozyme.

In these cases, the histiocytic lesions have been shown to share molecular genetic or cytogenetic features with the original leukaemia/lymphoma, leading to the suggestion that this may represent lineage plasticity or transdifferentiation. Recently, Brunner *et al.* (2014) described a case of histiocytic sarcoma in a patient previously treated for FL, which showed identical clonal immunoglobulin heavy chain rearrangement and the presence of t(14;18). In addition, however, the FL exhibited profound gains and losses of genetic material on array comparative genomic hybridization, whereas no aberrations were detected in the histiocytic sarcoma, suggesting that both neoplasms may have arisen separately from a common progenitor clone.

The presence of BRAF V600E mutations in histiocytic neoplasms raises the possibility of the BRAF inhibitor vemurafenib becoming a therapeutic option. The frequency of BRAF V600E mutation varies in different studies and has been found in up to 62.5 per cent of cases of histiocytic sarcoma, 25–38 per cent of Langerhans cell tumours (Haroche *et al.* 2012), and 18.5 per cent of cases of follicular dendritic cell (FDC) sarcoma, and has been reported in indeterminate cell tumour and an IDC sarcoma (O'Malley *et al.* 2015). Cases of blastic plasmacytoid dendritic cell neoplasm (BPDCN) or acute monocytic leukaemia did not harbour BRAF mutations (Go *et al.* 2014).

MACROPHAGE/HISTIOCYTIC NEOPLASMS

HISTIOCYTIC SARCOMA

In the past, large B-cell lymphomas, anaplastic large cell lymphoma and enteropathy-associated T-cell lymphoma were thought to be histiocyte-derived neoplasms. The endocrine pathologist Victor E. Gould, lamenting the lax use of the term 'dense core granule', once said, "'Dense core granule' is the third most abused term in the English language, coming after 'histiocyte' and 'I love you'." Since the advent of immunohistochemistry and molecular techniques for identifying B- or T-cell clonality, it has become apparent that neoplasms of histiocytes are uncommon. They can be differentiated from the monocytic leukaemias, particularly acute monoblastic leukaemias, which may present as myeloid sarcoma but which are usually more monomorphic and may show expression of CD34 or diffuse bone marrow involvement.

Histiocytic sarcoma occurs at all ages but is most common in adult life. There is a rare association with mediastinal germ cell tumour wherein the neoplastic histiocytes are thought to be derived from pluripotential germ cells. There is also an increased incidence of histiocytic sarcoma in patients with non-Hodgkin lymphoma (Bassarova A *et al.* 2009), and it has been described in association with Rosai–Dorfman disease in a patient with autoimmune lymphoproliferative syndrome (Venkataraman *et al.* 2010). Takahashi and Nakamura (2013) found that all patients with histiocytic sarcoma associated with ALL are male, whereas cases associated with mature B-cell neoplasms show no relation to gender.

Approximately one third of cases of histiocytic sarcoma present with lymphadenopathy, one third with skin infiltrates and one third at other extranodal sites. Systemic symptoms are common.

Lymph node involvement by histiocytic sarcoma may be diffuse or show a sinusoidal pattern. The tumour cells are large, usually four times or more the size of small lymphocytes. The nuclear morphology is typically rounded with varying degrees of pleomorphism, and there may be spindle cell areas. The cytoplasm is usually abundant and in some cases may show vacuolation. Phagocytosis, including erythrophagocytosis by the tumour cells, has been reported, although it is often difficult to determine whether it is the neoplastic cells or benign interspersed histiocytes that show this feature (Boxes 9.2 and 9.3 and Figure 9.1).

Immunohistochemistry

Histiocytic sarcomas are usually positive for CD45 and HLA-DR (Copie-Bergman *et al.* 1998). They express CD68; PG-M1 is the more specific stain, since KP1 also stains

BOX 9.2: Histiocytic sarcoma: Clinical features

- Occurs at all ages; most prevalent in adults
- Presenting features:
  - Lymphadenopathy
  - Skin tumours
  - Other extranodal tumours
- Systemic symptoms common
- Rare association with mediastinal germ cell tumours
- Association with lymphoid neoplasms

BOX 9.3: Histiocytic sarcoma: Morphology

- Large cells
- Variable degrees of pleomorphism
- Abundant cytoplasm, may be foamy
- Phagocytosis by tumour cells may be seen
- Often resemble large B- or T-cell lymphomas

Figure 9.1 Histiocytic sarcoma showing tumour cells with reniform and indented/grooved nuclei and abundant eosinophilic cytoplasm.

granulocytic cells (Table 9.2). It should be remembered, however, that PG-M1 also stains mast cells, synovial cells and some melanomas. CD163, a haemoglobin scavenger receptor protein, is a more specific marker of histiocyte lineage (Vos *et al.* 2005); however, reactive histiocytic proliferations express the same markers and the distinction between these entities rests on cytological features. CD43, CD45RO and CD4 are also expressed, which may lead to confusion with T-cell lymphoma unless more specific T-cell or histiocyte markers are used. Lysozyme (muramidase) is an excellent marker of histiocytic sarcoma, the cytoplasmic staining being finely granular with paranuclear accentuation (Box 9.4 and Figure 9.2).

BOX 9.4: Histiocytic sarcoma: Immunohistochemistry

- CD45+
- CD43+, CD45RO+, CD4+
- CD68+
- Lysozyme+

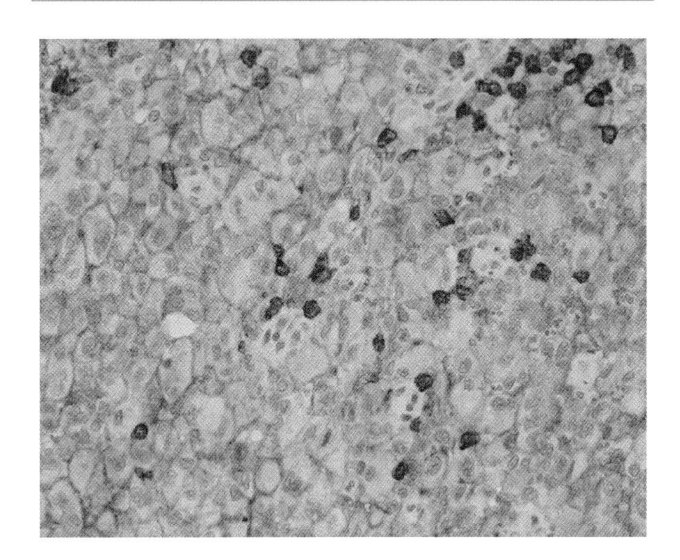

Figure 9.2 Histiocytic sarcoma immunostained for CD4. Note the membrane staining of the tumour cells. The darkly staining lymphocytes are helper T-cells.

Table 9.2 Immunohistochemistry of histiocytic and dendritic cell neoplasms

	Langerhans cell tumours	Interdigitating dendritic cell tumours	Follicular dendritic cell tumours	Histiocytic tumours
CD1a	+	–	–	–
Langerin	+	–	–	–
S100	+	+	+/–	+/–
CD68	+/–	+/–	+/–	+
CD21, CD23, CD35	–	–	+	–
Lysozyme	+/–	–	–	+
Podoplanin	–	–	+	–
Clusterin	–	–	+	–

Genetics

Histiocytic sarcomas usually lack clonal T-cell receptor and immunoglobulin gene rearrangements, but cases showing clonal rearrangements have been reported, as described earlier. Histiocytic sarcomas may show BRAF V600E mutation.

Differential diagnosis

The wide differential diagnosis of histiocytic sarcoma includes malignant melanoma, anaplastic carcinoma and large cell lymphomas. These can be separated by immunohistochemistry with a CD45+, CD68+, lysozyme+ profile confirming the diagnosis of histiocytic sarcoma. It should be noted that a high proportion of melanomas are CD68+.

DENDRITIC CELL NEOPLASMS

LANGERHANS CELL HISTIOCYTOSIS

Langerhans cells are specialized, antigen-presenting, dendritic cells in skin and mucosal surfaces. They migrate to lymph nodes and are closely related to IDCs.

LCH is an uncommon proliferation that encompasses the conditions previously designated as eosinophilic granuloma, Hand–Schuller–Christian disease and Letterer-Siwe disease, which were sometimes grouped together as histiocytosis X. Division into solitary and multicentric disease is now considered to be of greater clinical significance than the eponymous syndromes (Liebermann *et al.* 1996). Single or multiple bony lesions are most frequent and multisystem disease is usually an aggressive disease of young children.

Patients with lymphadenopathy show a spectrum of disease varying from solitary lymph node involvement by LCH (eosinophilic granuloma) to multisystem disease. Solitary lymph node involvement may be seen at any age, 3 months to 68 years in a large series (Edelweiss *et al.* 2007), but occurs most often in children and young adults. This nodal form of LCH is usually self-limiting and has a good prognosis.

With the exception of isolated pulmonary disease, LCH has been shown to be clonal and is considered to be neoplastic. It may be associated with a number of malignancies including acute leukaemia, both myeloid and lymphoblastic, Hodgkin lymphoma and carcinomas (Aricò *et al.* 2003). This is unlikely to be fortuitous and occasionally the LCH is of the aggressive, multisystem type. However, islands of Langerhans cells may be found in lymph nodes involved in Hodgkin or non-Hodgkin lymphomas. These, like the Langerhans cells seen in the lungs of heavy smokers, are probably reactive rather than neoplastic proliferations.

Typically, the Langerhans cells are found within the sinuses of the lymph node, in contrast with dermatopathic lymphadenopathy in which Langerhans cells and IDCs accumulate within the paracortex. However, this may not be apparent at a later stage when infiltration of the paracortex occurs, with partial or subtotal effacement of the node. Langerhans cells have complex nuclei, which in paraffin sections usually show one or more long nuclear folds or grooves. Nucleoli are inconspicuous. The cytoplasm is abundant and clear or palely eosinophilic. Giant cell forms with similar nuclear characteristics may or may not be present. Eosinophils are not always present in significant numbers. If they are abundant, they may undergo necrosis with the formation of eosinophilic abscesses and deposition of Charcot–Leyden crystals (Figures 9.3 and 9.4).

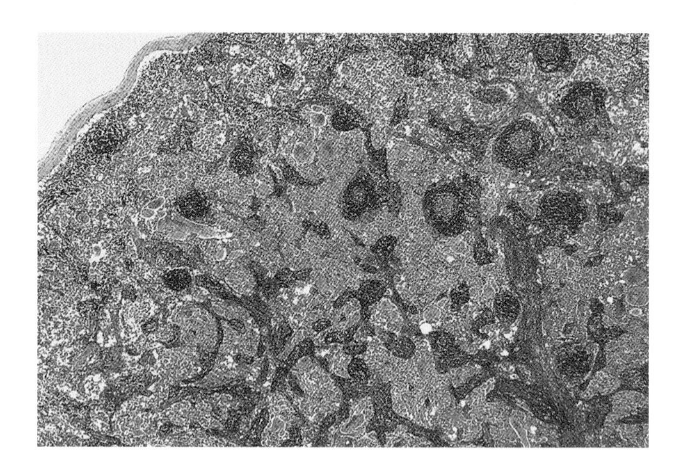

Figure 9.3 Langerhans cell histiocytosis of lymph node showing distension of the sinuses by Langerhans cells and giant cells.

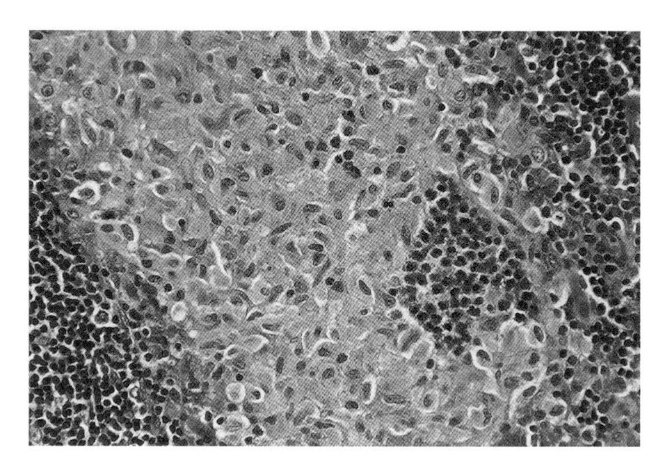

Figure 9.4 Langerhans cell histiocytosis of lymph node showing a sinus distended by Langerhans cells with elongated grooved nuclei and abundant pale eosinophilic cytoplasm. Eosinophils form a 'micro-abscess' in one area.

Immunohistochemistry

Like their normal counterparts, the cells of LCH express S100 protein, CDla and langerin, providing immunohistochemical confirmation of the diagnosis (see Table 9.2). This also highlights the lack of dendritic processes in the cells of LCH, an important distinction from normal Langerhans cells (Figures 9.5 and 9.6).

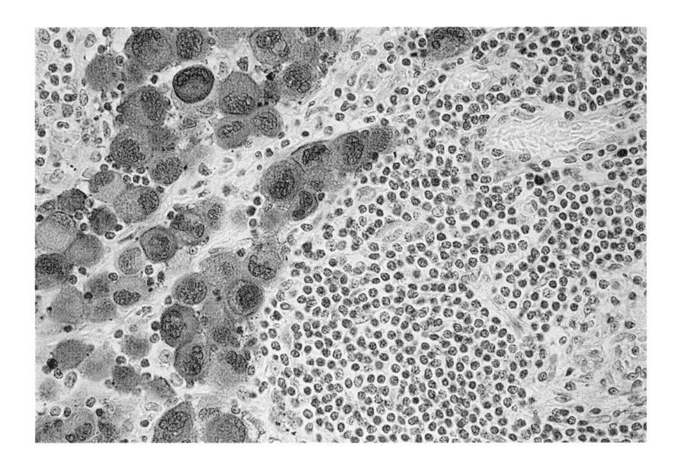

Figure 9.5 Langerhans histiocytosis of lymph node immunostained for S100 protein. The Langerhans cells, which have a sinusoidal distribution, are strongly stained.

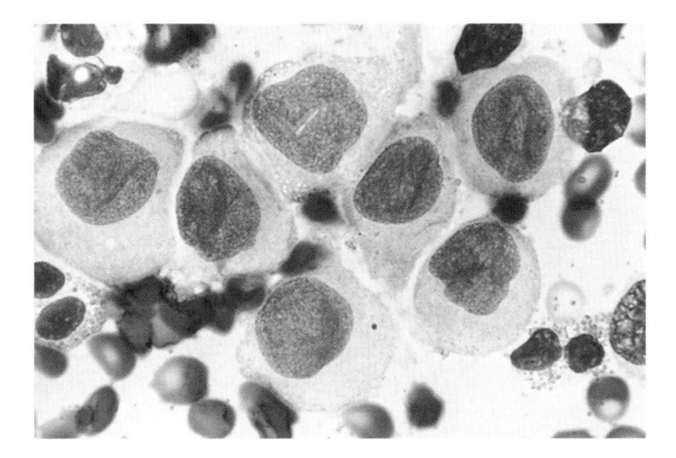

Figure 9.6 Imprint preparation of Langerhans cell histiocytosis of a lymph node. The Langerhans cells have grooved or folded nuclei and abundant pale-staining cytoplasm.

LANGERHANS CELL SARCOMA

Rare cases of malignant LCH or Langerhans cell sarcoma have been reported. This is usually seen in adults, predominantly women, as multifocal disease. It primarily involves extranodal sites and has a poor prognosis. The Langerhans cells, identified by the expression of S100, CD1a and langerin, or the presence of Birbeck granules on ultrastructural examination, appear frankly malignant. They show varying degrees of pleomorphism but some may preserve the characteristic nuclear grooving. Eosinophils are less frequent than in LCH (Figures 9.7 and 9.8).

INTERDIGITATING DENDRITIC CELL SARCOMA

Interdigitating reticulum cell sarcoma is a rare neoplasm presenting most commonly as solitary cervical lymphadenopathy. Involvement of the skin and other extranodal

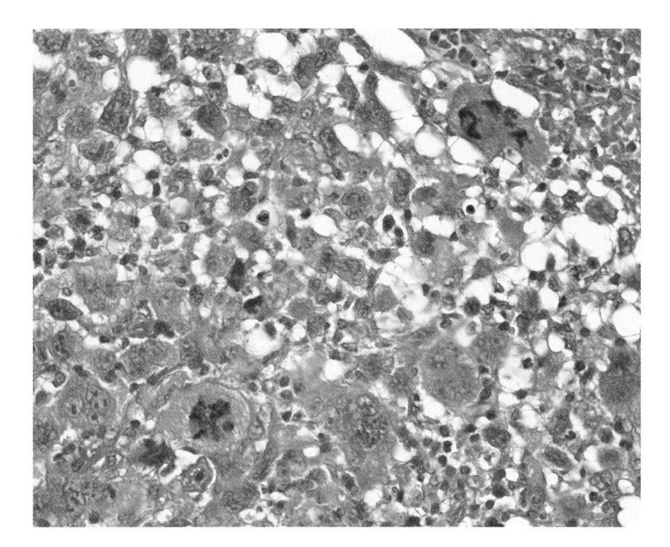

Figure 9.7 Langerhans cell sarcoma showing multinucleated cells and atypical mitoses with a background of more typical Langerhans cells.

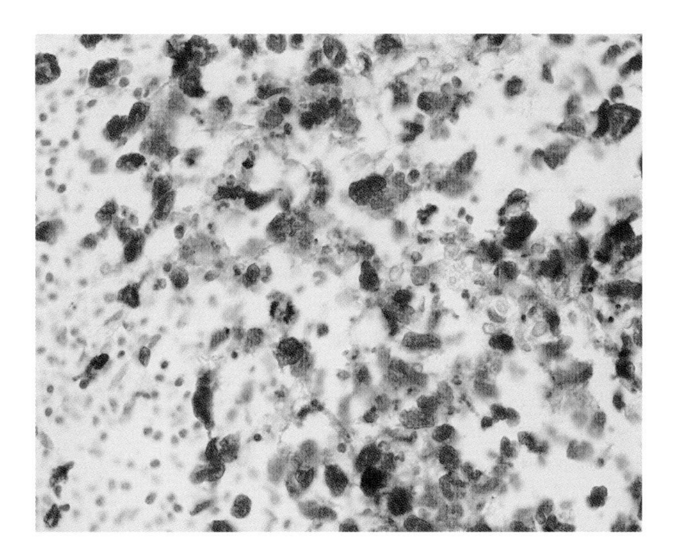

Figure 9.8 Langerhans cell sarcoma immunostained for S100 protein.

sites has been reported. The tumour often shows a paracortical distribution in the involved node with some residual follicles. Histologically, the tumour is composed of spindle cells arranged in fascicles and often forming a storiform pattern. More rounded cells and varying degrees of cytological atypia may be seen. The mitotic index is usually low. Small lymphocytes, mainly of T-cell phenotype, are present in variable numbers between the tumour cells (Figures 9.9 to 9.12).

Immunohistochemistry

The tumour cells express S100 protein, but are negative for CD1a, langerin and the follicular dendritic cell markers CD21, CD23 and CD35. CD68 and lysozyme expression may be seen in some cases.

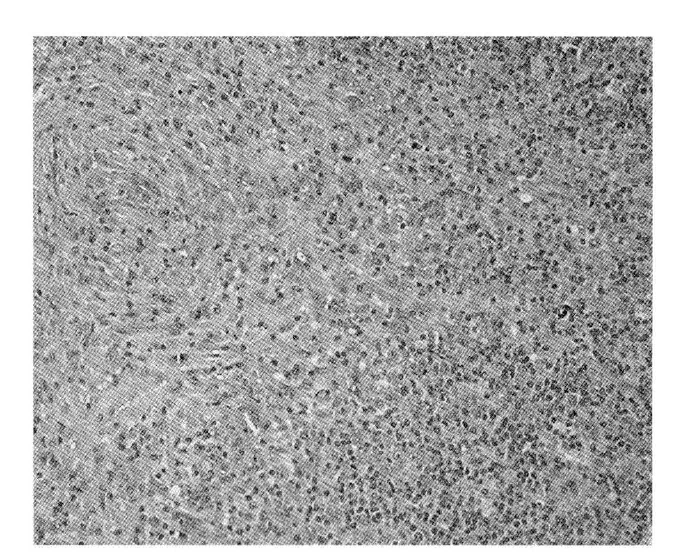

Figure 9.9 Interdigitating dendritic cell sarcoma in a lymph node involved by B-cell chronic lymphocytic leukaemia in transformation.

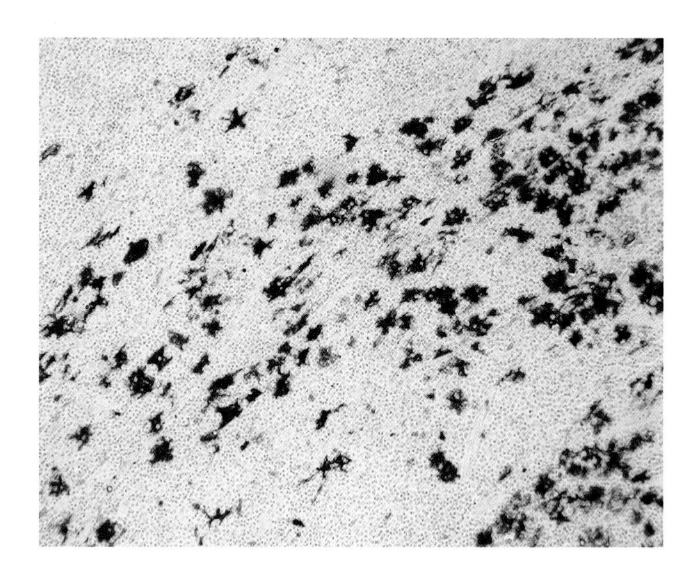

Figure 9.11 Interdigitating dendritic cell sarcoma in a lymph node involved by B-cell chronic lymphocytic leukaemia in transformation; S100 staining in the interdigitating dendritic cell sarcoma.

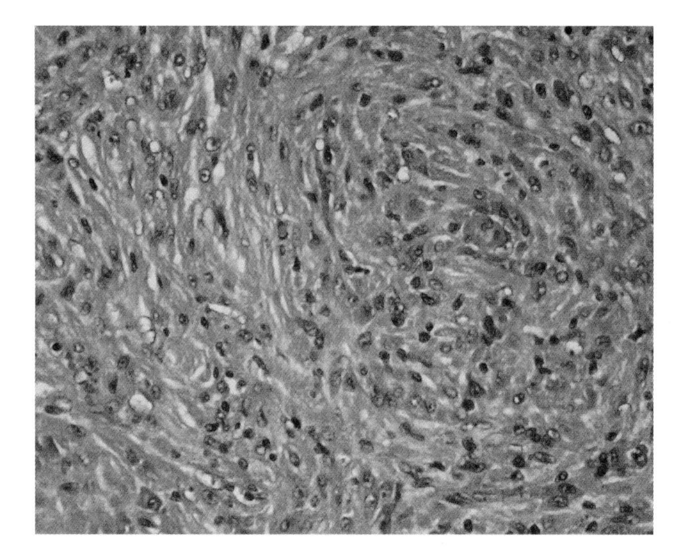

Figure 9.10 Interdigitating dendritic cell sarcoma in a lymph node involved by B-cell chronic lymphocytic leukaemia in transformation, high-power view of the spindle cell neoplasm.

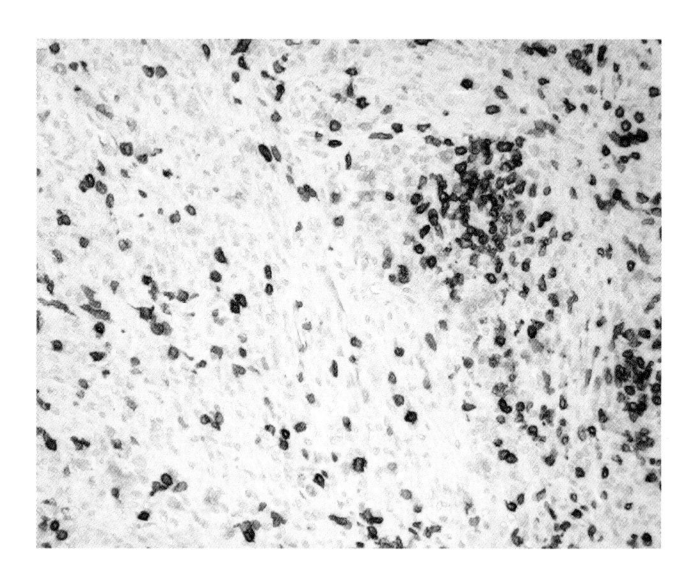

Figure 9.12 Interdigitating dendritic cell sarcoma in a lymph node involved by B-cell chronic lymphocytic leukaemia in transformation; the B-cell chronic lymphocytic leukaemia cells express CD5.

Interdigitating reticulum cell sarcoma may run a benign course. None of the four cases in the ILSG study died of disease, although Andriko *et al.* (1998) noted that, of 16 cases previously reported, seven patients died of widespread disease 1 week to 3 years after initial diagnosis. An association with B-CLL/SLL is described. The two neoplasms may be seen in the same lymph node and a clonal relationship has been demonstrated (Fraser *et al.* 2009). Other tumours occasionally associated with interdigitating reticulum cell sarcoma include FL, T-lymphoblastic lymphoma and carcinomas of colon, stomach, breast and liver (Cossu *et al.* 2006, Feldman *et al.* 2008).

FOLLICULAR DENDRITIC CELL SARCOMA

Follicular dendritic cell sarcoma is a rare neoplasm derived from follicular dendritic cells. The tumour occurs in adults, presenting most commonly with cervical lymphadenopathy, although a variety of extranodal sites are reported. There is an association with Castleman disease of the hyaline vascular type; the two diseases may occur synchronously or metachronously (Chan *et al.* 1994, Lin and Frizzera 1997). Extrafollicular clusters of atypical or 'dysplastic' follicular dendritic cells are occasionally observed in Castleman disease, suggesting a stage in the evolution of the neoplasm.

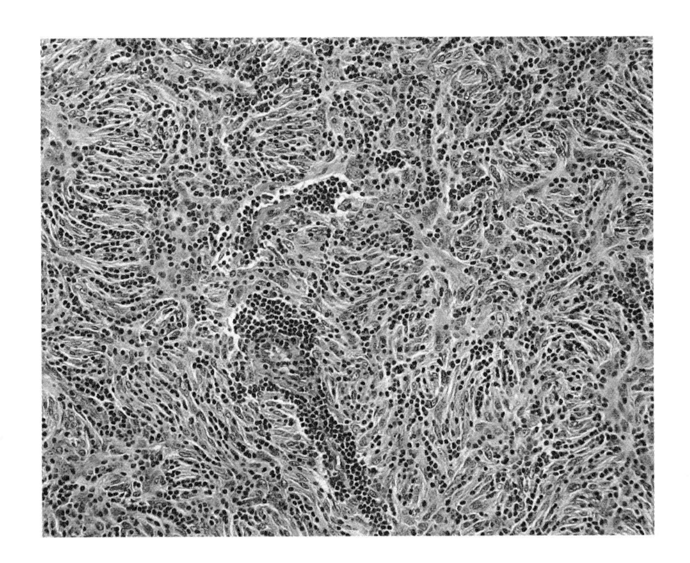

Figure 9.13 Follicular dendritic cell sarcoma of lymph node showing a storiform pattern of interlacing spindle cells.

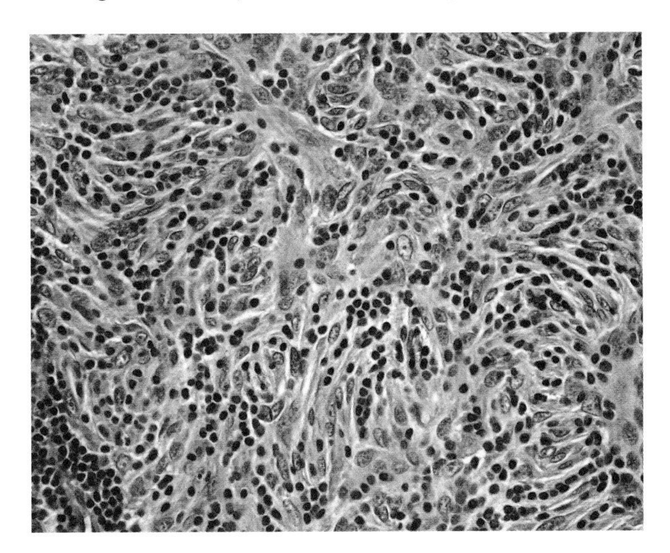

Figure 9.14 Follicular dendritic cell sarcoma of lymph node. In this area, the neoplastic spindle cells are interspersed with large numbers of lymphocytes.

Follicular dendritic cell sarcoma is a spindle cell tumour that forms whorls and fascicles and often has a storiform pattern. Varying degrees of cellular pleomorphism, sometimes associated with areas of necrosis, may be seen. These features are associated with more aggressive tumours (Perez-Ordonez *et al.* 1996, Perez-Ordonez and Rosai 1998). There is frequently a prominent infiltrate of small lymphocytes and plasma cells showing perivascular cuffing. The mitotic index and labelling index with Ki67 are usually low (Figures 9.13 and 9.14).

Immunohistochemistry

Immunohistochemical staining for one or more of the follicular dendritic cell markers, CD21, CD23 and CD35, is always present (see Table 9.2). A variable number of cases express CD68 and S100 protein. Pileri *et al.* (2002) found CD45 positivity in three of 13 cases. Newer antibodies include podoplanin and clusterin, both of which are positive with follicular dendritic cell sarcoma but negative with IDC sarcoma. Both, however, are also expressed by other types of spindle cell neoplasm (Grogg *et al.* 2005, Yu *et al.* 2007). Ultrastructurally, desmosomes are observed between the dendritic cell processes, and desmoplakin expression may be seen using immunohistochemistry on frozen sections.

Follicular dendritic cell sarcoma runs a variable course (Chan *et al.* 1997). Information available from 224 cases shows that it behaves like an intermediate-grade sarcoma with a substantial risk of local recurrence (28 per cent) and distant metastasis (27 per cent), and analysis of 50 of these cases showed that large tumour size (≥6 cm) and absence of lymphoplasmacytic response are associated with poor prognosis (Saygin *et al.* 2013).

An inflammatory pseudotumour-like variant is described, which usually involves liver or spleen. This has a female predominance and is frequently associated with systemic symptoms. It is consistently associated with Epstein–Barr virus (EBV) and, despite the presence of nuclear atypia, pursues an indolent course with late recurrence in some cases (Cheuk *et al.* 2001). The distinction between follicular dendritic cell sarcoma and nodal inflammatory pseudotumour is discussed in Chapter 3 (Figures 9.13 to 9.18).

OTHER RARE DENDRITIC CELL TUMOURS

These are very rare neoplasms that can be identified with confidence only if comprehensive immunohistochemistry and electron microscopy are performed. In the study by Andriko *et al.* (1998) one case was regarded as 'not otherwise specified' and three were classified as fibroblastic reticular cell tumour.

Fibroblastic reticular cells form a network that contributes to the microarchitecture of secondary lymphoid tissue and may be involved in regulating vascular growth.

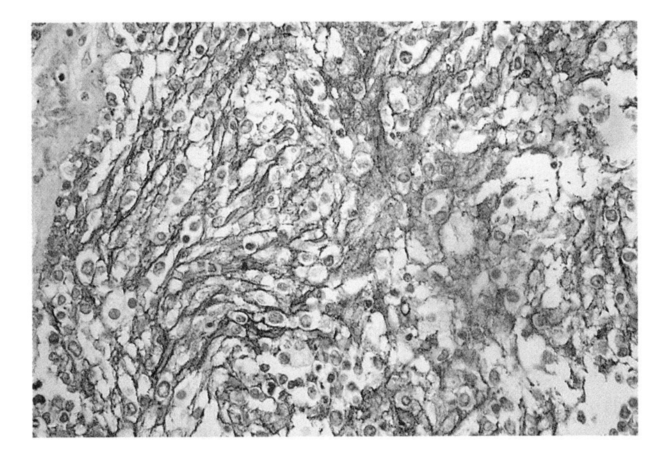

Figure 9.15 Follicular dendritic cell sarcoma immunostained for CD21, showing strong positivity of the spindle cells, which were also positive for CD23 and CD35.

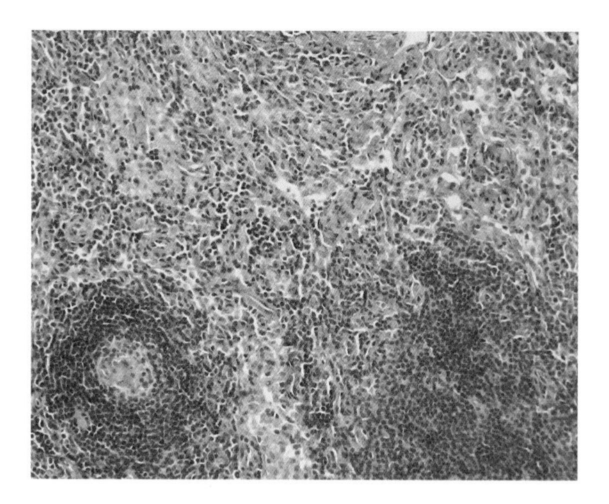

Figure 9.16 Follicular dendritic cell sarcoma complicating hyaline vascular Castleman disease. Castleman follicles are surrounded by the spindle cell neoplasm.

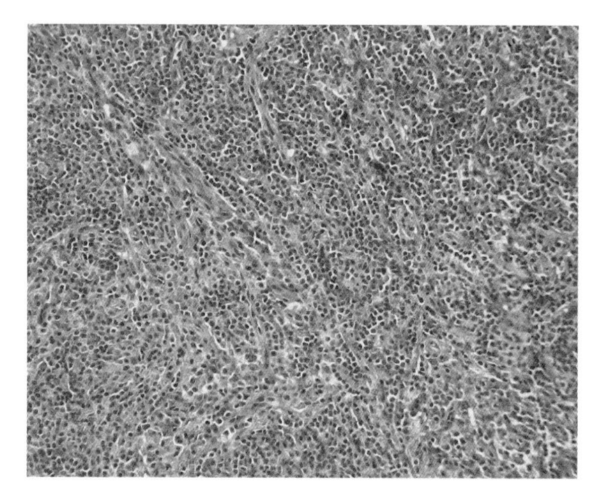

Figure 9.17 Follicular dendritic cell sarcoma complicating hyaline vascular Castleman disease showing spindle cell areas of follicular dendritic cell sarcoma.

Figure 9.18 Follicular dendritic cell sarcoma complicating hyaline vascular Castleman disease. CD21 staining shows the dendritic cell network.

Tumours may arise in lymph nodes, spleen or soft tissues and are similar in histological appearance to dendritic cell sarcoma and IDC sarcoma. However, they lack the typical immunohistochemical patterns of either tumour, showing only variable expression of smooth muscle actin, desmin and CD68. Some express cytokeratin, and it has been suggested that these arise from a separate subset of cytokeratin-positive interstitial reticular cells (Schuerfeld *et al.* 2003). Electron microscopy may be helpful in demonstrating features of myofibroblasts.

Indeterminate dendritic cell tumours are extremely rare neoplasms that usually arise in the dermis. They are composed of Langerhans-type cells that express CD1a and S100 but lack Birbeck granules and langerin.

BLASTIC PLASMACYTOID DENDRITIC CELL NEOPLASM

This is a rare aggressive neoplasm that has been previously designated as 'blastic NK-cell neoplasm', 'blastic natural killer cell lymphoma' or 'agranular CD4+ CD56+ haematodermic neoplasm'. It is now thought to be derived from precursors of plasmacytoid dendritic cells (PDC) or plasmacytoid monocytes (Petrella *et al.* 2005, Jegalian *et al.* 2009). BPDCNs are more frequent in males, patients usually presenting with skin plaques and nodules. About half of them also have lymphadenopathy. Most develop bone marrow involvement with a terminal leukaemic phase. This is often of myelomonocytic or myeloid type, in support of the concept that blastoid plasmacytoid dendritic cell neoplasms share a haemopoietic precursor cell lineage.

In lymph nodes the cells form a monomorphic infiltrate with a leukaemic pattern, involving the paracortex and medulla. They are small to intermediate in size with round to folded nuclear contours, fine chromatin, small nucleoli and a small to moderate amount of clear to eosinophilic agranular cytoplasm. Mitotic figures are frequent (Figure 9.19).

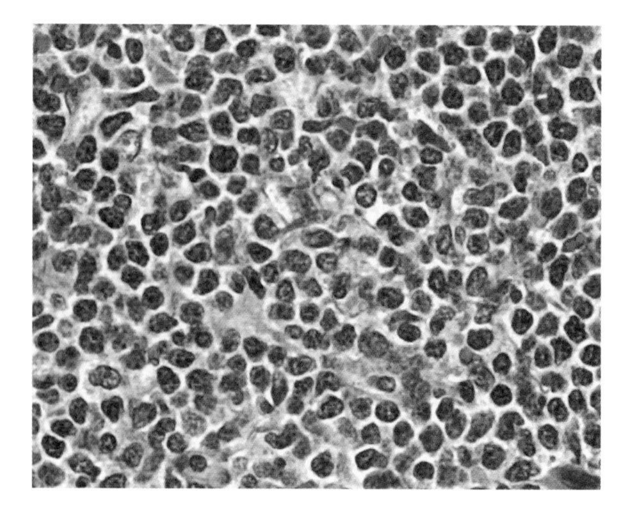

Figure 9.19 Blastic plasmacytoid dendritic cell neoplasm. The cells are small to intermediate in size with round or irregular nuclei, fine chromatin and small nucleoli.

Immunohistochemistry

A study of 91 cases of BPDCN (Julia *et al.* 2014) showed that the most characteristic markers of this entity are CD4, CD56, CD123 (interleukin-3 [IL-3] receptor), CD303 (blood dendritic cell antigen-2 [BDCA-2]) and T-cell leukaemia/lymphoma 1 protein (TCL1), which were expressed simultaneously in 46 per cent of cases. When four of these markers were expressed the diagnosis could still be reliably made. When only CD4, CD56 and CD123 are positive, the differential diagnosis includes acute myeloid leukaemia, which should be excluded by myeloid marker negativity. CD68 may be expressed as small cytoplasmic dots.

Terminal deoxynucleotidyl transferase (TdT) is positive in about one third of cases and expression of TdT and/or S100 correlates with varying degrees of maturation. TdT is a marker of immaturity, whereas S100 is a marker of dendritic cells, which are more mature than PDCs (indeterminate dendritic cells and Langerhans cells). However, in some cases both TdT and S100 were positive and double staining showed that these were expressed by different cells, suggesting different degrees of maturation within the same tumour.

The prognosis is poor. Statistical analysis showed that CD303 expression and high proliferative index (Ki67) were significantly associated with longer survival, suggesting that cases with a high proliferative index may have a better response to antimitotic chemotherapy, which may prolong survival.

MASTOCYTOSIS

In the WHO classification (Horny *et al.* 2008), mastocytosis is categorized as follows:

- Cutaneous mastocytosis.
- Indolent systemic mastocytosis.
- Systemic mastocytosis with clonal, haematological, non–mast-cell lineage disease.
- Aggressive systemic mastocytosis.
- Mast cell leukaemia.
- Mast cell sarcoma.
- Extracutaneous mastocytoma.

All of these conditions are uncommon, and significant lymphadenopathy is unusual. Lymph node involvement is most likely to be encountered in systemic mastocytosis. In 21 lymph node biopsies from patients with systemic mastocytosis, Horny and colleagues (1992) identified mast cell infiltrates in 80 per cent. Infiltration of the node begins in the medullary cords and sinuses and spreads to the paracortex, often leaving follicles intact. Germinal centres are frequently hyperplastic and high endothelial venules are prominent. The infiltrating mast cells may have rounded indented nuclei and abundant cytoplasm resembling histiocytes, or have spindle cell morphology. In haematoxylin and eosin–stained sections, the cytoplasm may appear granular or clear. If the cytoplasm is clear, the infiltrate may resemble hairy cell leukaemia or marginal zone lymphoma. Mast cell infiltrates are usually accompanied by large numbers of eosinophils, variable numbers of plasma cells and varying degrees of fibrosis, attributable to the release of mast cell mediators. Mast cell granules are better visualized in Giemsa-stained sections or imprint preparations. They are highlighted by metachromatic staining with toluidine blue. The naphthol ASD chloroacetate esterase (Leder) stain gives a positive reaction with mast cells but does not distinguish them from cells of the granulocyte lineage. Useful immunohistochemical stains include CD68, mast cell tryptase, CD117 (c-Kit or KIT), CD2 and CD25.

Mast cell disease is associated with activating point mutations within the KIT proto-oncogene, a transmembrane tyrosine kinase receptor for stem cell/mast cell growth factor. Systemic mastocytosis may be associated with other haematological neoplasms encompassed in the WHO classification as systemic mastocytosis with associated clonal haematological non–mast-cell lineage disease (SM-AHNMD). A study of 29 patients with SM-AHNMD found that these patients were generally older, had more systemic symptoms and had inferior survival compared with patients with pure systemic mastocytosis. The associated conditions include myelomonocytic leukaemia, acute myeloid leukaemia, myeloproliferative and myelodysplastic syndromes, low-grade B-cell lymphomas and plasma cell myeloma. The lymph node infiltrates in mastocytosis may therefore include other types of haemopoietic neoplasm (Wang *et al.* 2013). A small group of patients with 'lymphadenopathic mastocytosis with eosinophilia' have marked progressive lymphadenopathy with blood eosinophilia. Those with rearrangement of the PDGFRA gene are regarded as falling into a specific group of myeloid and lymphoid neoplasms with eosinophilia and abnormalities of PDGFRA, PDGFRB or FGFR1 (Bain *et al.* 2008), which are frequently associated with translocation and the formation of fusion genes encoding an aberrant tyrosine kinase. The importance of recognising these disorders is that tyrosine kinase inhibitors may be a therapeutic option (Figures 9.20 and 9.21).

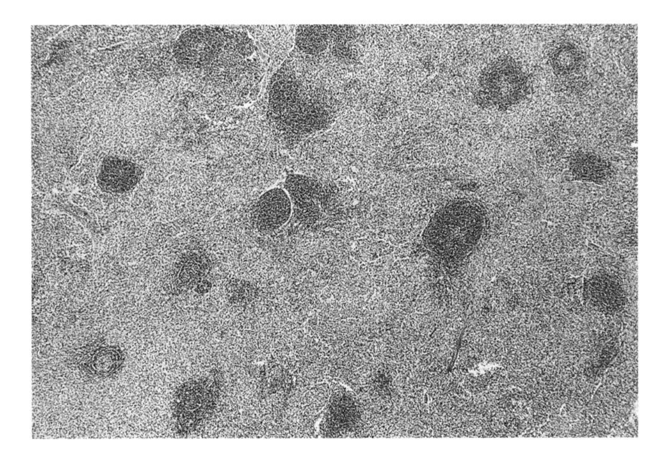

Figure 9.20 Systemic mastocytosis involving a lymph node and showing interfollicular spread of the mast cell infiltrate.

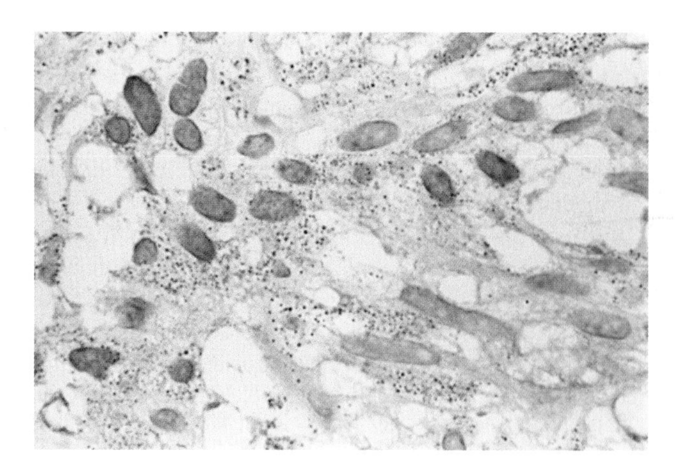

Figure 9.21 High-power view of systemic mastocytosis stained by toluidine blue. The mast cells have characteristic spindle cell morphology and show metachromatic granules.

MYELOID SARCOMA

The term 'myeloid sarcoma' is used in the WHO classification (Pileri *et al.* 2008) in preference to granulocytic sarcoma, because these neoplasms may show monocytic as well as granulocytic differentiation. The myeloperoxidase in tumours that show marked granulocytic differentiation gives the freshly sliced tumour a green colouration, a feature that gave rise to the name chloroma for these neoplasms.

Myeloid sarcoma is a relatively uncommon tumour, presenting most frequently in the skeleton and lymph nodes. Tumours showing monoblastic differentiation frequently involve the skin and oral cavity.

Myeloid sarcoma is associated with acute and chronic myeloid leukaemia, myeloproliferative disorders and myelodysplasia. It may occur before there is any haematological evidence of leukaemia and, in such cases, is reported to often show long survival after local therapy. Even without therapy, there may be a long interval between the diagnosis of myeloid sarcoma and evidence of leukaemia. If myeloid sarcoma develops in patients previously treated for acute myeloid leukaemia (AML), it is evidence of relapse.

Morphologically, myeloid sarcoma may show varying degrees of granulocytic differentiation. In lymph node biopsies, the cells form a diffuse infiltrate that initially involves the paracortex with follicular sparing. The appearances are most often misinterpreted as non-Hodgkin lymphoma, diffuse large B-cell lymphoma (DLBCL) in particular. Clues to the diagnosis may be given by preservation of some of the underlying lymph node structure in the form of sinuses outlined by reticulin and residual follicles. The tumour cells have rounded or indented nuclei that, in the less differentiated cases, are usually more uniform and have more delicate chromatin and smaller nucleoli than most DLBCLs. The finding of granular cells, particularly eosinophil myelocytes, provides more positive identification. Histochemical stains for myeloperoxidase and chloroacetate esterase will identify granular cells and may be of diagnostic value (Boxes 9.5 and 9.6 and Figures 9.22 to 9.24).

IMMUNOHISTOCHEMISTRY

Unexpected immunohistochemical staining will often be the first clue to the diagnosis of myeloid sarcoma. The tumour cells do not stain with restricted B- or T-cell lineage markers. Myeloid sarcomas showing granulocytic differentiation are the easiest to recognize. The granular cells will give a positive reaction with CD15 and antibodies to

BOX 9.5: Myeloid sarcoma: Clinical features

- Occurs at all ages, including childhood
- Lymphadenopathy, usually solitary
- May occur together with or precede acute or chronic myeloid leukaemia, myeloproliferative disease, or myelodysplasia
- Myeloid sarcoma in patients treated for acute myeloid leukaemia indicates relapse

BOX 9.6: Myeloid sarcoma: Morphology

- Effacement of underlying nodal architecture but may show some preservation
- Tumour cells have relatively uniform cytology, often resemble diffuse large B-cell lymphoma
- Nuclear chromatin fine, nucleoli small in less differentiated cases
- Granulocytic differentiation may be present—eosinophil myelocytes

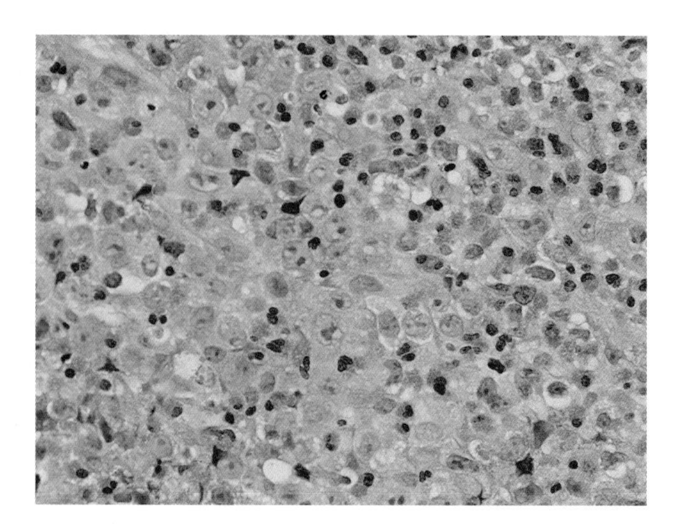

Figure 9.22 Myeloid neoplasm with PDGFRA gene mutation showing blasts admixed with numerous eosinophils.

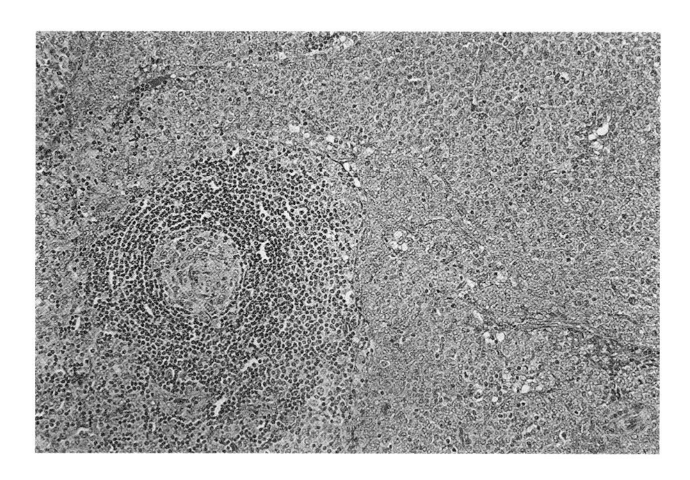

Figure 9.23 Myeloid sarcoma involving a lymph node showing a residual reactive follicle surrounded by a uniform infiltrate of blast cells.

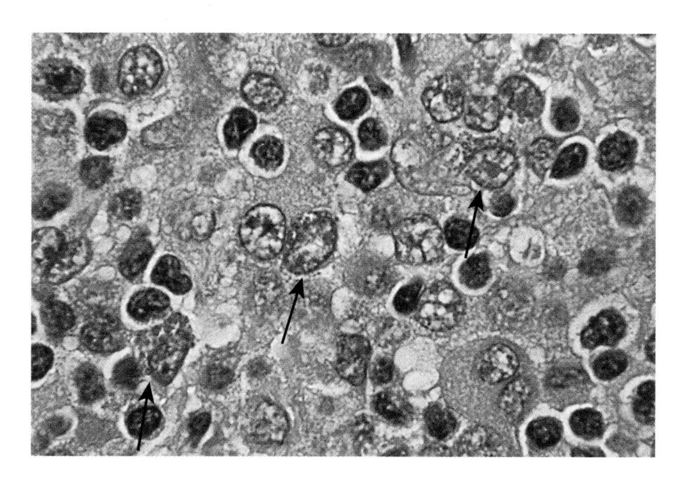

Figure 9.24 High-power view of myeloid sarcoma. The neoplastic cells have rounded nuclei with fine nuclear chromatin. Three cells in this photograph show eosinophilic cytoplasmic granules (arrows).

neutrophil elastase and myeloperoxidase. Many myeloid sarcomas stain positively for lysozyme. Of the two antibodies to CD68, KP1 stains most examples of myeloid sarcoma, whereas PG-M1 staining suggests a monocytic/monoblastic lineage. CD43 is usually positive in myeloid sarcomas, even the least differentiated forms. The finding of a tumour that is CD43+ and CD3− should raise suspicions of myeloid sarcoma. Additional useful antibodies include CD117 (c-KIT), which stains about 80 per cent of myeloid sarcomas, CD99 (54 per cent), CD34 (43 per cent) and TdT (31 per cent). Occasional tumours express CD56, and even fewer CD30 or CD4 (Pileri *et al.* 2007) (Box 9.7 and Figures 9.25 and 9.26).

GENETICS

Myeloid sarcoma is associated with chromosomal abnormalities in about 50 per cent of cases (Pileri *et al.* 2007). In children there is an association with acute myeloid leukaemia exhibiting t(8;21)(p22;q22). Among the most frequent abnormalities are those involving chromosome 16, including inv(16) and monosomy 16. For example, acute myeloid leukaemia with abnormal eosinophils exhibits inv(16)(p13;q22) or t(16;16)(p13;q22). Acute myeloid

BOX 9.7: Myeloid sarcoma: Immunohistochemistry

- No reactivity with B- or T-cell lineage markers
- CD43 positivity, even in undifferentiated tumours
- CD15, myeloperoxidase and neutrophil elastase identify granulocytic differentiation
- CD68 (KP1) stains the majority of tumours
- Lysozyme and CD68 (PG-M1) identify histiocytic differentiation
- Majority CD117 (c-KIT receptor)+
- Minority CD34+

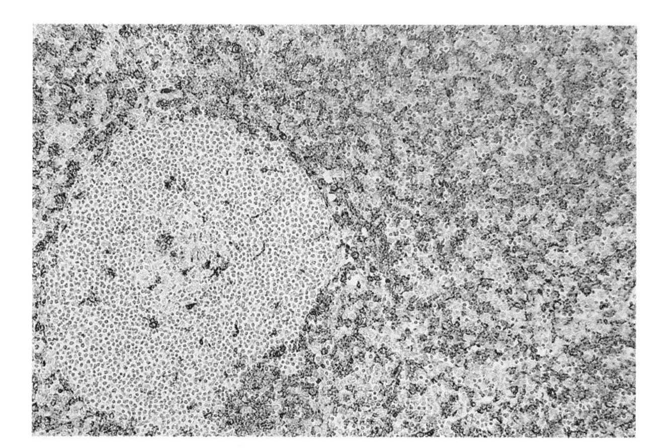

Figure 9.25 Myeloid sarcoma of lymph node immunostained for CD68 (PG-M1) showing positivity of many of the myeloid cells surrounding a reactive follicle.

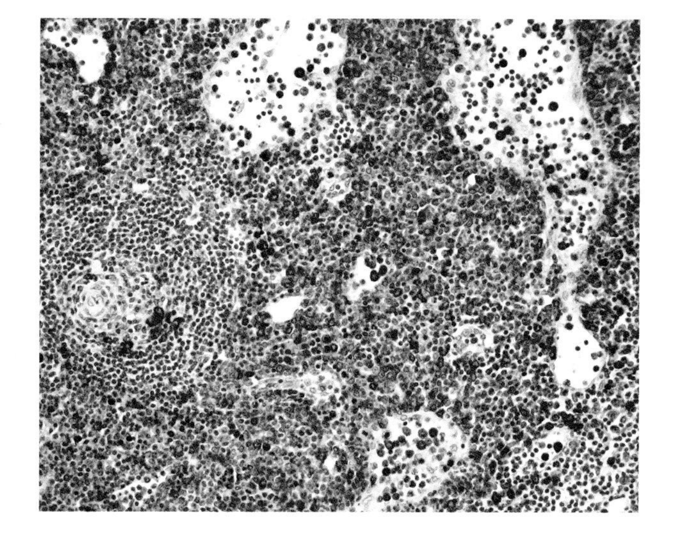

Figure 9.26 Myeloid sarcoma immunostained for neutrophil elastase. Many of the cells show granulocytic differentiation.

leukaemia with 11q23 abnormalities usually shows monocytic features and is associated with monocytic sarcoma.

One third of acute myeloid leukaemia patients show a mutation of the nucleophosmin (NPM1) gene, and about 18 per cent of myeloid sarcomas show a similar mutation. Chromosomal microarray analysis (CMA) using formalin-fixed paraffin-embedded (FFPE) tissues successfully identified genomic aberrations in six cases of myeloid sarcoma and confirmed the previously reported frequent occurrence of Fms-related ryrosine kinase 3 (FLT3) and NPM mutations as well as identifying multiple mutations in each case in a wide range of other AML-associated genes (Mirza *et al.* 2014). The factors involved in extramedullary localisation remain to be determined.

A rare variant of myeloid sarcoma has been described in which granulocytic precursors (usually eosinophilic) are found in a background of T-lymphoblastic lymphoma. This syndrome is associated with translocations involving fibroblast growth factor receptor 1 (FGFR1) on the chromosome region 8p11 (Inhorn *et al.* 1995) now classified as myeloid and lymphoid neoplasms with FGFR1 abnormalities. Most patients progress to acute myeloid leukaemia (see also Chapter 4).

REFERENCES

Andriko JA, Kaldjian EP, Tsokos M, *et al.* 1998 Reticulum cell neoplasms of lymph nodes. A clinicopathological study of 11 cases with recognition of a new subtype derived from fibroblastic reticulum cells. *American Journal of Surgical Pathology* **22:** 1048–1058.

Aricò M, Girschikofsky M, Généreau T, *et al.* 2003 Langerhans cell histiocytosis in adults. Report from the International Registry of the Histiocyte Society. *European Journal of Cancer* **39:** 2341–2348.

Bain BJ, Gilliland DG, Horny HP, *et al.* 2008 Myeloid and lymphoid neoplasms with eosinophilia and abnormalities of PDGFRA, PDGFRB or FGFR1. In Swerdlow SH, Campo E, Harris NL, *et al.* (Eds) *WHO Classification of Tumours of Haematopoietic and Lymphoid Tissues.* Lyon: IARC Press.

Bassarova A, Troen G, Fossa A, *et al.* 2009 Transformation of B cell lymphoma to histiocytic sarcoma: Somatic mutations of PAX-5 gene with loss of expression cannot explain transdifferentiation. *Journal of Hematopathology* **2:** 135–141.

Brunner P, Rufle A, Dirnhofer S, *et al.* 2014 Follicular lymphoma transformation into histiocytic sarcoma: Indications for a common neoplastic progenitor. *Leukemia* **28:** 1937–1940.

Castro EC, Blazquez C, Boyd J, *et al.* 2010 Clinicopathologic features of histiocytic lesions following ALL, with a review of the literature. *Pediatric and Developmental Pathology* **13:** 225–337.

Chan JK, Fletcher CD, Nayler SJ, Cooper K 1997 Follicular dendritic cell sarcoma: Clinicopathological analysis of 17 cases suggesting a malignant potential higher than currently recognized. *Cancer* **79:** 294–313.

Chan JK, Tsang WY, Ng CS 1994 Follicular dendritic cell tumor and vascular neoplasm complicating hyaline-vascular Castleman's disease. *American Journal of Surgical Pathology* **18:** 517–525.

Cheuk W, Chan JK, Shek TW, *et al.* 2001 Inflammatory pseudotumor-like follicular dendritic cell tumor: A distinctive low-grade malignant intra-abdominal neoplasm with consistent Epstein–Barr virus association. *American Journal of Surgical Pathology* **25:** 721–731.

Copie-Bergman C, Wotherspoon AC, Norton AJ, *et al.* 1998 True histiocytic lymphoma. A morphological, immunohistochemical and molecular genetic study of 13 cases. *American Journal of Surgical Pathology* **22:** 1386–1392.

Cossu A, Deiana A, Lissia A, *et al.* 2006 Synchronous interdigitating dendritic cell sarcoma and B-cell small lymphocytic lymphoma in a lymph node. *Archives of Pathology and Laboratory Medicine* **130:** 544–547.

Edelweiss M, Medeiros LJ, Suster S, *et al.* 2007 Lymph node involvement by Langerhans cell histiocytosis: A clinicopathologic and immunohistochemical study of 20 cases. *Human Pathology* **38:** 1463–1469.

Feldman AL, Arber DA, Pittaluga S, *et al.* 2008 Clonally related follicular lymphomas and histiocytic/dendritic cell sarcomas: Evidence for transdifferentiation of the follicular lymphoma clone. *Blood* **111:** 5433–5439.

Fraser CR, Wang W, Gomez M, *et al.* 2009 Transformation of chronic lymphocytic leukemia/small lymphocytic lymphoma to interdigitating dendritic cell sarcoma: Evidence for transdifferentiation of the lymphoma clone. *American Journal of Clinical Pathology* **132:** 928–939.

Go H, Jeon YK, Huh J, *et al.* 2014 Frequent detection of BRAF(V600E) mutations in histiocytic and dendritic cell neoplasms. *Histopathology* **65:** 261–272.

Grogg KL, Macon WR, Kurtin PJ, *et al.* 2005 A survey of clusterin and fascin expression in sarcomas and spindle cell neoplasms: Strong clusterin immunostaining is highly specific for follicular dendritic cell tumor. *Modern Pathology* **18:** 260–266.

Haroche J, Charlotte F, Arnaud L, *et al.* 2012 High prevalence of BRAF V600E mutations in Erdheim-Chester disease but not in other non-Langerhans cell histiocytoses. *Blood* **120:** 2700–2703.

Horny HP, Kaiserling E, Parwaresch MR, *et al.* 1992 Lymph node findings in generalized mastocytosis. *Histopathology* **21:** 439–446.

Horny HP, Metcalfe DD, Bennett JM, *et al.* 2008 Mastocytosis. In Swerdlow SH, Campo E, Harris NL, *et al.* (Eds) *WHO Classification of Tumours of Haematopoietic and Lymphoid Tissues.* Lyon: IARC Press.

Inhorn RC, Aster JC, Roach SA, *et al.* 1995 A syndrome of lymphoblastic lymphoma, eosinophilia and myeloid hyperplasia/malignancy associated with a t(8;13)(p11;q11): description of a distinctive clinicopathological entity. *Blood* **85:** 1881–1887.

Jegalian AG, Facchetti F, Jaffe ES 2009 Plasmacytoid dendritic cells: Physiologic roles and pathologic states. *Advances in Anatomical Pathology* **16:** 392–404.

Julia F, Dalle S, Duru G, *et al.* 2014 Blastic plasmacytoid dendritic cell neoplasms: Clinico-immunohistochemical correlations in a series of 91 patients. *American Journal of Surgical Pathology* **38:** 673–680.

Krautler NJ, Kana V, Kranich J, *et al.* 2012 Follicular dendritic cells emerge from ubiquitous perivascular precursors. *Cell* **150:** 194–206.

Liebermann PH, Jones CR, Steinman RM, *et al.* 1996 Langerhans cell (eosinophilic) granulomatosis. A clinicopathological study encompassing 50 years. *American Journal of Surgical Pathology* **20:** 519–552.

Lin O, Frizzera G 1997 Angiomyoid and follicular dendritic cell proliferative lesions in Castleman's disease of hyaline vascular type: A study of 10 cases. *American Journal of Surgical Pathology* **21:** 1295–1306.

Mirza MK, Sukhanova M, Stölzel F, *et al.* 2014 Genomic aberrations in myeloid sarcoma without blood or bone marrow involvement: Characterization of formalin-fixed paraffin-embedded samples by chromosomal microarrays. *Leukemia Research* **38:** 1091–1096.

O'Malley DP, Agrawal R, Grimm KE, *et al.* 2015 Evidence of BRAF V600E in indeterminate cell tumor and interdigitating dendritic cell sarcoma. *Annals of Diagnostic Pathology* **19:** 113–116.

Perez-Ordonez B, Erlandson RA, Rosai J 1996 Follicular dendritic cell tumor. Report of 13 additional cases of a distinctive entity. *American Journal of Surgical Pathology* **20:** 944–955.

Perez-Ordonez B, Rosai J 1998 Follicular dendritic cell tumor: Review of the entity. *Seminars in Diagnostic Pathology* **15:** 144–154.

Petrella T, Bagot M, Willemza R, *et al.* 2005 Blastic NK-cell lymphomas (agranular CD4+CD56+ hematodermic neoplasms): A review. *American Journal of Clinical Pathology* **123:** 662–675.

Pileri SA, Ascani S, Cox MC, *et al.* 2007 Myeloid sarcoma: Clinico-pathologic, phenotypic and cytogenetic analysis of 92 adult patients. *Leukaemia* **21:** 340–350.

Pileri SA, Grogan TM, Harris NL 2002 Tumours of histiocytes and accessory dendritic cells: An immunohistochemical approach to classification from the International Lymphoma Study Group based on 61 cases. *Histopathology* **41:** 1–29.

Pileri SA, Orazi A, Falini B 2008 Myeloid sarcoma. In Swerdlow SH, Campo E, Harris NL, *et al.* (Eds) *WHO Classification of Tumours of Haematopoietic and Lymphoid Tissues.* Lyon: IARC Press.

Saygin C, Uzunaslan D, Ozguroglu M, *et al.* 2013 Dendritic cell sarcoma: A pooled analysis including 462 cases with presentation of our case series. *Critical Reviews in Oncology Hematology* **88:** 253–271.

Schuerfeld K, Lazzi S, De Santi MM 2003 Cytokeratin-positive interstitial cell neoplasm: A case report and classification issues. *Histopathology* **43:** 491–494.

Shao H, Xi L, Raffeld M, *et al.* 2011 Clonally related histiocytic/dendritic cell sarcoma and chronic lymphocytic leukemia/small lymphocytic lymphoma: A study of seven cases. *Modern Pathology* **24:** 1421–1432.

Takahashi E, Nakamura S 2013 Histiocytic sarcoma: An updated literature review based on the 2008 WHO classification. *Journal of Clinical and Experimental Hematopathology* **53:** 1–8.

Swerdlow SH, Campo E, Harris NL, *et al.* (Eds) 2008 *WHO Classification of Tumours of Haematopoietic and Lymphoid Tissues.* Lyon: IARC Press.

Venkataraman G, McClain KL, Pittaluga S, *et al.* 2010 Development of disseminated histiocytic sarcoma in a patient with autoimmune lymphoproliferative syndrome and associated Rosai–Dorfman disease. *American Journal of Surgical Pathology* **34:** 589–594.

Vos J, Abbondanzo SL, Barekman C, *et al.* 2005 Histiocytic sarcoma: A study of five cases including the histiocyte marker CD163. *Modern Pathology* **18:** 693–704.

Wang SA, Hutchinson L, Tang G, *et al.* 2013 Systemic mastocytosis with associated clonal hematological non-mast cell lineage disease (SM-AHNMD): Clinical significance and comparison of chromosomal abnormalities in SM and AHNMD Components. *American Journal Hematology* **88:** 219–224.

Yu H, Gibson JA, Pinkus GS, *et al.* 2007 Podoplanin (D2–40) is a novel marker for follicular dendritic cell tumors. *American Journal of Clinical Pathology* **128:** 776–782.

10

Non-haematolymphoid tumours of lymph nodes

METASTATIC CARCINOMAS AND MELANOMAS

A number of non-haematolymphoid proliferations cause lymphadenopathy that may clinically simulate malignant lymphoma. The commonest of these are metastatic anaplastic carcinomas and melanomas, which may mimic large cell lymphomas and occasionally Hodgkin lymphoma. Carcinomas usually have a more cohesive growth pattern than malignant lymphomas and, in their early stages, may be predominantly sinusoidal in their distribution. However, sinusoidal distribution is also seen in some cases of anaplastic large cell lymphoma and diffuse large B-cell lymphomas.

Although an important and sometimes difficult differential diagnosis in the past, the distinction between metastatic tumours and malignant lymphomas is now readily made using immunohistochemistry. Now that lymphoma subtypes have been more precisely identified and their morphological features defined, it is usually possible to differentiate between metastatic tumours and malignant lymphomas on cytomorphology alone in good-quality histological sections.

SPINDLE CELL NEOPLASMS/ PROLIFERATIONS

Rare spindle cell neoplasms/proliferations may cause lymphadenopathy (usually solitary). These are readily distinguished histologically from malignant lymphomas, although they may bear some resemblance to dendritic cell neoplasms.

INTRANODAL PALISADED MYOFIBROBLASTOMA

Intranodal palisaded myofibroblastoma or palisaded spindle cell tumour with amianthoid fibres is an uncommon benign neoplasm with only about 50 reported cases (Suster and Rosai 1989, Nguyen and Eltorky 2007) (Figures 10.1 to 10.3). Although occasionally described in other sites, it occurs almost invariably in inguinal lymph nodes in adults, usually in the 45–55 year age group, with a male to female ratio of 2:1. There is low recurrence rate with no reports of metastases. The nodes are replaced by interlacing fascicles of spindle cells with areas of nuclear palisading. These are interspersed with stellate areas of thick collagen fibres, called amianthoid fibres because their crystalline appearance is thought to resemble asbestos. Areas of haemorrhage are present with free red cells and haemosiderin deposition. The tumour is surrounded by a compressed rim of residual lymph node.

Immunohistochemistry is required to exclude other spindle cell tumours, such as Kaposi sarcoma and dendritic cell neoplasms. The tumour cells are positive for smooth muscle actin and myosin and show nuclear expression of cyclin D1. Smooth muscle actin staining also demonstrates both intracellular and extracellular bodies. These are not readily identified by routine haematoxylin and eosin staining but are red (fuchsinophilic) with trichrome stains. Electron microscopy shows features of myofibroblasts and smooth muscle cells.

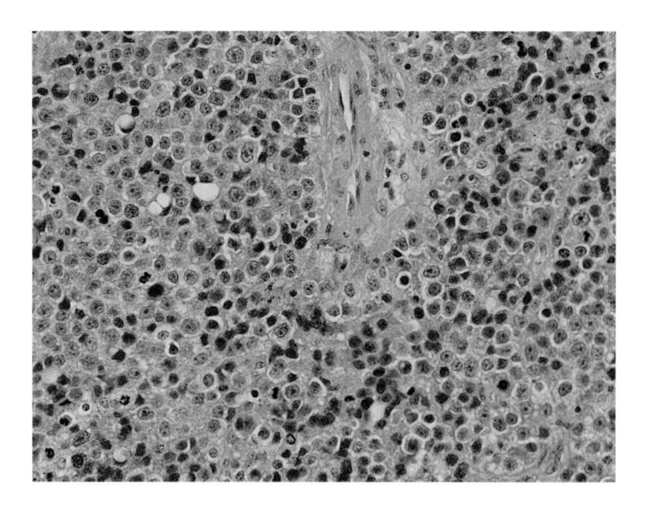

Figure 10.1 Poorly differentiated tumour in a lymph node. The tumour is composed of a diffuse infiltrate of large discohesive cells. Fine needle aspiration cytology had suggested this was a high-grade lymphoma.

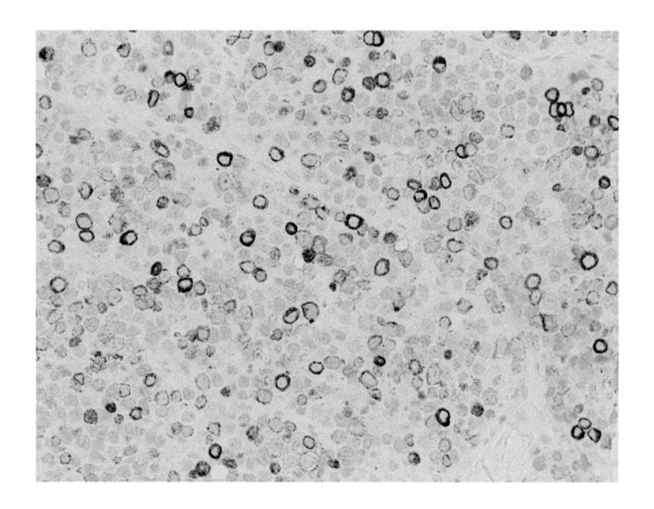

Figure 10.2 The tumour cells were negative for all lymphoid markers and showed expression of cytokeratin, confirming a poorly differentiated carcinoma.

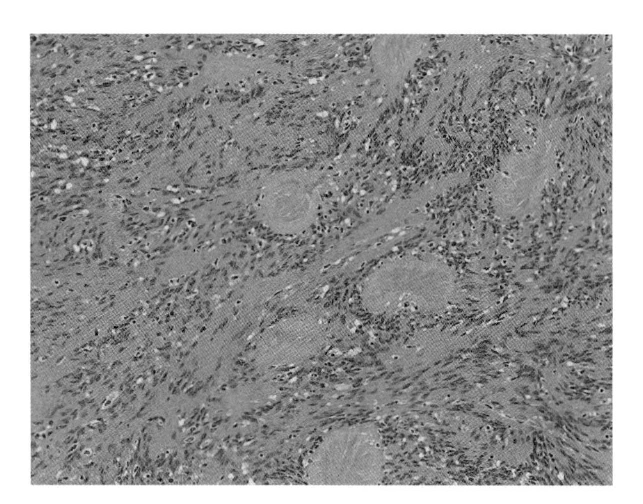

Figure 10.3 Lymph node showing an intranodal palisaded myofibroblastoma with palisaded spindle cells and amianthoid fibres.

LYMPHANGIOMYOMATOSIS

Lymphangiomyomatosis may involve lymph nodes, usually intrathoracic or para-aortic, but pulmonary involvement and other manifestations of the condition will usually be clinically dominant. The smooth muscle cells are positive for HMB45 and variably positive for smooth muscle actin.

LEIOMYOMATOSIS

Leiomyomatosis is rarely encountered in the pelvic lymph nodes of females. These smooth muscle tumours probably arise within the node, although metastasis from benign fibroids has been suggested.

INFLAMMATORY PSEUDOTUMOUR OF LYMPH NODE

This is discussed in Chapter 3.

MYCOBACTERIAL SPINDLE CELL PSEUDOTUMOUR

Mycobacterial spindle cell pseudotumour is composed of sheets of plump spindle cell macrophages filled with *Mycobacterium avium-intracellulare*. Periodic acid–Schiff and Ziehl–Neelsen staining demonstrates these organisms. The condition is seen most commonly in patients with acquired immune deficiency syndrome (AIDS) (see also Chapter 3).

VASCULAR TUMOURS/ PROLIFERATIONS

HAEMANGIOMAS

Haemangiomas may occur in lymph nodes, where they are most frequently situated in the hilum. The so-called angiomyomatous hamartoma is a rare variant most frequently involving inguinal nodes with irregular thick-walled vessels extending into the node from the hilum.

Epithelioid haemangioma (angiolymphoid hyperplasia with eosinophilia) has been described as a cause of lymphadenopathy. It is not always easy to determine, however, whether a lesion with many lymphoid follicles has evolved *in situ* or was originally a lymph node. The lack of sinus structure in most cases would favour evolution in an extranodal site.

BACILLARY ANGIOMATOSIS

Bacillary angiomatosis is a vascular proliferation, usually seen in patients with AIDS, caused by infection with *Bartonella henselae*, the causative organism of cat scratch disease. It most frequently presents as skin lesions but can affect other organs including lymph nodes. Clusters of vessels, some ectatic, are lined by plump endothelial cells. These are surrounded by amorphous eosinophilic or amphophilic material and neutrophils. With the Warthin–Starry stain this is seen to consist of large numbers of aggregated bacteria (Chan *et al.* 1991).

KAPOSI SARCOMA

Kaposi sarcoma may cause multiple lymphadenopathy in children, sometimes associated with malignant lymphoma or tuberculosis in the same lymph nodes (Figures 10.4 and 10.5). Most such cases have been reported from southern Europe or Africa. Adult disease is seen most commonly in patients with AIDS. It occurs most frequently in the capsular or subcapsular region of the node and may be confused with vascular transformation of the sinuses. The distinguishing features of Kaposi sarcoma are fascicles of spindle cells that often contain eosinophilic hyaline globules. The tumour shows varying degrees of vascularity. Red cells are seen in non–endothelial-lined clefts between the spindle cells. Mitoses are usually easily found among the spindle cells. Human herpesvirus 8 (HHV-8) is now recognized as a major aetiological factor in all variants of Kaposi sarcoma. Immunostaining for HHV-8 using an antibody to latency-associated nuclear antigen (LANA) is a sensitive and specific method for diagnosing Kaposi sarcoma in paraffin-embedded sections. Patients with human immunodeficiency virus (HIV) infection have a higher incidence of two HHV-8–related diseases: Kaposi sarcoma and multicentric Castleman disease (MCD). Careful examination of lymph nodes from patients with MCD frequently reveals evidence of Kaposi sarcoma (Naresh *et al.* 2008).

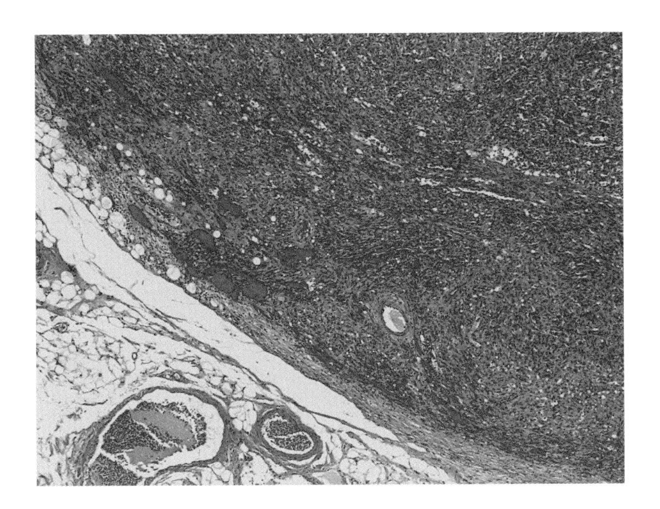

Figure 10.4 Low-power view of Kaposi sarcoma showing a vascular proliferation with fascicles of spindle cells.

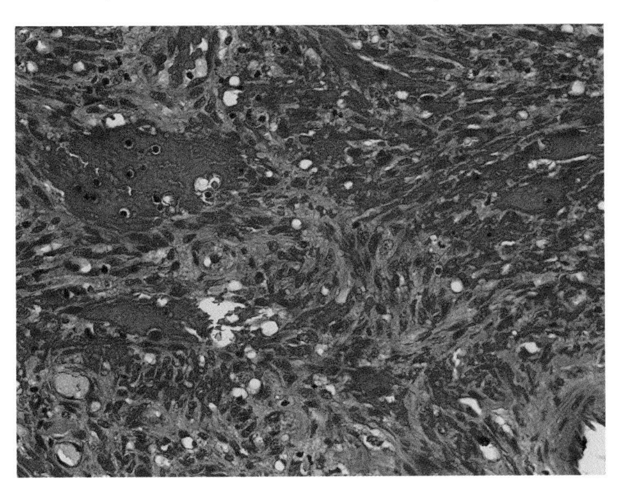

Figure 10.5 High-power view of Kaposi sarcoma demonstrating the atypical spindle cells and red cells in clefts between the spindle cells.

EPITHELIAL AND NEURAL INCLUSIONS

MÜLLERIAN INCLUSIONS

These may be seen in pelvic lymph nodes in females (Figure 10.6). These glandular structures should not be misinterpreted as metastatic adenocarcinoma.

ENDOMETRIOSIS

Endometriosis of pelvic lymph nodes is identified by the presence of both glandular and stromal elements.

DECIDUALIZATION

Decidualization of lymph nodes is occasionally encountered in lymph node biopsies from pregnant women. This usually appears to be a result of decidualization of resident nodal cells but in some cases may be superimposed on preexisting endometriosis.

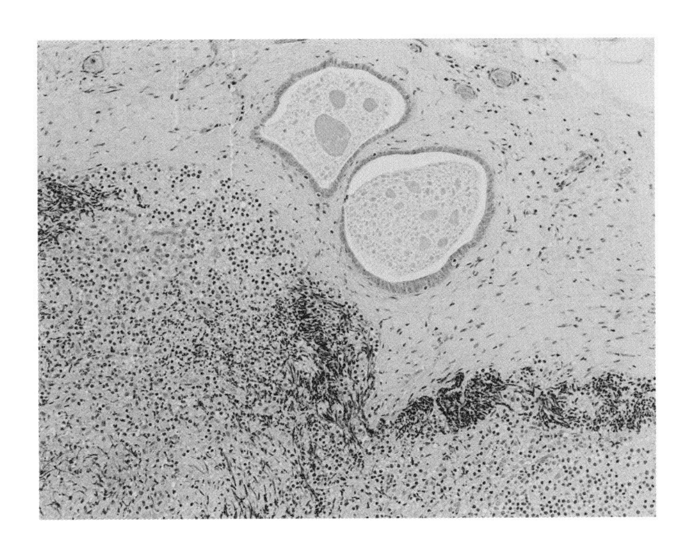

Figure 10.6 Pelvic lymph node showing benign müllerian inclusions within the capsule.

SALIVARY DUCTS

Salivary ducts are frequently seen within intraparotid lymph nodes and sometimes in periparotid nodes. In the presence of sialadenitis, the identification of the lymph node capsule may be necessary to distinguish between the node and the surrounding salivary tissue.

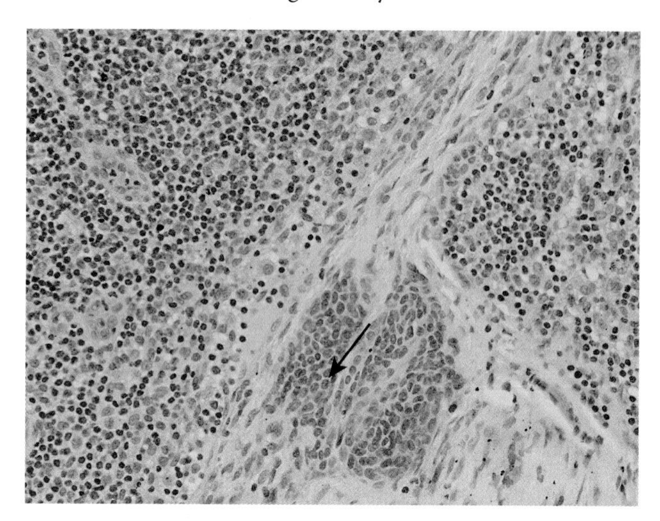

Figure 10.7 Lymph node showing a benign subcapsular inclusion of naevus cells. Note the bland nuclear morphology (arrow).

NEURAL INCLUSIONS

Various neural inclusions may be seen in lymph nodes. The most common is the presence of clusters of naevus cells in the capsule of the node (Figure 10.7). These are readily identified by morphology and immunohistochemistry.

REFERENCES

Chan JK, Lewin KJ, Lombard CM, *et al.* 1991 Histopathology of bacillary angiomatosis of lymph node. *American Journal of Surgical Pathology* **15:** 430–437.

Naresh KN, Rice AJ, Bower M 2008 Lymph nodes involved by multicentric Castleman disease among HIV-positive individuals are often involved by Kaposi sarcoma. *American Journal of Surgical Pathology* **32:** 1006–1012.

Nguyen T, Eltorky MA 2007 Intranodal palisaded myofibroblastoma. *Archives of Pathology and Laboratory Medicine* **131:** 306–310.

Suster S, Rosai J 1989 Intranodal hemorrhagic spindle-cell tumor with 'amianthoid' fibers. Report of six cases of a distinctive mesenchymal neoplasm of the inguinal region that simulates Kaposi's sarcoma. *American Journal of Surgical Pathology* **13:** 347–357.

11

Needle (trephine) core biopsies of lymph nodes

Since the publication of the previous editions of this book we have seen a further increase in the use of needle core biopsies of lymph nodes for diagnostic purposes. This has been made possible by the expansion and improvement of immunohistochemistry. Biopsies are often taken by radiologists using computed tomography (CT) guidance and are thus targeted at the node that appears most pathological rather than the most accessible one. CT-guided needle biopsies have been of particular value in the diagnosis of deep-seated lesions, such as retroperitoneal tumours, that would previously have required laparoscopy or laparotomy. The biopsy can be relatively easily repeated should the first one be unsatisfactory or unrepresentative. If the diagnosis is of a lymphoma, the patient can begin chemotherapy immediately without having to await recovery from abdominal surgery. We are being further challenged by the increasing use of endoscopic ultrasound (EUS) to sample mediastinal and deep abdominal nodes using fine needle aspiration (FNA). Ancillary techniques such as flow cytometry and cell blocks of FNA material may allow a firm diagnosis in a reasonable proportion of these samples.

Although needle core biopsies are clearly of advantage for deep-seated lesions and cases in which the patient is debilitated, we have been of the opinion that superficial lymph nodes in patients who are reasonably fit should be biopsied by open surgery. However, we are now seeing increasing numbers of such patients undergoing needle core biopsy, and in most instances a definitive diagnosis can be made (Goldschmidt *et al.* 2003, Hu *et al.* 2013). In view of the ease and speed of needle core biopsies, it is inevitable that their use will increase, with open biopsy being reserved for those lymph nodes in which the core biopsy is inconclusive.

We have therefore set out in this chapter our approach to the diagnosis of needle core biopsies with illustrations of the common, and a few unusual, lymphomas encountered. These are illustrated as instructive clinical cases in Figures 11.6 to 11.16.

HOW TO HANDLE NEEDLE CORE BIOPSIES

It is important that sufficient sections be cut initially for morphology and immunohistochemistry because re-cutting into the block usually results in a loss of tissue. Levels should not be cut on the block. A single haematoxylin and eosin (H&E) sample is cut, followed by spares for immunohistochemistry and an additional H&E sample if desired. If the biopsy consists of multiple cores it is worth separating these into individual blocks to conserve tissue; the most representative block is chosen for immunohistochemistry. Duplicate stains on both blocks should not be requested. The second block may be needed for additional immunohistochemistry, molecular analysis, or fluorescence *in-situ* hybridization (FISH), so it is important to preserve as much tissue as possible. Other than in exceptional circumstances it is wise to make a careful morphological assessment before ordering targeted immunohistochemistry. It is not uncommon to receive cases in which multiple immunohistochemical stains have been performed, with leukocyte common antigen being the only one to give a positive result, but with insufficient tissue remaining to complete the diagnosis.

The rapid fixation of needle biopsy specimens usually results in cells appearing smaller than they do in specimens from open biopsies and resections. This applies, in particular, to the cells of diffuse large B-cell lymphomas and to Hodgkin/Reed–Sternberg cells. At first sight this may make the distinction between small and large cell lymphomas appear difficult. However, this distinction can usually be made by the presence of apoptotic cells and cell debris in large cell lymphomas, a feature that is rarely seen in small cell lymphomas. A high proliferation fraction, as detected by Ki67 staining and seen in most large cell lymphomas, will further aid this distinction. Clinical correlation is vital in the interpretation of needle core biopsies, and knowing the clinical scenario and reasons for the core biopsy may help establish a firm diagnosis in cases in which there may otherwise have been some uncertainty. It is generally more difficult to make a primary diagnosis of lymphoma than to confirm recurrence in patients with known disease. For example, a diagnosis of persistent disease in a residual mass after treatment in a patient with diffuse large B-cell lymphoma could be made more confidently from a tiny biopsy specimen than a de novo diagnosis of diffuse large B-cell lymphoma.

LIMITATIONS, PROBLEMS AND PITFALLS

The disadvantages of needle core biopsies are generally related to the size of the biopsy. Encouraging the use of as large a core as possible would maximize the likelihood of reaching a diagnosis. Technical artefacts are more frequent in core biopsies than in excision biopsies of lymph nodes. The nodal tissue is fragile and poor handling of the tissue before fixation will result in crush artefact. Biopsy of sclerotic tissue is particularly prone to causing traction artefact in the adjacent tissue (Figure 11.1, crush artefact). Sponge artefact is another change that is easy to recognize as triangular spaces in the tissue. These are caused by incompletely fixed cores being placed between two sponges in a tissue cassette. This may be avoided by wrapping the core in biopsy paper rather than between sponges (Figure 11.2, sponge artefact).

For most lymphomas the histological features in core biopsies are the same as those in excision biopsies although the lack of architectural features may present some diagnostic challenges. However, there are some instances where the nature of the biopsy proves to be a limitation. It is easier to make a positive diagnosis of lymphoma on a core biopsy than to rule out the presence of disease. A negative biopsy may result from sampling the wrong node Figure 11.3) or

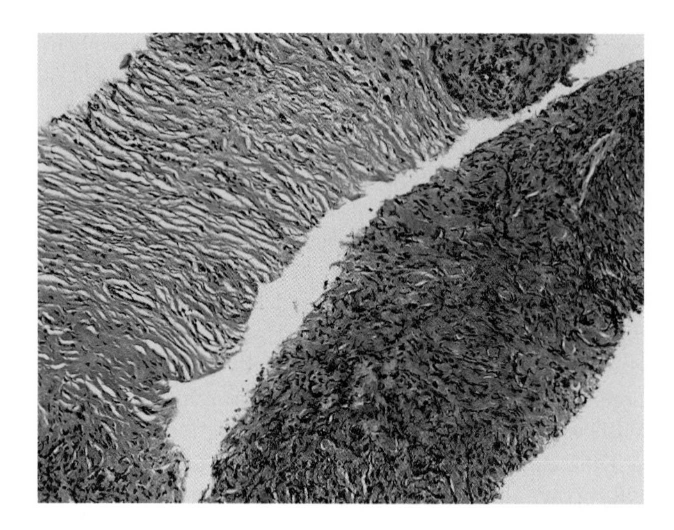

Figure 11.1 Core biopsy showing typical crush artefact.

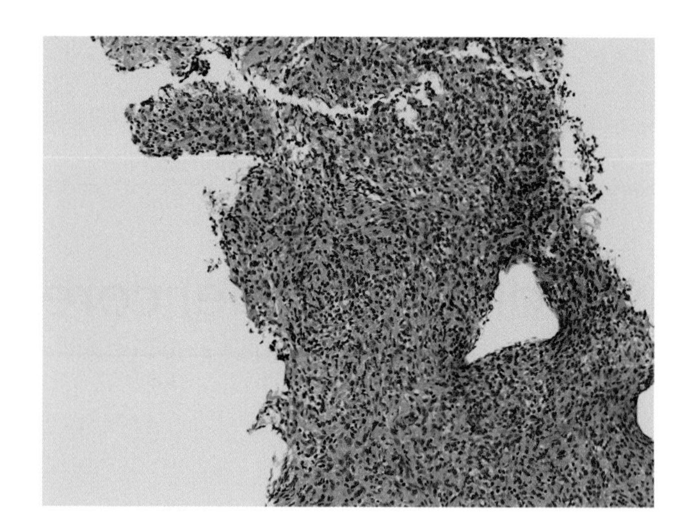

Figure 11.2 Core biopsy showing sponge artefact.

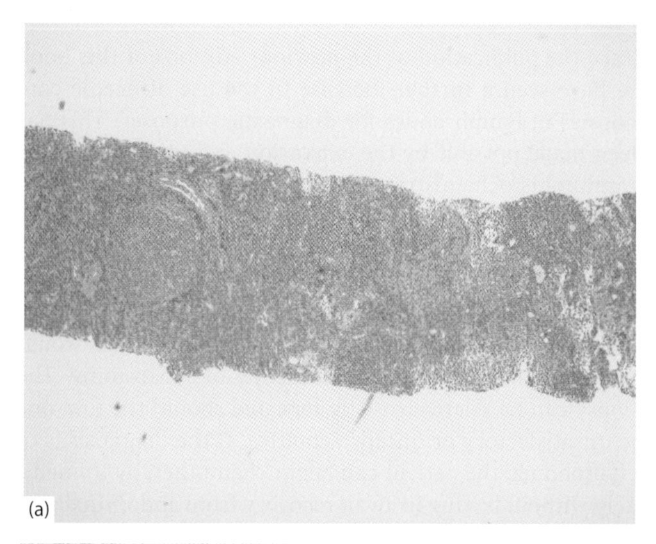

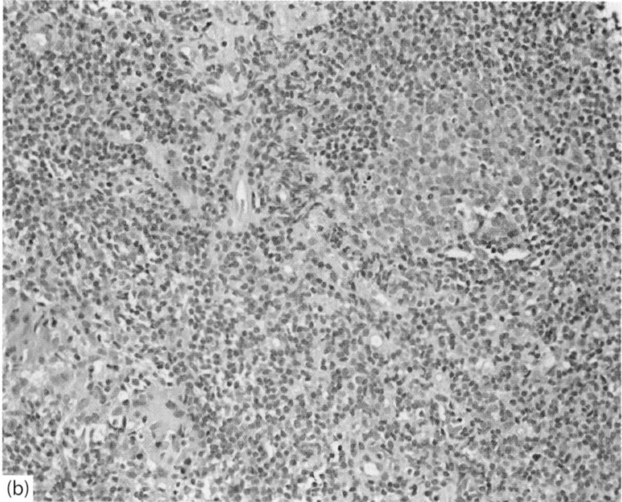

Figure 11.3 Unrepresentative core biopsy. (a) Core biopsy of a retroperitoneal mass. Low-power view showing a well-defined follicle to the left and open sinuses to the right. (b) Higher power view showing a reactive-looking follicle (right) and a loose collection of epithelioid histiocytes (low left). *(Continued)*

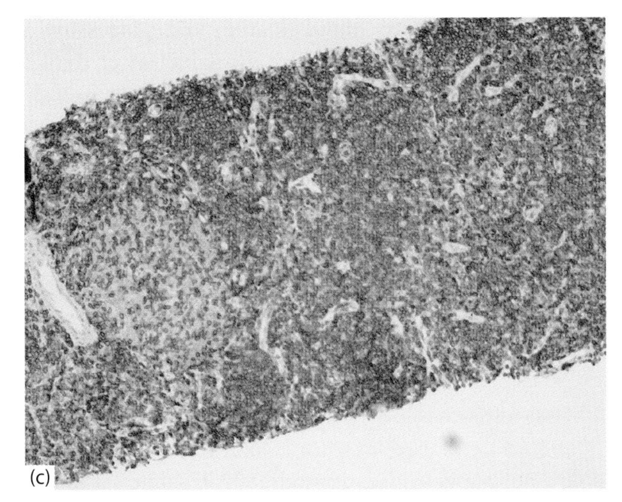

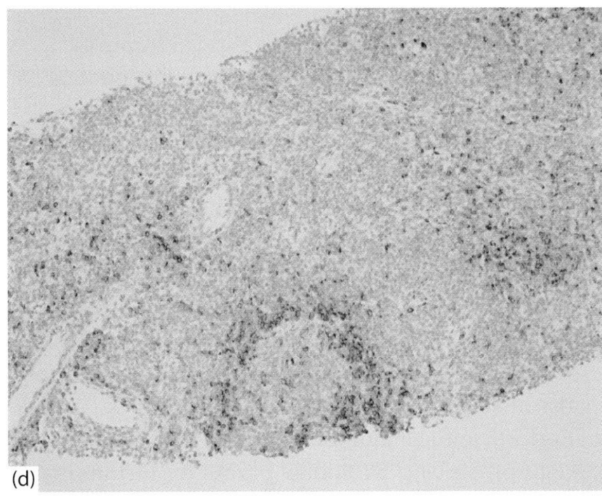

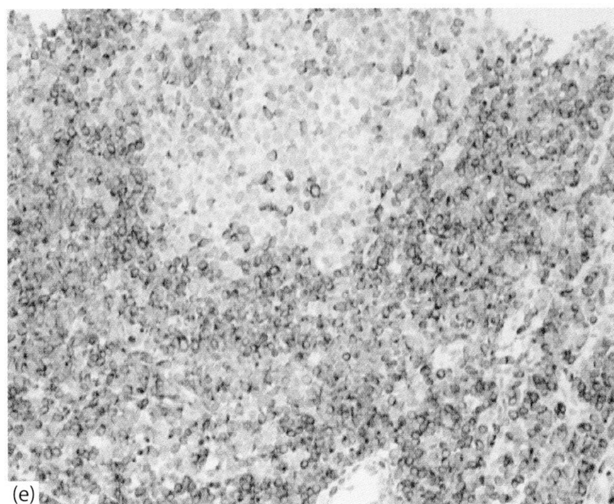

Figure 11.3 (Continued) Unrepresentative core biopsy. (c) Section stained for CD3 showing large numbers of T-cells, including many within a reactive follicle (left). (d) Section stained for CD79a showing scattered B-lymphocytes and plasma cells. The mantle around a reactive follicle (centre) is well defined. (e) Section stained for BCL-2 showing positive staining of small B- and T-cells. The follicle centre cells (upper centre) are negative.

Comments and conclusion: The features indicate that this is a reactive lymph node and that it is presumably not representative of the retroperitoneal mass.

from a focal lesion within a node that has not been sampled. The small size of the biopsy may make diagnosis difficult in the case of a non-homogeneous lesion or may result in failure to detect variability in tumour (e.g. areas of high-grade transformation may be missed). This is demonstrated in Figure 11.4, which shows a reactive core from a retroperitoneal mass biopsy and an excision of the same node that revealed interfollicular Hodgkin lymphoma that was missed by the core.

In some instances it may not be possible to make a firm diagnosis. One example would be the distinction between grade 3b follicular lymphoma and diffuse large B-cell lymphoma, which may be impossible based on a tiny core. Uncommon lymphomas present a diagnostic challenge in an excision biopsy and it may not be possible to make a firm diagnosis from a core biopsy. Insufficient tissue in the core presents a problem; unless care is taken to conserve tissue, the biopsy may be cut through before all necessary immunohistochemical stains have been completed. Careful thought has to be given as to how to maximize the use of sparse tissue.

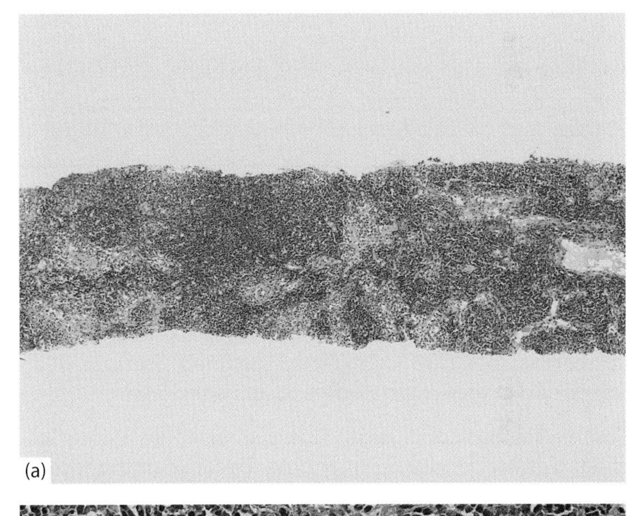

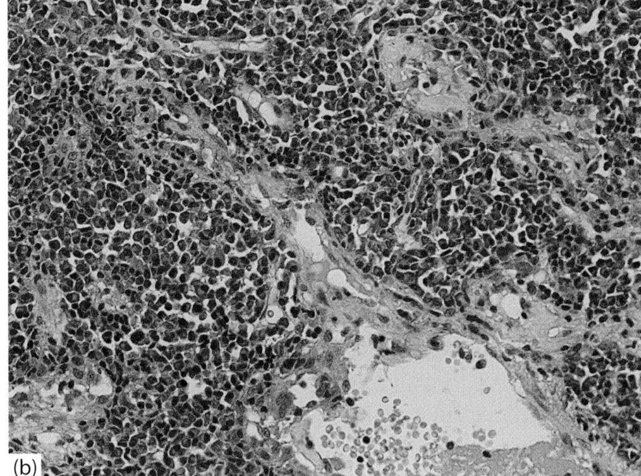

Figure 11.4 Paired core biopsy and excision biopsy of node. (a) Low-power view of the core showing paracortex with open sinuses. (b) Prominent plasma cells were present but no atypical cells were identified and the plasma cells were polytypic. *(Continued)*

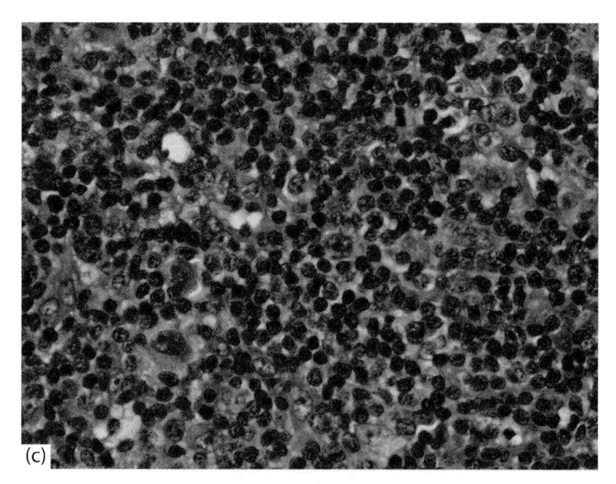

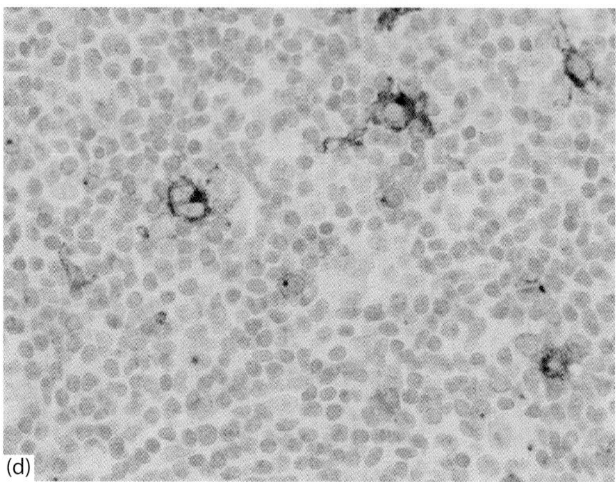

Figure 11.4 (Continued) Paired core biopsy and excision biopsy of node. (c) High-power view of the excision biopsy demonstrating occasional Reed–Sternberg cells within the paracortex. (d) CD30 highlighting the Reed–Sternberg cells found in the lymph node excision specimen.

Comments and conclusion: This is an example of interfollicular classical Hodgkin lymphoma. Because of the focal distribution of the Reed-Sternberg cells, these were not sampled in the core biopsy.

ADVANTAGES OF NEEDLE CORE BIOPSIES

There are many advantages to the use of needle core biopsies, for the patient, oncologist and pathologist. These biopsies are usually performed as an outpatient or day-case procedure and do not require inpatient admission and are generally quicker to arrange than an excision biopsy. The majority of biopsies are performed under ultrasound or CT guidance by a radiologist. Multidisciplinary meetings allow discussion of the most appropriate site before the biopsy is taken. This ensures that the most clinically worrying area is sampled. These biopsies are very well tolerated by patients and have a very low complication rate and therefore are suitable for patients who may not be fit enough for a surgical procedure. Additional cores can be taken to obtain tissue for cytogenetics or molecular analysis. Because of the small size of the biopsy, rapid fixation and processing are possible and diagnostic slides are available faster than for excision biopsies. Core biopsies are particularly useful in staging and in identifying recurrent disease and for investigating persistent masses after treatment (Soudack *et al.* 2008). Because the morbidity from core biopsies is low, they can be repeated if the first biopsy is unsatisfactory or unrepresentative (Boxes 11.1 and 11.2).

BOX 11.1: Needle biopsies

- Tissue for histology and immunohistochemistry should be placed in isotonic fixative immediately. Do not allow to dry; do not leave in saline for long periods.
- Allow at least 12 hours for fixation in formalin.
- Additional cores may be taken for cytogenetics, molecular analysis, etc.
- Put each core in a separate cassette to preserve tissue for immunohistochemistry.
- Cut a skim haematoxylin and eosin (H&E) sample only; do not cut levels; a further H&E specimen may be cut after taking spares for immunohistochemistry.
- Cut multiple spare sections for immunohistochemistry on one block at the same time as the sections for routine stains are cut. This avoids the waste of tissue associated with re-cutting the block.
- Perform initial immunohistochemistry on one block only, preserving additional blocks for further markers if required.

BOX 11.2: Advantages and disadvantages of needle core biopsies

Advantages	Disadvantages
• Outpatient or day-case procedure.	• Small size of the biopsy may make diagnosis difficult.
• Low morbidity.	• Small size of the biopsy may result in failure to detect variability in tumour; high-grade transformation may be missed.
• Good for disease staging and identifying recurrent disease.	• Require care and technical skill to conserve tissue.
• Can be repeated if the first biopsy is unsatisfactory or unrepresentative.	• There may be insufficient tissue to perform all immunohistochemical stains required.
• Additional cores can be taken for cytogenetics, molecular analysis.	• Traction artefact can cause problems.
• Small size of the biopsy allows rapid fixation and processing.	

ENDOSCOPIC ULTRASOUND BIOPSIES

EUS techniques are being increasingly used to sample mediastinal and deep abdominal lymph nodes that are not amenable to traditional core sampling. These are FNA biopsies, and the use of ancillary techniques such as flow cytometry and cell block preparations from aspirate material does allow a diagnosis of lymphoma in some cases (Amador-Ortiz *et al.* 2011; Ribeiro *et al.* 2010). The same limitations apply to these samples as to traditional cores, albeit the material is even more limited (Stacchini *et al.* 2012). Typically the cell block preparation contains blood with a few fragments of lymphoid tissue. Figure 11.5 demonstrates a typical EUS cell block preparation.

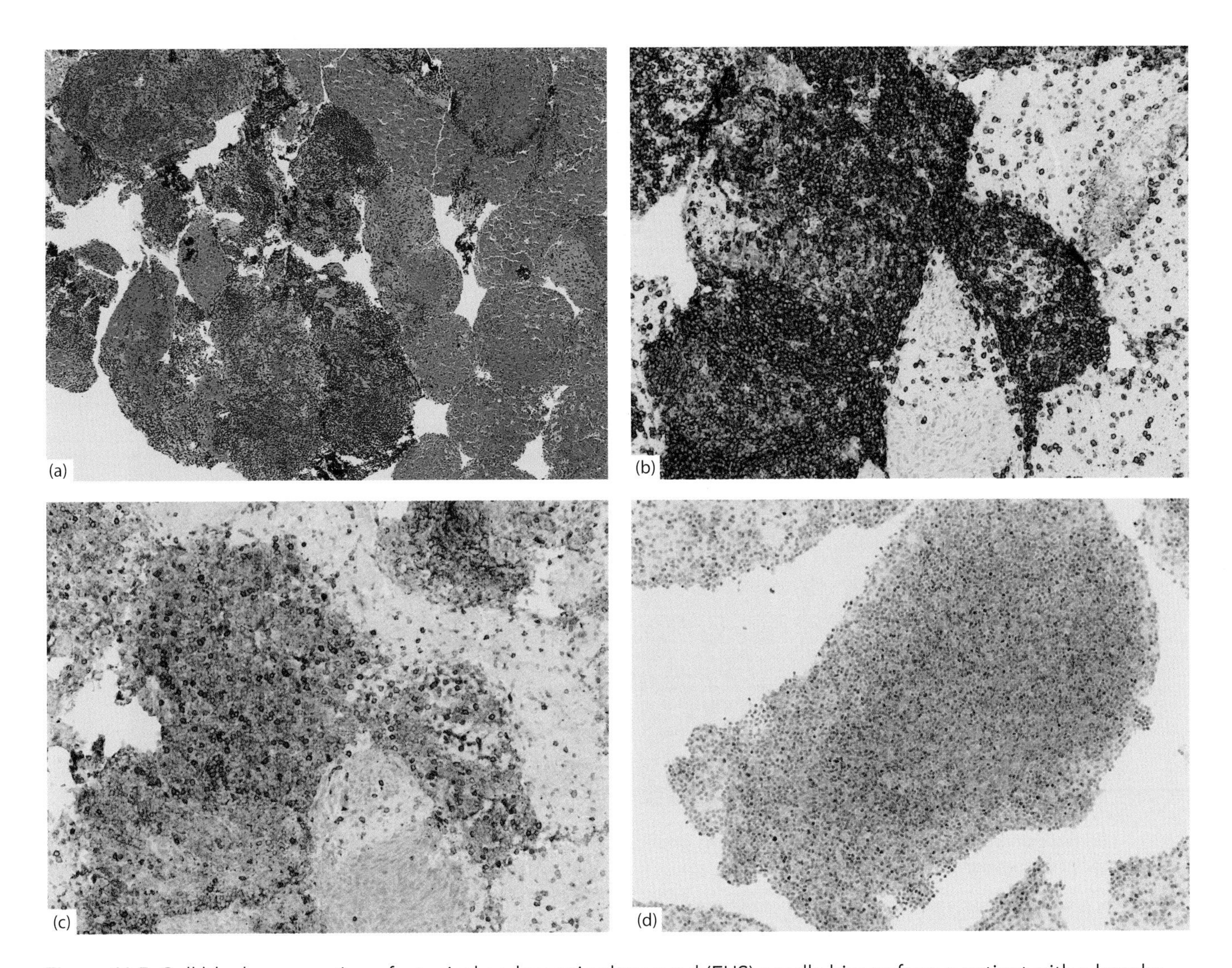

Figure 11.5 Cell block preparation of a typical endoscopic ultrasound (EUS) needle biopsy from a patient with a lymphocytosis. (a) There are small fragments of lymphoid tissue with much blood. (b) Section stained for CD20 showing strong positivity of the small lymphocytes. (c) Section stained for CD5 showing positivity of the B-cells. The small numbers of very darkly staining cells are reactive T-cells. (d) Cyclin D1 staining demonstrates positive nuclear expression.

Comments and conclusion: There was no positivity for CD23 and CD3 highlighted only scattered T-cells. A diagnosis of mantle cell lymphoma was made based on the CD5 and cyclin D1 expression. The blood lymphocytes had the same phenotype as the EUS biopsy.

CASE 1: B-CELL CHRONIC LYMPHOCYTIC LEUKAEMIA/SMALL LYMPHOCYTIC LYMPHOMA (Figure 11.6)

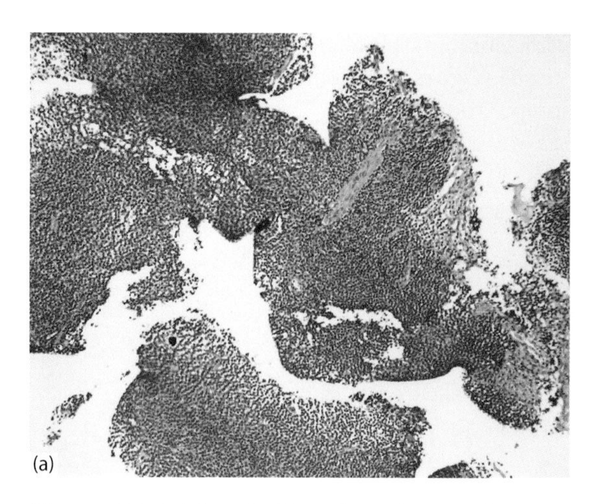

Figure 11.6 (a) Cervical lymph node core. Low-power view showing a fragmented core composed of regular small lymphoid cells.

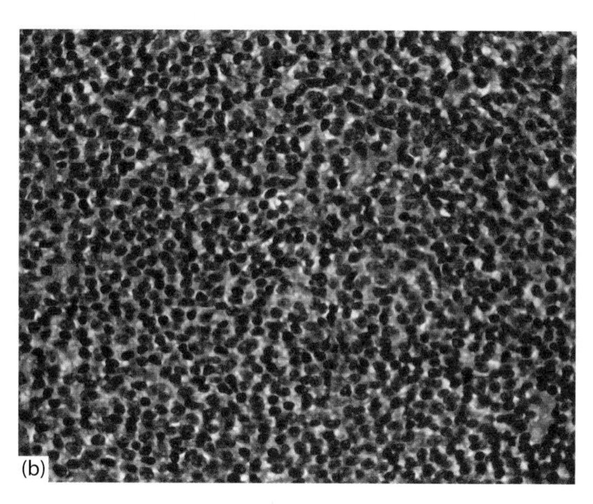

(b) Higher power view showing regular small lymphoid cells with rounded nuclei and clumped chromatin. Occasional cells have more open chromatin and prominent central nucleoli (paraimmunoblasts).

(c) Section stained for CD20 showing strong positivity of tumour cells.

(d) Section stained for CD5 showing positivity of tumour cells. The cells staining very darkly are reactive T-cells.

(e) Section stained for CD23. The tumour cells are strongly positive.

Comments and conclusion: The morphological appearances and histochemical profile indicate that this is B-cell chronic lymphocytic leukaemia/small lymphocytic lymphoma (B-CLL/SLL).

CASE 2: TRANSFORMATION OF KNOWN B-CELL CHRONIC LYMPHOCYTIC LEUKAEMIA (Figure 11.7)

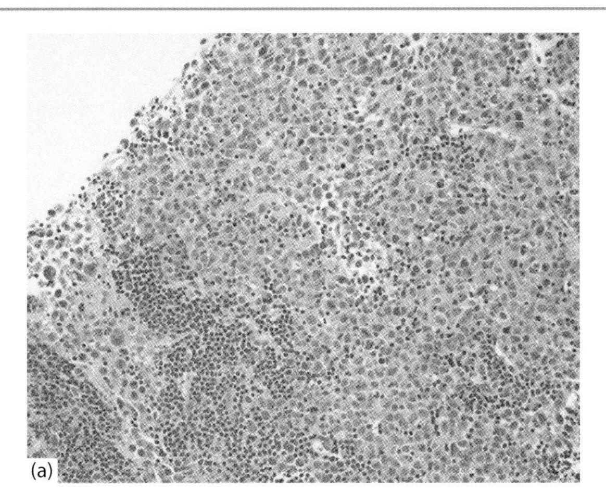

Figure 11.7 Axillary core biopsy from a 75-year-old woman with known chronic lymphocytic leukaemia. (a) The biopsy shows predominantly small B-cell chronic lymphocytic leukaemia/small lymphocytic lymphoma (B-CLL/SLL) cells to the left with blast cells of varying sizes mainly to the right.

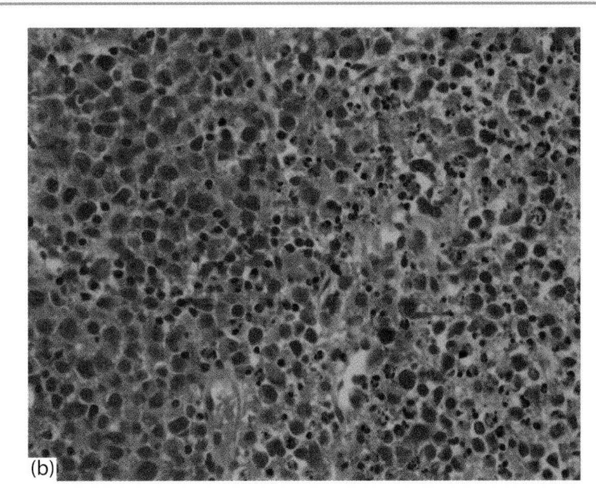

(b) High-power view showing extensive apoptosis among the transformed blast cells.

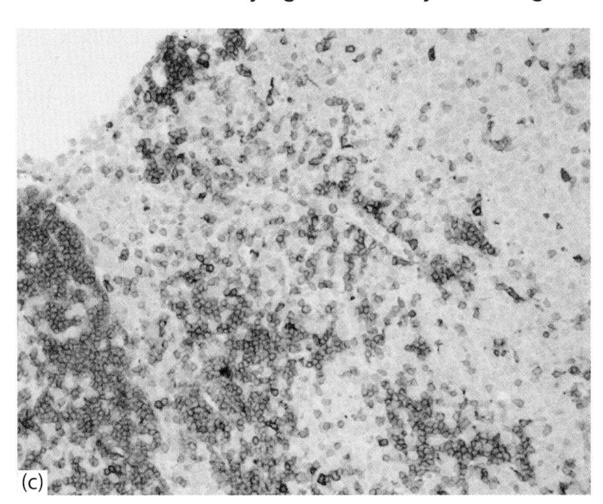

(c) Section stained for CD5 showing positive staining of the small cells (mainly to the left of the picture). Many of the large transformed cells are unstained.

(d) Section stained for CD23 showing positivity of most of the small cells to the left and negativity of most of the large cells to the right.

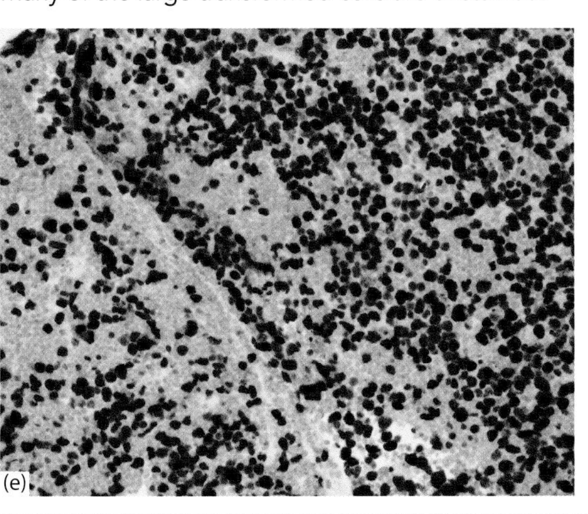

(e) Section stained for Ki67 showing a high labelling fraction among the large transformed cells. The identifiable small B-cell chronic lymphocytic leukaemia/small lymphocytic lymphoma cells show a low labelling fraction.

Comments and conclusion: Core biopsies are often performed to determine whether cases of B-CLL have undergone transformation. Not all biopsies are as successful as this one, which clearly shows transformation to a high-grade lymphoma. The transformed cells were positive for CD20 and CD79a but many have lost reactivity for CD5 and CD23. Transformed B-CLL/SLL cells are often more pleomorphic than those shown in this case. Transformation to Hodgkin lymphoma may occur.

CASE 3: MANTLE CELL LYMPHOMA (Figure 11.8)

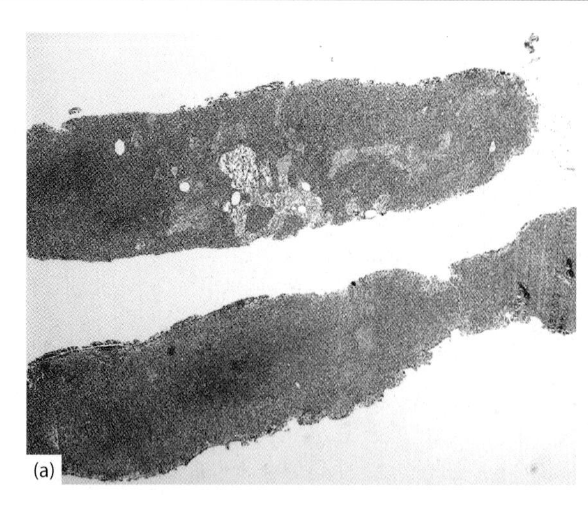

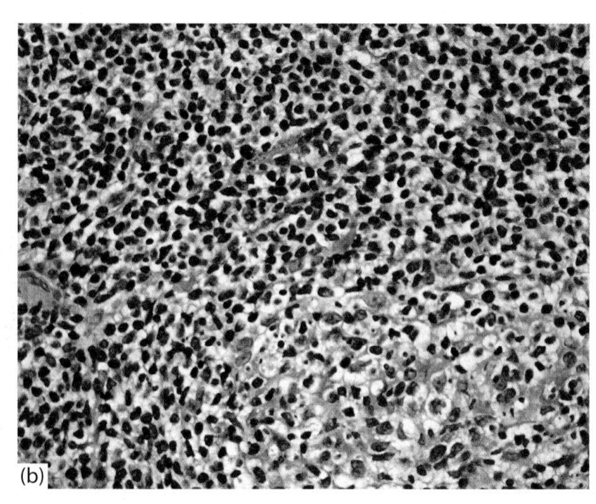

Figure 11.8 Axillary core biopsy from a man aged 63. (a) Low-power view showing uniform dense cellularity, although the presence of open sinuses raises the possibility that this is a reactive process.

(b) High-power view showing a 'regressed' germinal centre. The surrounding small lymphoid cells have slightly angulated nuclei.

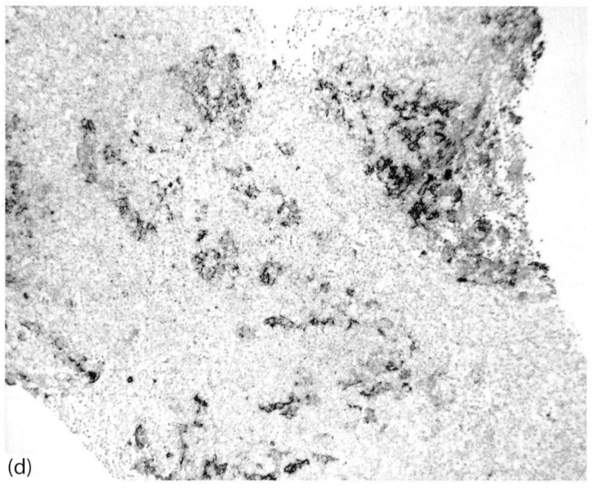

(c) Section stained for CD5 showing positivity of all the small lymphoid cells. The minority darkly staining cells are reactive T-cells. The majority of the cells were also positive for CD20 and CD79a.

(d) Section stained for CD23, showing dispersed follicular dendritic cells. The small lymphoid cells are negative.

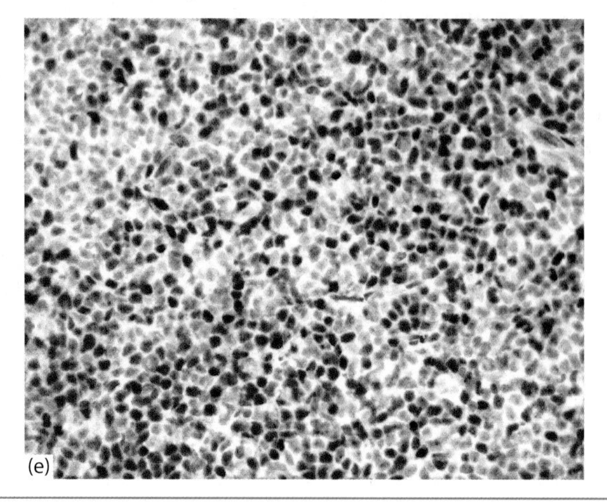

(e) Section stained for cyclin D1 (BCL-1) showing strong nuclear positivity of the tumour cells.

Comments and conclusion: The morphological features and immunohistochemistry indicate partial involvement of the node by mantle cell lymphoma.

CASE 4: FOLLICULAR LYMPHOMA (Figure 11.9)

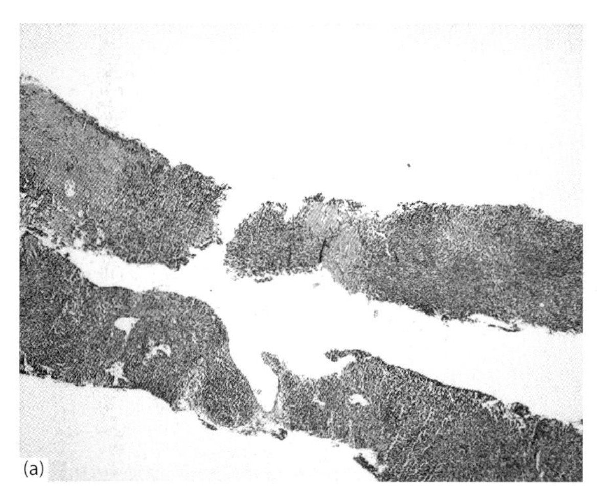

Figure 11.9 Core biopsy of abdominal lymph node from a 55-year-old woman. (a) The low power demonstrates areas of sclerosis (characteristic of follicular lymphomas).

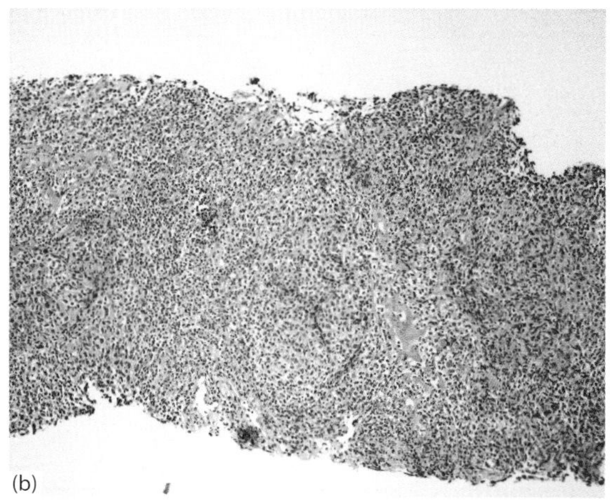

(b) Higher power view showing neoplastic follicles.

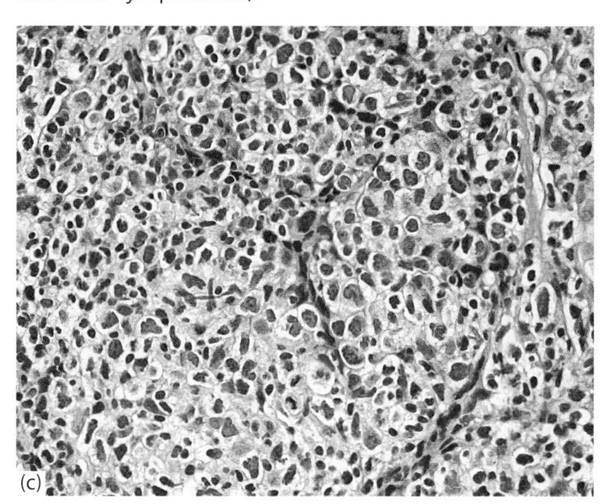

(c) High-power view of neoplastic follicle. The follicle centre cells have pale-staining cytoplasm and irregular nuclei. A few cells have visible nucleoli. This is probably best categorized as grade 2.

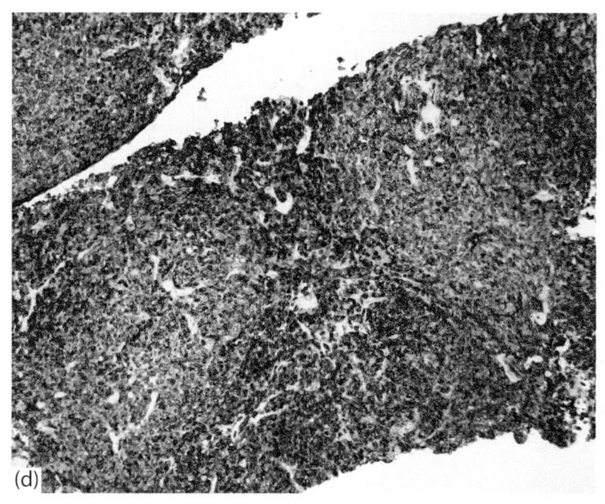

(d) Section stained for BCL-2. The follicle centre cells are BCL-2 positive but stain more weakly than the surrounding small lymphocytes.

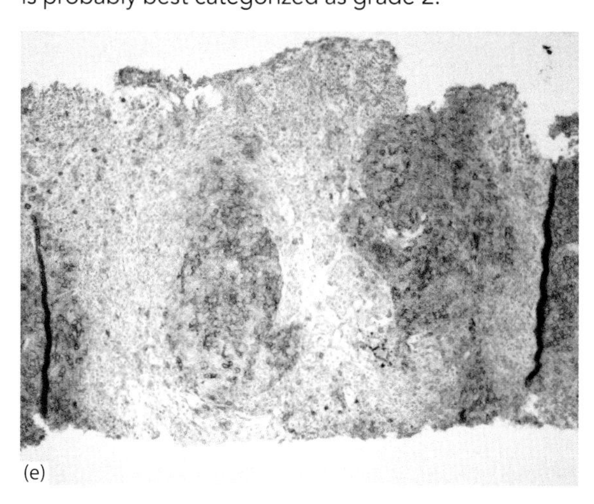

(e) Section stained for CD10, highlighting the neoplastic follicles.

(Continued)

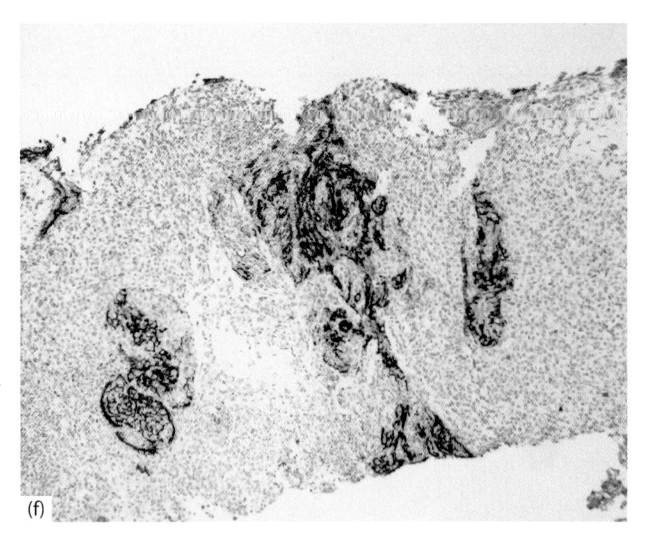

(f) Section stained for CD23, showing residual dendritic cell networks in the neoplastic follicles.

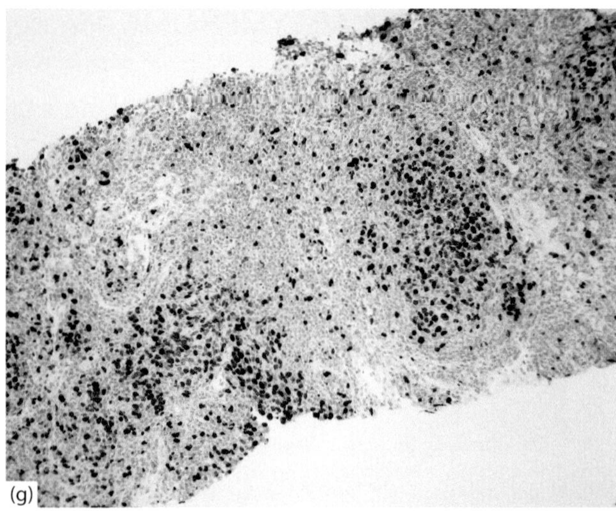

(g) Section stained for Ki67 showing a high labelling fraction in the neoplastic follicles.

Comments and conclusion: This is a follicular lymphoma. Most of the tumour cells have irregular nuclear outlines and lack visible nucleoli. They show a high proliferation index on Ki67 staining. The follicle centre cells do not have the morphology of centroblasts. This tumour should probably be best categorized as grade 2.

CASE 5: DIFFUSE LARGE B-CELL LYMPHOMA/NECROTIC CORE (Figure 11.10)

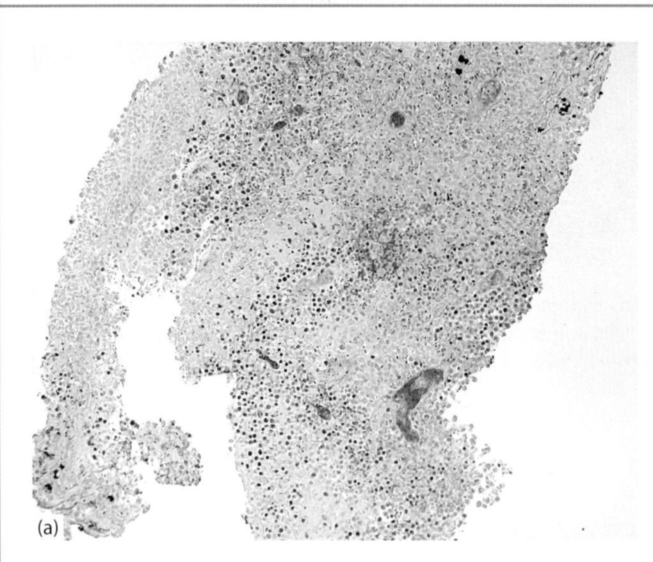

Figure 11.10 Core biopsy from a mesenteric mass in an elderly man. (a) Needle core biopsy showing almost complete necrosis with some surviving nuclei. Note the absence of inflammatory cells.

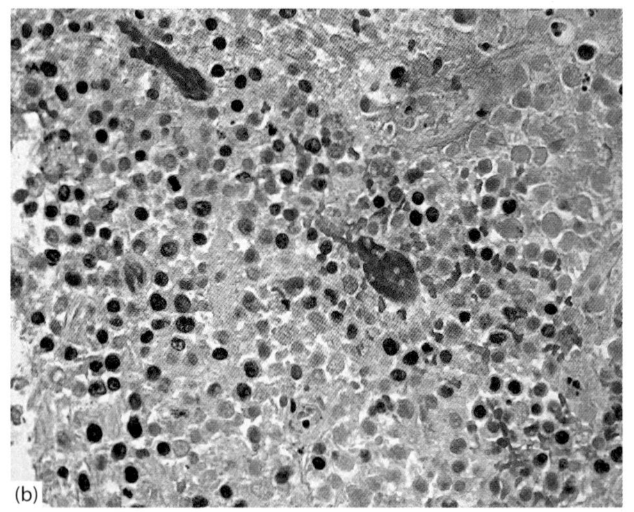

(b) High-power view showing a small number of apparently 'viable' cells among the majority of necrotic tumour cells.

(Continued)

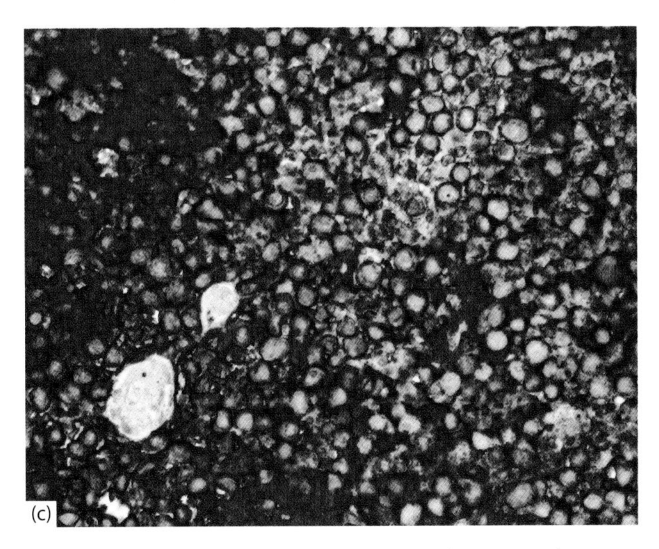

(c) High power view of section stained for CD20 showing positive membrane staining of tumour cells (to the right) and strong positivity of necrotic areas (to the left).

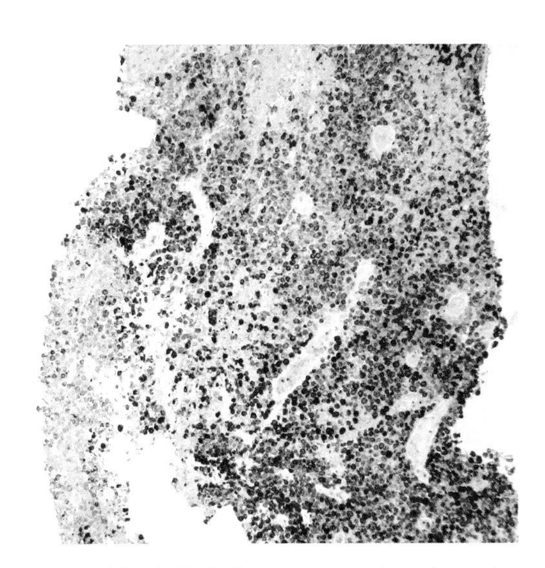

(d) Section stained for BCL-2 showing cytoplasmic positivity of many of the tumour cells.

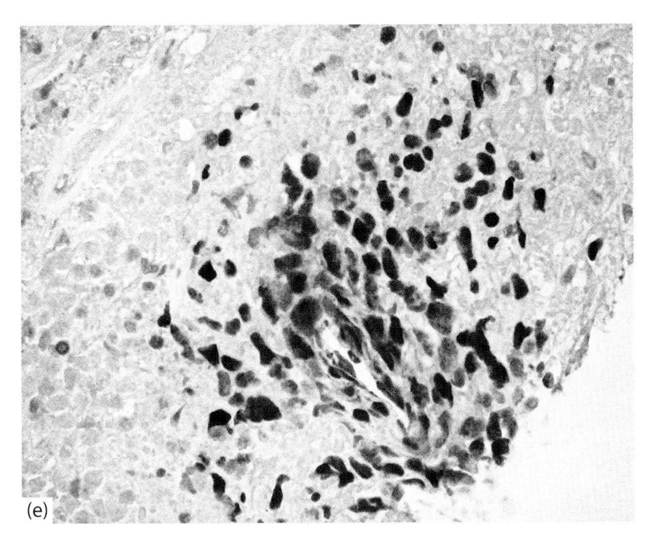

(e) In one 'corner' of the biopsy a small group of tumour cells shows nuclear MUM1 positivity.

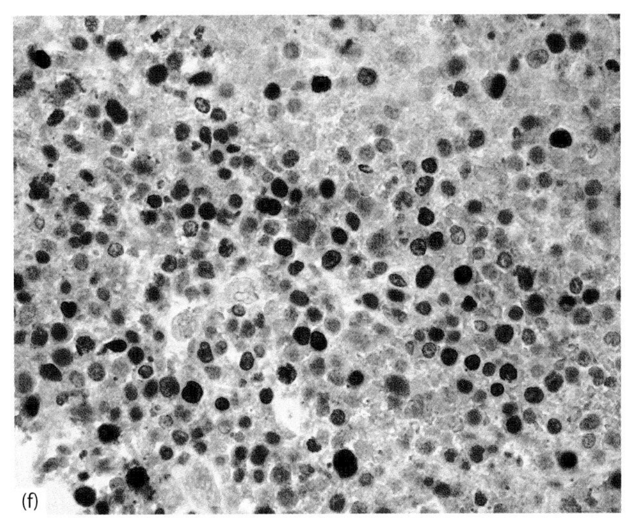

(f) High-power view of section stained for Ki67. A high proportion of the least necrotic looking cells show nuclear staining.

Comments and conclusion: Before the advent of immunohistochemistry this biopsy would probably have been reported as 'not diagnostic', although the appearances are characteristic of diffuse large B-cell lymphoma (DLBCL) with necrosis. CD20 is a very robust antigen, surviving for some time in necrotic tissue. In this case it shows the B-cell phenotype of the tumour. Ki67 is a less persistent antigen but in this case labels more than 40 per cent of tumour cells in a few areas, supporting the diagnosis of diffuse large B-cell lymphoma (DLBCL). The tumour could also be shown to be positive for BCL-2 and for MUM1 (CD10 and BCL-6 appeared to be negative). With a reasonable degree of confidence one could report this 'undiagnosable' biopsy as a DLBCL of activated B-cell type.

CASE 6: CLASSICAL HODGKIN LYMPHOMA (Figure 11.11)

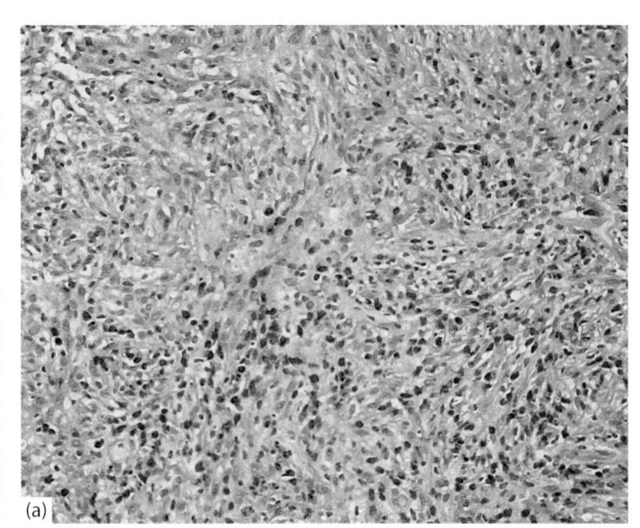

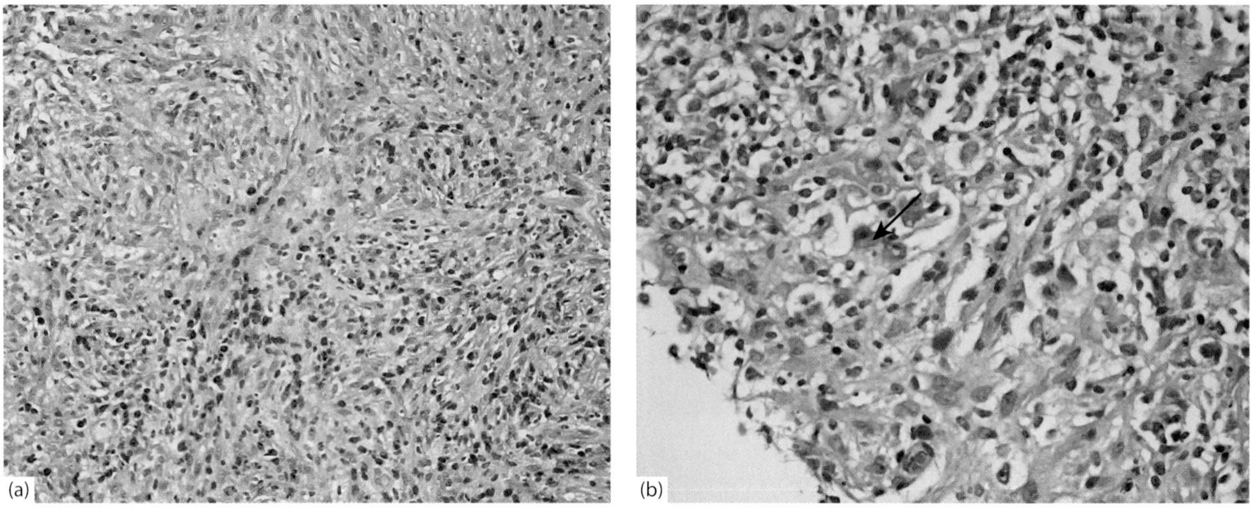

(b)

Figure 11.11 Core biopsy from a mediastinal mass in a child. (a) Low-power view of core biopsy showing extensive fibrosis with a scanty infiltrate of small lymphocytes and large numbers of eosinophils.

(b) High-power view showing mononuclear Hodgkin cells (arrow).

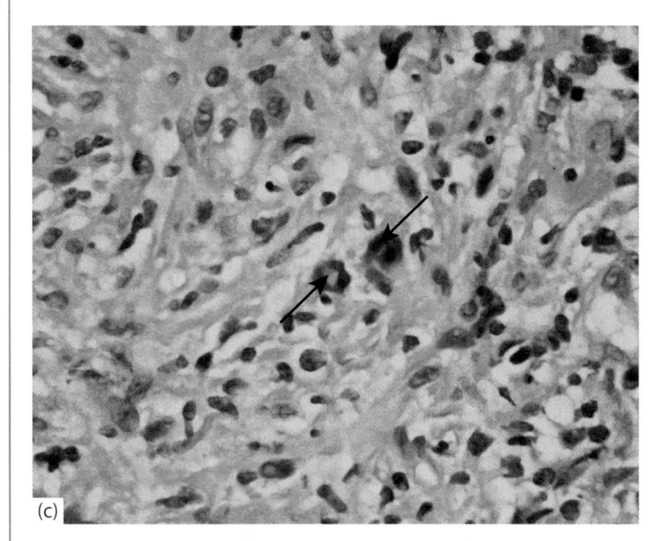

(c) High-power view showing more characteristic Reed–Sternberg cells (arrows). Note that these cells appear smaller than equivalent cells seen in standard sections.

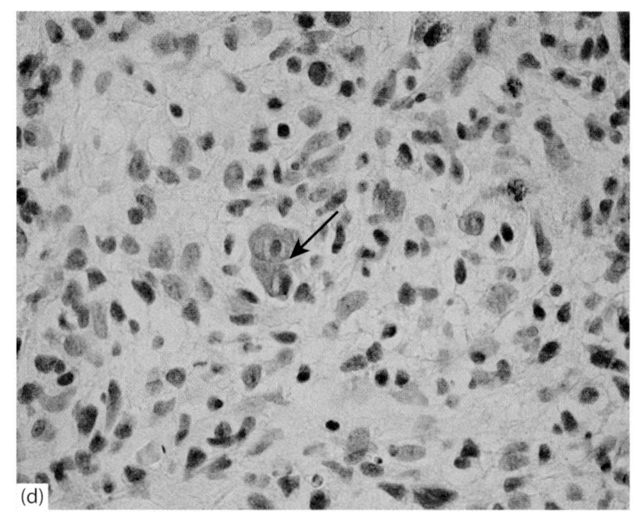

(d) Section stained for CD15 showing a characteristic Reed–Sternberg cell arrow, (unstained). When the amount of tissue is limited it is worthwhile looking at morphology in slides used for immunohistochemistry.

(Continued)

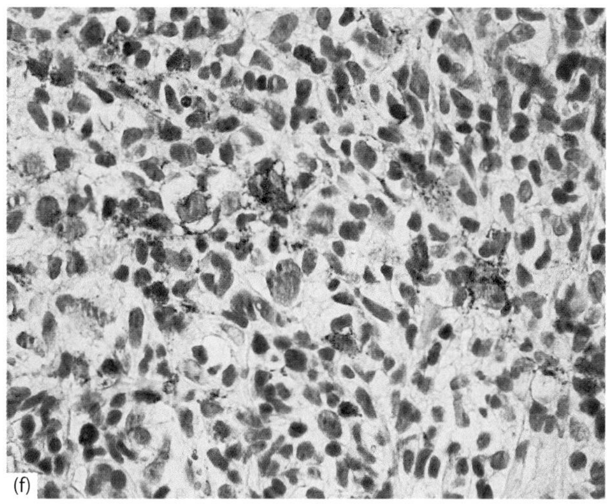

(e) Section stained for CD30 showing several positive mononuclear and binucleate cells. The background smearing of CD30 positivity is a result of traction artefact in this fibrotic tissue.

(f) Section stained for CD15 showing rather granular membrane staining of several large atypical cells.

Comments and conclusion: The differential diagnosis of a mediastinal mass in an 11-year-old boy would include T-lymphoblastic lymphoma, Hodgkin lymphoma and mediastinal large B-cell lymphoma. The low-power histology of this case immediately points to Hodgkin lymphoma with fibrosis and large numbers of eosinophils. The Hodgkin/Reed–Sternberg (H/RS) cells appear smaller than in standard biopsies (possibly because of rapid fixation). Although not 'textbook' the atypical cells express CD15 and CD30, supporting a diagnosis of classical Hodgkin lymphoma. Although the site and the sclerosis favour the diagnosis of nodular sclerosis Hodgkin lymphoma, it does not fall within the World Health Organization definition of this category since we have not seen collagen bands surrounding at least one nodule and the H/RS cells do not have lacunar morphology.

CASE 7: LYMPHOBLASTIC LYMPHOMA (Figure 11.12)

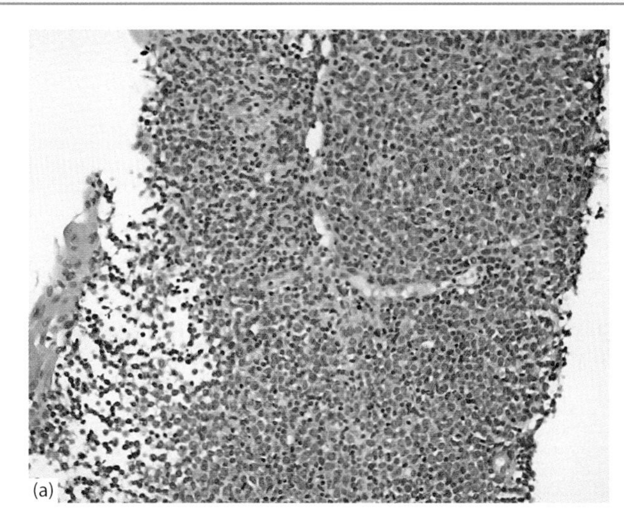

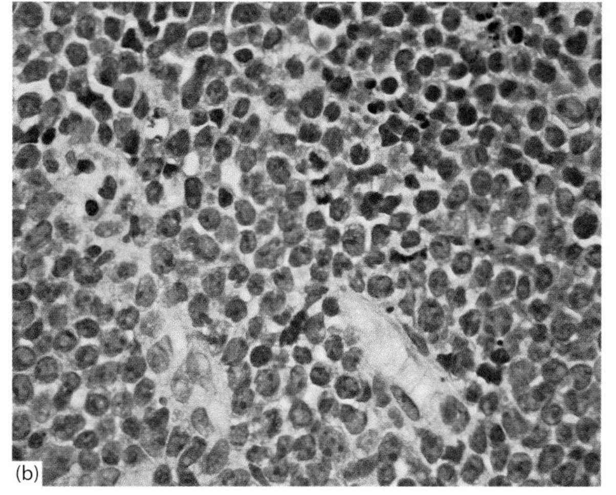

Figure 11.12 Core biopsy of a mediastinal mass in a 29-year-old man. (a) Medium-power view of core biopsy showing a uniform lymphoid population. The tumour cells are two to three times the size of the scattered small lymphocytes and have relatively pale smooth nuclear chromatin. The strip of squamous epithelium to the left of the picture indicates that the biopsy has come from the thymus.

(b) High-power view showing regular lymphoblastoid cells with rounded nuclei, smooth nuclear chromatin and one or more small eosinophilic nuclei. The cytoplasm of these cells is weakly stained and ill defined.

(Continued)

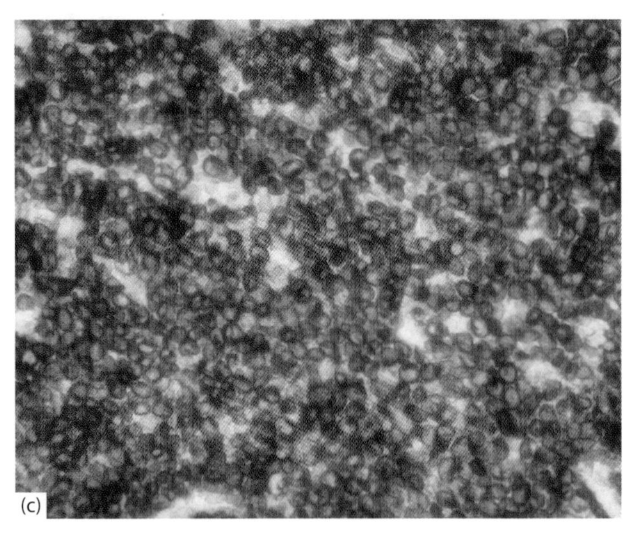

(c) Section stained for CD3 showing cytoplasmic positivity of most of the cells present. The cytoplasmic staining highlights nuclear clefts in many of the cells.

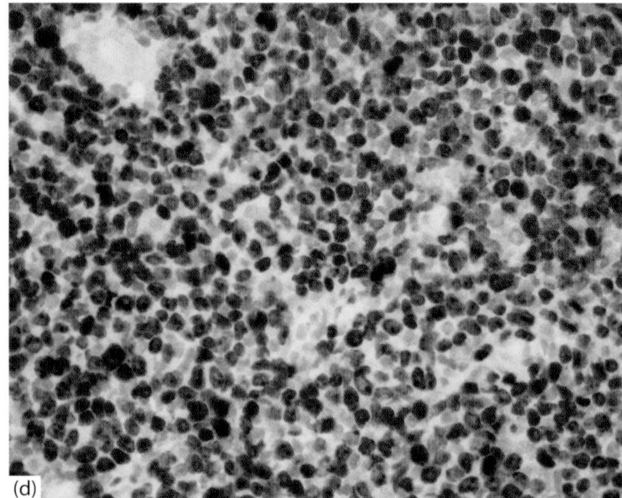

(d) Section stained for Ki67 showing a high labelling fraction.

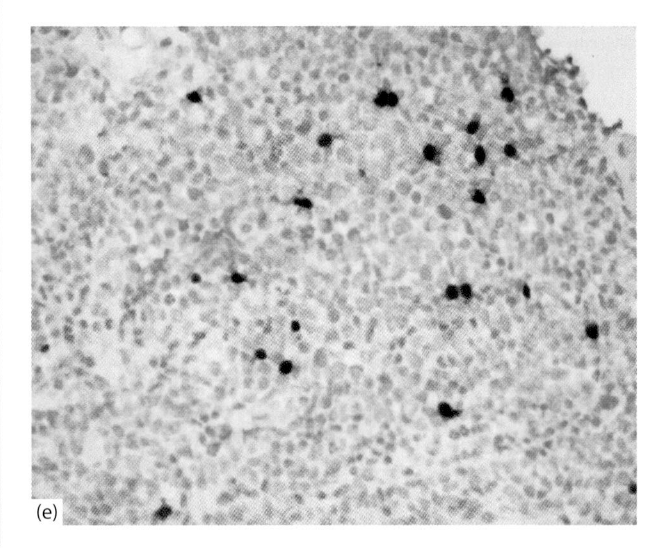

(e) Section stained for terminal deoxynucleotidyl transferase (TdT), showing very few strongly staining nuclei.

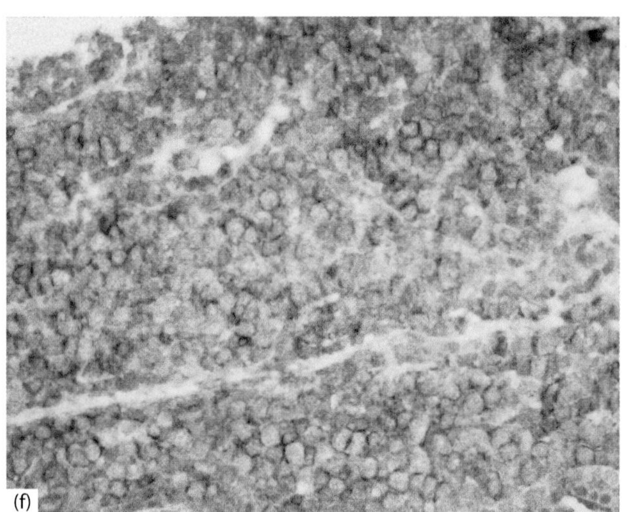

(f) Section stained for CD99 showing membrane positivity of the tumour cells.

Comments and conclusion: The morphology of the tumour is that of a lymphoblastic lymphoma. The site in the thymus and strong CD3 positivity of the tumour cells confirm its T-cell phenotype. The almost complete lack of staining for TdT might appear to be a discordant feature, but approximately 20 per cent of T-lymphoblastic lymphomas are TdT negative (Lewis *et al.* 2006). In the absence of TdT staining CD99 is the best marker of the precursor nature of these tumour cells (Borowitz and Chan 2008).

CASE 8: BURKITT LYMPHOMA (Figure 11.13)

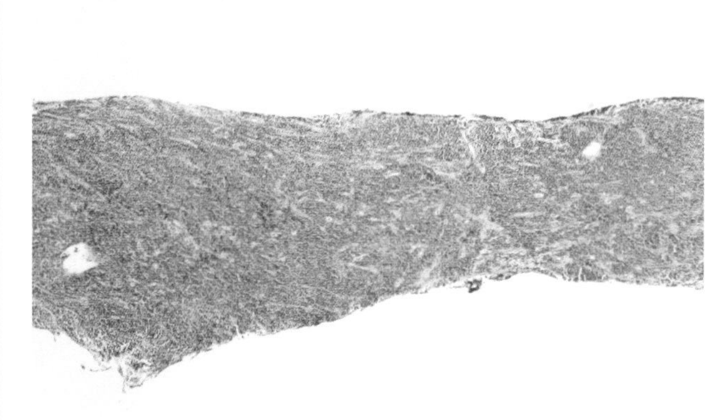

Figure 11.13 Abdominal mass core biopsy from a 24-year-old man. (a) Low-power view showing a core of highly cellular tumour with a diffuse growth pattern.

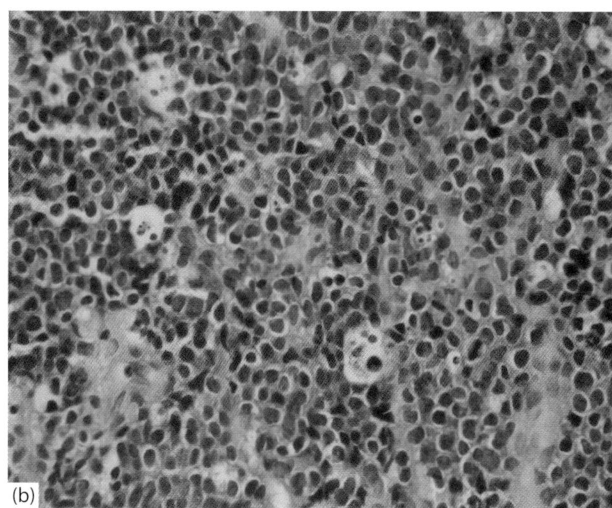

(b) High-power view shows a diffuse neoplasm of medium-sized cells with smooth nuclear chromatin and inconspicuous nucleoli. Apoptotic bodies are present between the tumour cells and within 'starry-sky' macrophages.

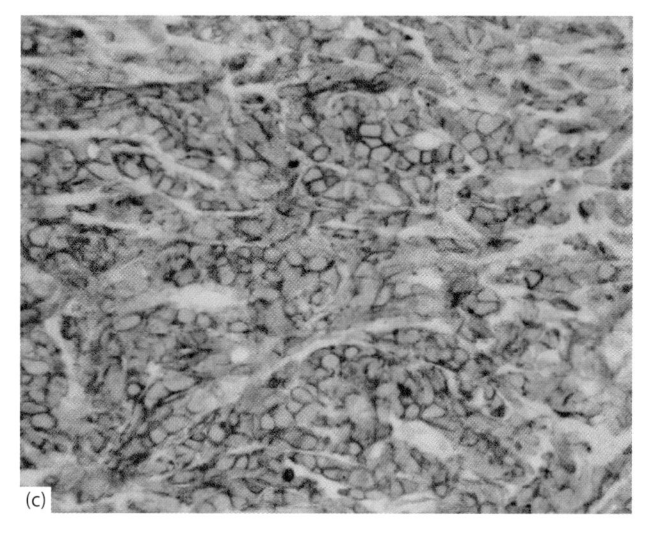

(c) Section stained for CD10 showing positivity of the tumour cells. They were also positive for CD20 and CD79a.

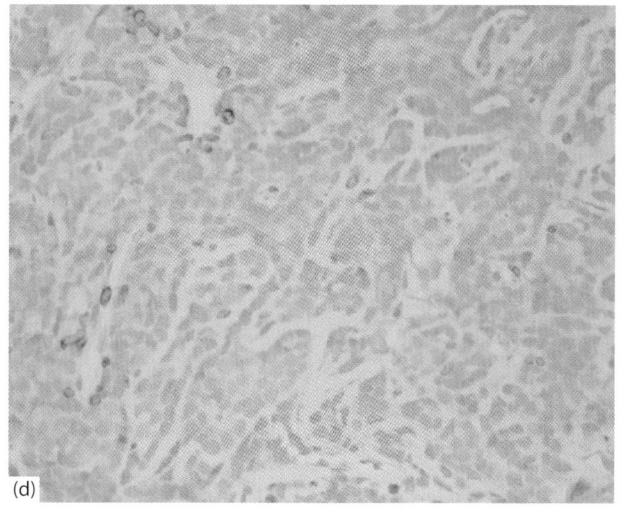

(d) Section stained for BCL-2. The tumour cells are negative. Occasional reactive T-cells show cytoplasmic positivity.

(Continued)

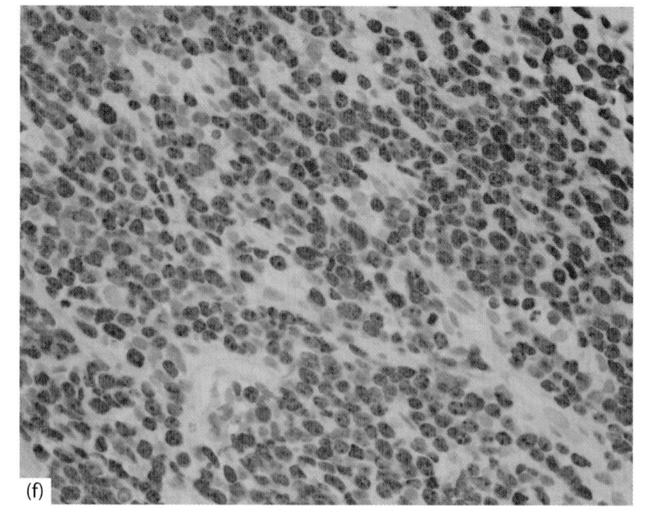

(e) Section stained for BCL-6 showing strong nuclear positivity of the tumour cells.

(f) Section stained for Ki67 showing a labelling fraction in the region of 100 per cent.

Comments and conclusion: The morphology and immunohistochemistry support a diagnosis of Burkitt lymphoma.

CASE 9: PERIPHERAL T-CELL LYMPHOMA NOS (Figure 11.14)

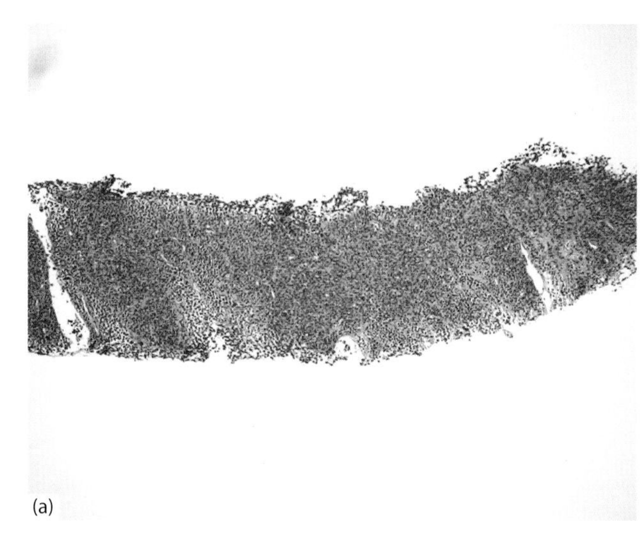

Figure 11.14 Inguinal lymph node core from a 45-year-old man. (a) Low-power view of the core biopsy showing a highly cellular diffuse neoplasm.

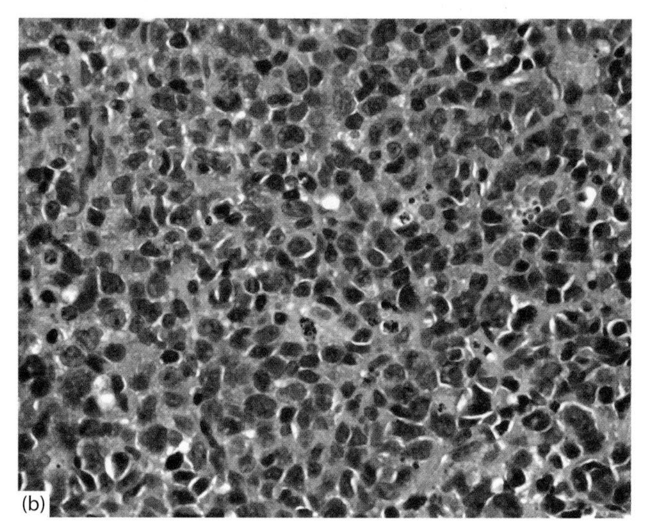

(b) High-power view showing a large cell lymphoma. Many of the tumour cells have indented or folded nuclei with fine chromatin and one or more eosinophilic nuclei. Numerous apoptotic bodies are seen.

(Continued)

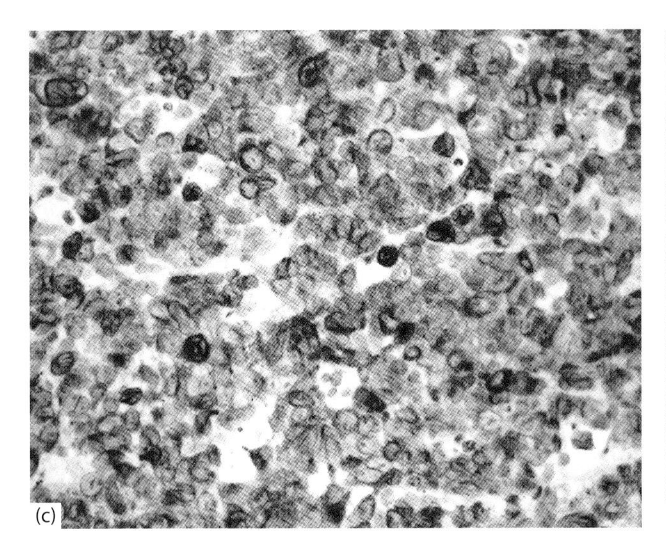

(c) Section stained for CD3 showing strong cytoplasmic positivity of the tumour cells.

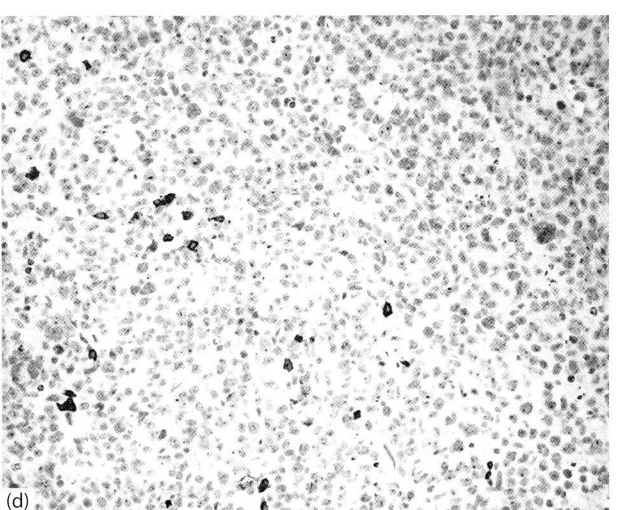

(d) Section stained for CD20 showing occasional residual small B-cells. The tumour cells are negative, although some show nucleolar staining.

(e) Section stained for CD30 showing strong positivity of the tumour cells.

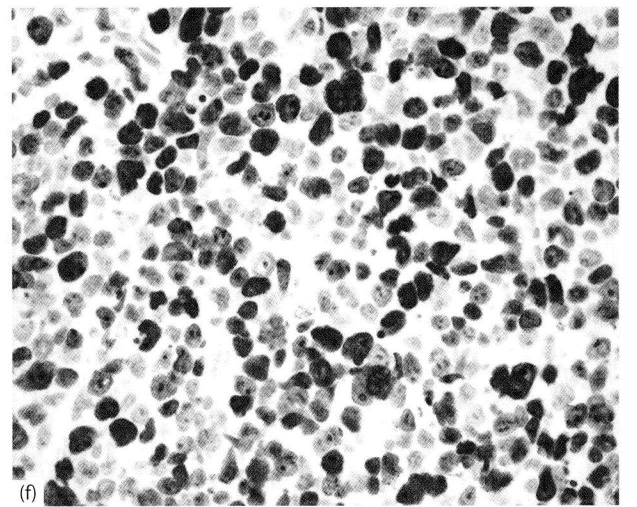

(f) Section stained for Ki67 showing a high labelling fraction.

Comments and conclusion: The morphology and immunohistochemistry support the diagnosis of a peripheral T-cell lymphoma not otherwise specified (NOS). The tumour is CD30 positive but does not have the morphology of an anaplastic large cell lymphoma and was anaplastic lymphoma kinase 1 (ALK-1) negative.

CASE 10: ANGIOIMMUNOBLASTIC T-CELL LYMPHOMA (Figure 11.15)

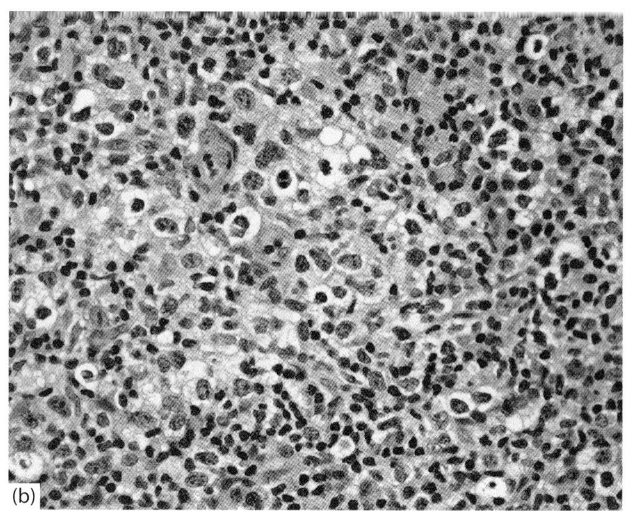

Figure 11.15 Cervical lymph node core from a 62-year-old woman. (a) Core biopsy of a lymph node with a dense lymphoid infiltrate. Clusters of cells with clear cytoplasm can be seen.

(b) High-power view showing large cells with abundant clear cytoplasm.

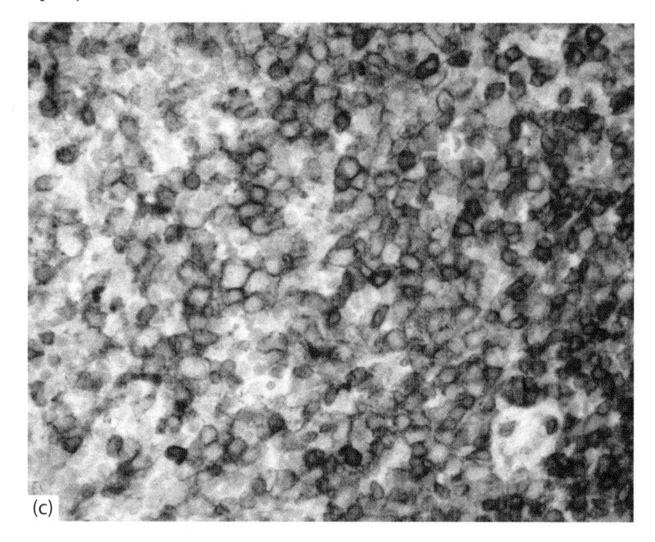

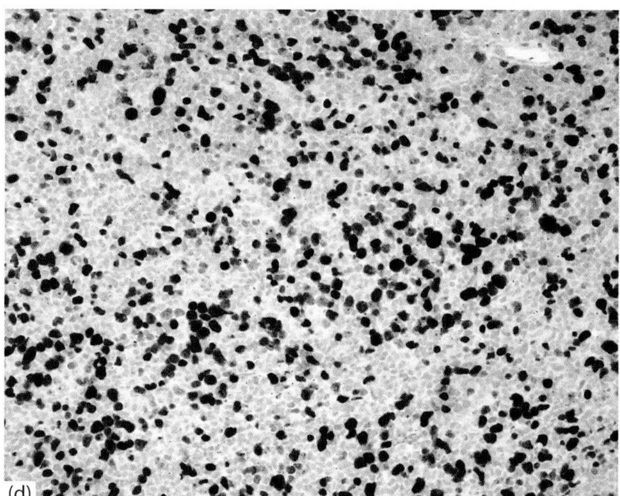

(c) Section stained for CD3 showing positive small T-cells (right) with positive staining of large tumour cells (centre and left).

(d) Ki67-stained section showing positive labelling of large cell nuclei.

(Continued)

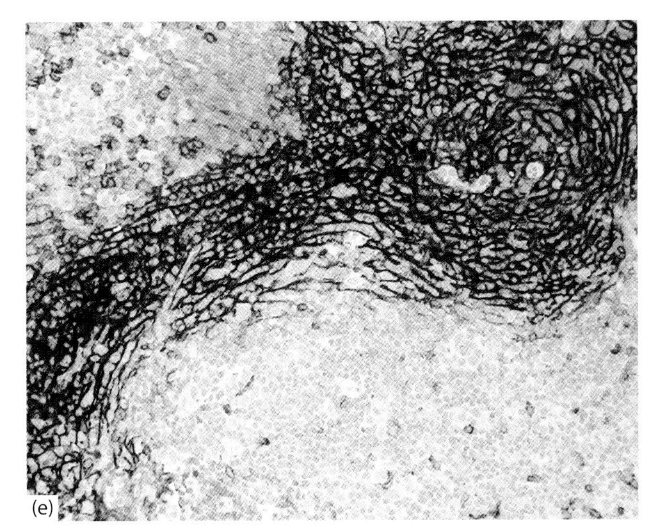

(e) Section stained for CD21 showing an expanded dendritic cell network at one end of the core biopsy. Scattered CD21+ small B-cells are also seen.

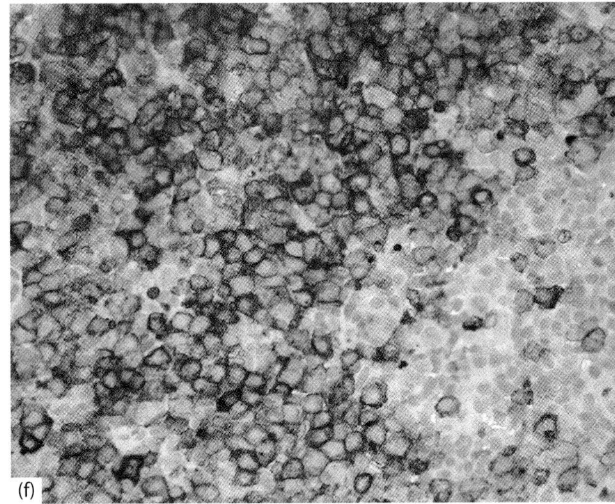

(f) Section stained for CD10 showing positive staining of the large tumour cells.

Comments and conclusion: This core biopsy has captured many of the diagnostic features of angioimmunoblastic T-cell lymphoma (AITL) (large cells with clear cytoplasm often surrounding blood vessels; immunophenotype of T-cells [CD3+, CD4+, CD10+]; expanded network of dendritic cells). The CD10 positivity of the large T-cells imparts a degree of specificity in favour of the diagnosis of AITL.

CASE 11: NODULAR LYMPHOCYTE PREDOMINANT HODGKIN LYMPHOMA (Figure 11.16)

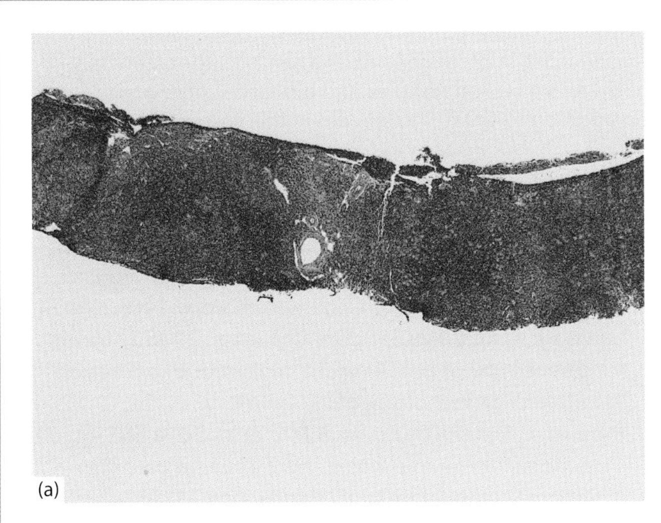

Figure 11.16 Core biopsy from a cervical node in a 27-year-old man. (a) Low-power view of core biopsy showing a vaguely nodular appearance with many small lymphocytes.

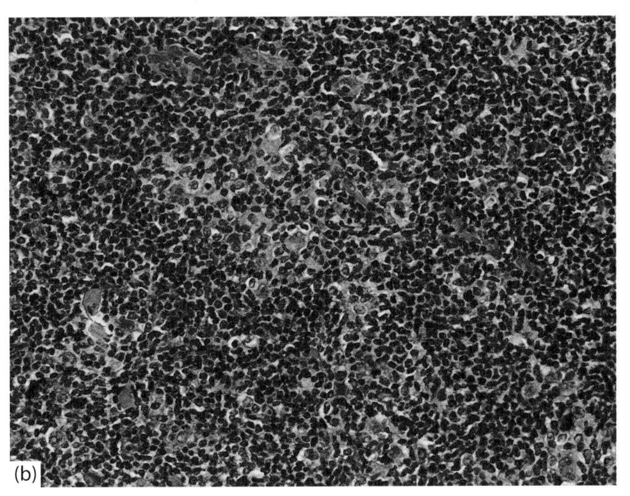

(b) High-power view showing several LP cells. The thick section hampers the morphological assessment.

(Continued)

(c) Section stained for CD21 highlighting the nodular architecture.

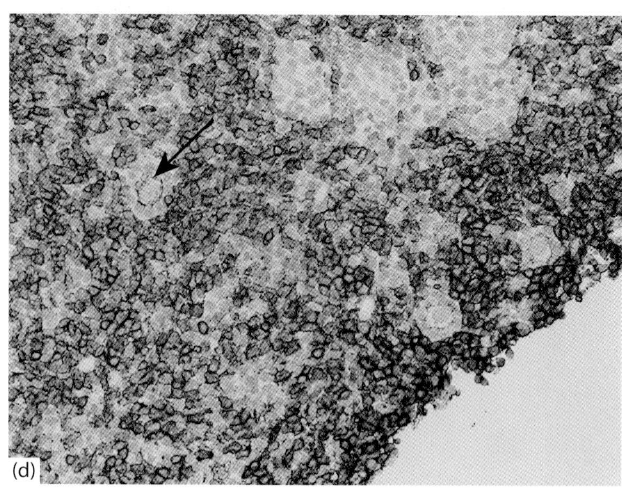

(d) Section stained for CD20 showing a characteristic LP cell surrounded by a rosette of negative small lymphocytes (arrow).

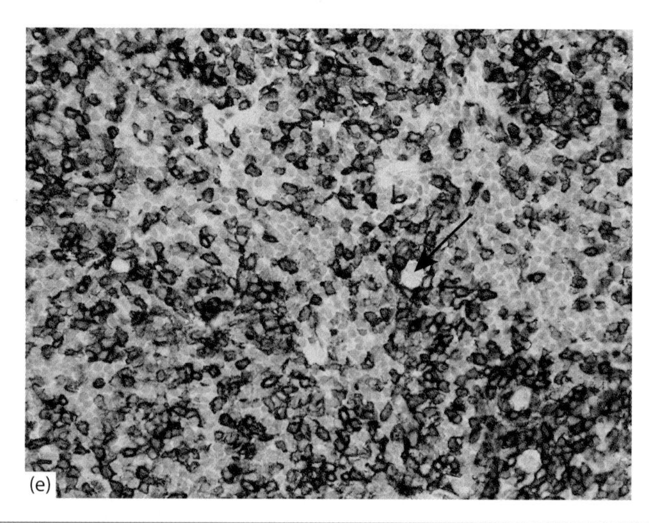

(e) Section stained for CD3 highlighting the T-cell rosettes (arrow).

Comments and conclusion: The diagnosis of nodular lymphocyte-predominant Hodgkin lymphoma is challenging in core biopsies and can be suggested only if there are typical nodules and a classical phenotype. In addition to CD20 the LP cells in this case expressed CD79a, BCL-6, PAX5 and CD45. There was no expression of CD15 or CD30.

REFERENCES

Amador-Ortiz C, Chen L, Hassan A, *et al.* 2011 Combined core needle biopsy and fine-needle aspiration with ancillary studies correlate highly with traditional techniques in the diagnosis of nodal-based lymphoma. *American Journal of Clinical Pathology* **135:** 516–524.

Borowitz MJ, Chan JK 2008 T lymphoblastic leukaemia/lymphoma. In Swerdlow SH, Campo E, Harris NL, *et al.* (Eds) *WHO Classification of Tumours of Haematopoietic and Lymphoid Tissues*. Lyon: IARC Press.

Goldschmidt N, Libson E, Bloom A, *et al.* 2003 Clinical utility of computed tomography–guided core needle biopsy in the diagnostic re-evaluation of patients with lymphoproliferative disorders and suspected disease progression. *Annals of Oncology* **14:** 1438–1441.

Hu Q, Naushad H, Xie Q, *et al.* 2013 Needle-core biopsy in the pathologic diagnosis of malignant lymphoma showing high reproducibility among pathologists. *American Journal of Clinical Pathology* **140:** 238–247.

Lewis RE, Cruse JM, Sanders CM, *et al.* 2006 The immunophenotype of preTALL/LBL revisited. *Experimental and Molecular Pathology* **81:** 162–165.

Ribeiro A, Pereira D, Escalón MP, *et al.* 2010 EUS-guided biopsy for the diagnosis and classification of lymphoma. *Gastrointestinal Endoscopy* **71:** 851–855.

Soudack M, Shalom RB, Israel O, *et al.* 2008 Utility of sonographically guided biopsy in metabolically suspected recurrent lymphoma. *Journal Ultrasound Medicine* **27:** 225–231.

Stacchini A, Carucci P, Pacchioni D, *et al.* 2012 Diagnosis of deep-seated lymphomas by endoscopic ultrasound-guided fine needle aspiration combined with flow cytometry. *Cytopathology* **23:** 50–56.

Appendix 1: Immunohistochemical markers used in diagnosis of lymphoproliferative conditions*

Annexin A1: Encoded by the ANXA1 gene, upregulated in HCL, in which it shows membranous or sometimes cytoplasmic expression. It is currently considered the most sensitive and most specific marker for diagnosis of HCL. By contrast, B-cell lymphomas other than HCL are negative. Importantly, splenic MZL, HCv and diffuse red pulp small B-cell lymphoma are all negative for this marker. Assessment of expression in bone marrow can be difficult owing to strong background staining of the myeloid cells.

BCL-2: Expressed in many normal T- and B-cells and many lymphomas. It is not expressed in reactive germinal centres and therefore BCL-2 immunostaining is useful in distinguishing follicular hyperplasia from FL. Care must be taken not to misinterpret reactive germinal centre T-cells, which are normally BCL-2 positive and may be present in abundance for positive germinal centre B-cells. In systemic, nodal FL, expression of BCL-2 is a consequence of t(14;18); however, in most other small cell B-cell lymphomas BCL-2 is positive as a result of epigenetic control in the absence of the translocation. Therefore BCL-2 cannot be used to distinguish FL from other systemic small lymphoid cell B-cell lymphomas. Primary cutaneous follicle centre lymphoma and paediatric FLs do not express BCL-2 and show no evidence of t(14;18). A significant proportion of grade 3B FLs are also negative for this marker and lack the translocation. In DLBCL, expression of BCL-2 in conjunction with MYC positivity represents a significant adverse prognostic factor.

BCL-6: Nuclear antigen expressed in germinal centre cells and FL. BCL-6 also reflects cellular activation and is often seen in B-cell lymphoid blasts unrelated to their germinal centre origin. This could be seen in CLL/SLL, MZLs and a proportion of DLBCLs of non–germinal centre derivation. H/RS cells of cHL are usually negative whereas L&H cells of NLPHL express BCL-6. This marker is also positive in T-cells of germinal centre derivation and hence in AITL as well.

BOB1: See OCT2, BOB1.

CD1a: Identifies Langerhans cells in the skin and in lymph nodes. It may be significantly increased in dermatopathic lymphadenopathy. Langerhans cell histiocytosis is also positive. Most T-lymphoblastic lymphomas are positive.

CD2, CD3: Robust T-cell markers expressed by normal T-cells and most T-cell lymphomas. T-cell lymphomas may show loss of pan-T-cell markers so application of a number of markers for general T-cell differentiation is advised. Some myeloid tumours express these antigens. In particular, systemic mastocytosis is characterised by co-expression of CD2 and CD25.

CD4, CD8: Markers of T-cell subsets are useful in the differential diagnosis of T-cell proliferations and in T-cell lymphoma typing. CD4 is abundantly expressed in histiocytes which may hamper assessment of expression in presumed T-cells. PDC neoplasm is also positive for CD4.

CD5, CD7: Pan-T-cell markers. CD5 is also expressed by some B-cell lymphomas. Expression may be lost in some T-cell lymphomas and CD7 may also be downregulated in inflammatory conditions.

CD11c: An integrin, a member of the cellular adhesion molecule family. It is strongly expressed in HCL and, when co-expressed with CD22, was proposed as a unique marker of this lymphoma. It is also variably expressed in acute myeloid leukaemia with monocytic/monoblastic differentiation and some cases of CLL. It is also one of the best markers used for identification of macrophages/histiocytes.

CD15: Useful in the diagnosis of Hodgkin lymphoma. Available antibodies to CD15, a carbohydrate antigen, are all IgM clones and their detection by reagents optimised for the more usual IgG monoclonal antibodies is sometimes suboptimal. In addition, CD15 is fixation sensitive; overall, expression of CD15 is demonstrable in approximately 80% of cHL; it is expressed in 15–20% of ALCL. Co-expression of CD15 with CD30 does not mandate diagnosis of cHL. Diagnosis should rely on the assessment of a range of parameters including morphology and clinical presentation and should not rely on co-expression of CD30 and CD15 alone.

* Adapted from Appendix E, Royal College of Pathologists Standards for specialist laboratory integration and Dataset for the histopathological reporting of lymphomas. Permission given to reproduce Appendix E. See explanation of abbreviations on page xvii.

CD16: Expressed by NK cells, some T-cells, NK neoplasms and a subset of LGL leukaemias.

CD20: B-cell marker in widespread CD20 expression starts in late pro-B-cells and is lost at the late postfollicular stage of plasma cell differentiation. Expression of CD20 may be lost in relapsed or persisting B-cell lymphomas after rituximab treatment.

CD21, CD23 and CD35: These identify FDCs; they show different patterns of staining, reflecting functional variation within the FDC population. Their use can help in the identification of follicular growth patterns and in the diagnosis of AITL. CD23 is normally expressed by many FDCs and a minor subpopulation of mantle B-cells. CD23 is also expressed by 93% of B-CLL, in occasional cases of other small B-cell lymphomas (particularly in FL) and in some LBCLs (particularly mediastinal LBCL). They are robust markers for FDC tumours.

CD25: The interleukin-2 receptor widely expressed on activated T-cells, B-cells and macrophages and in both non-Hodgkin and Hodgkin lymphomas. CD25 is expressed in ATLL associated with HTLV, some peripheral T- cell lymphoma NOS, mycosis fungoides, especially in the large cell transformation, and in some ALCLs. It is also a widely used marker of HCL and is expressed in some cases of splenic MZLs. Neoplastic mast cells show CD25 immunoreactivity together with CD2. This pattern of co-expression is highly specific for diagnosis of systemic mastocytosis (reactive mast cells in the marrow are CD25/CD2 negative). CD25 can be expressed in H/RS cells as well as by reactive T-cells in the tumour microenvironment.

CD30: Useful in the diagnosis of Hodgkin lymphoma and ALCL. Interpretation of CD30 depends on the detection system used; the more sensitive the technique, the more reactive B-cell blasts will be detected. Diagnosis should rely on the assessment of a range of parameters including morphology and clinical presentation and should not rely on co-expression of CD30 and CD15 alone.

CD35: See CD21, CD23 and CD35.

CD38: Marker of plasma cells also expressed on mature circulating B-cells and may be expressed on lymphomas with plasma cell and plasmablastic differentiation. In addition, CD38 is positive in a proportion of B-CLL/SLL where its expression to an extent correlates with the unmutated genotype.

CD43: One of the major glycoproteins of thymocytes and T-lymphocytes, used as a pan-T-cell marker. It is also expressed by myeloid cells and macrophages. In addition, it is aberrantly expressed in MCL B-CLL/SLL, a proportion of MZL and some DLBCL. FL and HCL do not express this marker. In this context, CD43 may be useful in resolving differential diagnosis between FL and DLBCL or MZL.

CD45 (LCA), CD45RA, CD45RO: These are a number of antibodies against tyrosine phosphatases present on the surface of almost all haematolymphoid cells. CD45 (LCA) recognises all the isoforms. The usefulness of this antibody is in confirming haematolymphoid origin of proliferations. Negativity for this marker is also of significant diagnostic value. The negative haematolymphoid neoplasms include a proportion of ALCLs, ALK+ LBCL, cHL and plasma cell neoplasms. There are several antibodies against CD45RA including 4KB5, MB1, KiB3 and MT2. The most commonly used MB1 and KiB3 stain most B-cell lymphomas. MT2 in reactive tissues and FL show a pattern of staining of germinal centres similar to that obtained with BCL-2 immunostaining. MT2 stains a proportion of T-cell lymphomas. Antibodies against CD45RO include UCHL1, A6 and OPD4 which are used in confirming T-cell derivation.

CD56, CD57: NK and NK-like T-cell markers, essential for the diagnosis of malignancies derived from these cells. CD56 is expressed in NK/T-cell lymphomas of nasal type. CD57 is expressed by germinal centre T-cells and is also a marker of LGLs; it is useful in the diagnosis of NLPHL. CD56 is also expressed in many cases of plasma cell neoplasia (but not in normal plasma cells) and in nonlymphoid tumours.

CDw75 (LN1): A neuraminidase-sensitive sialoprotein, present on cell membrane and cytoplasm of germinal centre B-cells and derived lymphomas. LN1 reacts with erythroid precursors, ductal and ciliated epithelial cells of kidney, breast, prostate, pancreas, lung, and with glioblastomas, astrocytomas, and L&H cells of NLPHL. LN1 is shown to be a reliable antibody for ascribing a B-cell phenotype in known lymphoid tissues.

CD79a: B-cell marker in widespread use. Its spectrum of staining is slightly different from that of CD20. CD79a expression starts earlier in B-cell development than expression of CD20 and CD20 expression is lost at the late post-follicular stage of plasma cell differentiation.

CD103: Alpha-E integrin expressed by intestinal intraepithelial T-lymphocytes, mucosal B-cells, and HCL cells. In HCL the staining pattern is predominantly membranous. It is useful in differentiating HCL and HCv from other small cell lymphomas including CLL/SLL, MCL, FL, LPL and MZL. Intestinal lymphomas including EATL and FL are positive.

CD123: Interleukin-3 receptor α-chain regarded as a reliable marker of PDCs. These represent one of the three subsets of normal dendritic cells, originate from CD34+ bone marrow progenitor cells and have been identified in the thymus and lymphoid tissues, including tonsil, bone marrow, peripheral blood and spleen. It is a useful marker in the diagnosis of blastic PDC neoplasms (formerly known as 'blastic NK-cell lymphoma' or 'CD4-positive/CD56-positive neoplasm') and reactive proliferations of PDCs including Kikuchi lymphadenitis.

CD138: Marker of plasma cells. CD138 is not expressed on mature circulating B-cells but may be expressed on lymphomas with plasma cell and plasmablastic differentiation. CD138 is also a robust epithelial marker, wherein its expression to an extent correlates with the unmutated genotype.

CD246 (ALK-1, 5A4): These antibodies enable visualisation of the nucleophosmin–anaplastic lymphoma kinase (NPM-ALK) fusion protein associated with t(2:5) and variant translocations involving the ALK gene. Positive staining identifies a subgroup of ALCL with good prognosis and is currently the defining feature of this entity within WHO 2008. A rare variant of LBCL is also characterised by nuclear ALK expression.

Cyclin D1: Not normally expressed by lymphoid cells. It is expressed in MCL displaying nuclear positivity, reflecting t(11;14) translocation. However, up to 20% of MCLs may be negative and require alternative immunostaining (e.g. with SOX11, cyclin D2) or FISH for diagnostic confirmation. Cyclin D1–positive DLBCL has also been described and positivity is seen in HCL and a proportion of plasma cell myelomas. In reactive lymphoid tissue and bone marrow, normal macrophages and histiocytes show strong nuclear expression.

Cytotoxic molecules: TIA-1, granzyme B, perforin: TIA-1, granzyme B and perforin are cytotoxic molecules stored in cytoplasmic granules. TIA-1 is present in all cytotoxic cells whereas granzyme B and perforin expression depends on the activation status. The expression of cytotoxic molecules is helpful in typing of T-cell and NK-cell neoplasms. Aggressive NK-cell leukaemia, extranodal NK/T-cell lymphoma of nasal type, subcutaneous panniculitis-like T-cell lymphoma and EATL express TIA-1, granzyme B and perforin. Hepatosplenic T-cell lymphoma, T-cell large granular lymphocytic leukaemia and T-cell PLL usually express TIA-1 but not granzyme/perforin. HTLV-associated ATLL does not express cytotoxic molecules. These are also expressed by CD8+ PTCL.

D2-40 (podoplanin): Directed against the M2A antigen, a sialoglycoprotein found on the cell surface of testicular gonocytes, germ cell tumours, lymphatic endothelium and mesothelial cells. Using a human podoplanin-Fc fusion protein, it has been shown that the commercially available mouse monoclonal antibody D2-40 specifically recognised human podoplanin. This antibody was initially described as specific and sensitive for diagnosis of FDC tumours; however, a large spectrum of tumours has been shown to express this marker including mesothelioma, a range of vascular tumours, carcinomas and benign epithelial tumours.

EBV: Anti-LMP1 identifies EBV in about 20–30% of infected lymphoma cells of various types. It should be emphasised that immunostaining for LMP1 alone does not exclude EBV positivity and involvement of this virus in pathogenesis of certain lymphomas. LMP1 is expressed only in latency types II and III, which include cHL and a range of other lymphomas, many associated with underlying severe immunosuppression (latency III). LMP1 does not stain tumours that belong to latency I such as extranodal T/NK lymphoma of nasal type, lymphomatoid granulomatosis and Burkitt lymphoma. Therefore the gold standard for assessment of EBV is *in-situ* hybridization for EBER, which is present in all types of latency.

EMA: This is one of several glycoproteins found in HMFGP. Because HMFGPs are packaged in the Golgi apparatus, dot-type reactivity in the Golgi zone may be seen. The glycoprotein identified with EMA is now known to be one of a series of glycoproteins or mucins designated as MUC1. This marker is found on a wide range of epithelial and soft tissue tumours but also in a range of normal haematopoietic cells, lymphocytes and plasma cells. In the context of diagnosis of haematolymphoid malignancies, EMA is seen expressed in myelomas and plasmacytomas, all subtypes of ALCL, ALK-1+ LBCL and NLPHL. Importantly, it is not seen in RS cells of cHL or TCRBCL.

Fascin: Actin-binding protein that is specifically expressed by some dendritic cells and a high percentage of H/RS cells of cHL.

FOXP1: Transcription factor essential for transcriptional regulation of B-cell development. By gene expression profiling it has been shown to be highly expressed by activated B-cell type DLBCL. Expression is nuclear. GCET1 and FOXP1 are part of the Choi algorithm for typing of DLBCL with CD10, MUM1 and BCL-6.

FOXP3: The protein encoded by FOXP3 is a member of the forkhead/winged-helix family of transcriptional regulators. It is a marker of T-regulatory cells (Tregs) and is expressed in adult ATLL. In FL and cHL, high numbers of FOXP3+ Tregs correlate with better overall survival. However, in both FL and cHL, this is not part of routine investigations or the basis for specific therapeutic decisions.

GCET1: Highly expressed in normal germinal centre B-cells and B-cell lymphomas of germinal centre derivation including FL and germinal centre type DLBCL. Expression is cytoplasmic and membranous.

Granzyme B: See Cytotoxic molecules.

HHV8: An antibody reacting against the latent nuclear antigen of the Kaposi sarcoma virus (HHV8). The immunostaining is nuclear and detected in Kaposi sarcomas of HIV-positive and other immunosuppressed and elderly patients. In addition, this antibody helps diagnosis of primary effusion lymphoma and plasmablastic variant of Castleman disease of HIV-positive patients.

ICOS: The ICOS protein is a member of the CD28 co-stimulatory receptor family and identifies TFH cells, a specialised subpopulation of T-helper cells residing primarily in the germinal centre. ICOS expression appears to be restricted to certain subsets of T-cells, with the highest expression on CD4-positive T-cells and moderate expression on Tregs. The expression of ICOS has been described in AITL, follicular variant of PTCL, primary cutaneous CD4+ lymphoma, and cases of PTCL, NOS with borderline AITL features. Expression of ICOS is variably associated with other markers of follicular T-helper cell differentiation including CD10, BCL-6, CXCL13 and PD-1. In the context of AIL, ICOS is at present considered most sensitive.

Immunoglobulins: Kappa and lambda light chains, IgG, IgG4, IgM, IgA, IgD: Light chain immunostains are very useful because light chain restriction may be used as a proxy for implying clonality but are technically difficult and must be interpreted with caution. Some laboratories prefer to use *in-situ* hybridization for kappa and lambda mRNA, at least when there is visible plasma cell maturation. Use of *in-situ* hybridization for the assessment of light chain restriction of lymphocytic B-cell population is problematic. Heavy chain staining can also be useful in identifying some lymphoma subtypes and assessing suspected plasma cell myeloma. IgG4 has become important in identifying systemic IgG4-related disease.

IRF4/MUM1: Nuclear positivity for MUM1 is a consequence of activity of the NF-κB pathway. This is seen in late post-follicular B-cells including plasma cells. However, its expression should not be interpreted as a definite indicator of plasmacytic differentiation. MUM1 is positive in non–germinal centre type of DLBCL and is part of the Hans and Choi algorithms for typing of DLBCL. H/RS cells of cHL are usually positive but not L&H cells of NLPHL. This marker is also expressed in a range of T-cell lymphomas and has recently been suggested as a useful discriminator between cutaneous ALCL and lymphomatoid papulosis.

IRTA1: IRTA1 is selectively associated with normal (Peyer patches) and acquired MALT. It is expressed in both extranodal (93%) and nodal MZLs (73%) but is not seen in other small B-cell NHLs and is therefore regarded as a specific marker for MZLs. It is an important aid to diagnosis of MZLs, highlighting tumour cells in colonised follicles. Expression is membranous.

Ki67: Nuclear antigen expressed in nuclei of cells in cycle but not in G0. It can help in identifying highly proliferative lymphomas such as Burkitt lymphoma, in which the Ki67-positive fraction approaches 100% and the majority of cells show uniform and very strong expression. Whereas Burkitt lymphoma generally shows proliferation of 100%, other aggressive B-cell lymphomas do as well and this is no longer a marker that helps to distinguish between Burkitt and non-Burkitt B-cell NHLs. Proliferation assessment in MCL is of significant prognostic value. Ki67 is also useful in distinguishing reactive follicular hyperplasia from follicular and other forms of B-cell lymphomas with nodular growth patterns.

Langerin: A type II transmembrane C-type lectin specific to normal Langerhans cells, Langerhans cell histiocytosis and Langerhans cell sarcoma.

LEF1: Nuclear expression of LEF1 has been shown in all cases of CLL/SLL, even those that are CD5 negative. This marker is not expressed in MZL, MCL or FL and is regarded as a robust marker of CLL/SLL.

MUM1: See IRF4/MUM1.

MYC: Participates in the regulation of gene transcription and cell cycle. The antibody displays nuclear expression of MYC, which could be seen in normal cells in cycle. It must be emphasised that evidence of expression of MYC does not equate with the translocational status of the MYC gene because the protein could be overexpressed in the absence of the translocation as a result of epigenetic regulation. DLBCLs, which co-express MYC and BCL-2 regardless of the presence or absence of the MYC gene translocation, pursue an aggressive course. In this context, the cut-off point for positivity is 40% nuclei positive for MYC.

OCT2, BOB1: These two markers are used in conjunction. They aid transcription of immunoglobulin genes in B-cells and play a role in germinal centre formation and differentiation of B-cells into plasma cells. Expression of both these markers is nuclear and is uniformly distributed in reactive, non-neoplastic B-cells. Most non-Hodgkin B-cell lymphomas and NLPHL show strong nuclear expression. Most cases of cHL are negative for both or at least one of these markers.

p24: HIV p24-gag viral capsid protein can be demonstrated by immunohistochemistry. The staining is localised to FDCs. Cutaneous Langerhans cells are also immunoreactive for p24-gag in early stages of HIV infection, even before seroconversion. Positivity in various parenchymal cells of a variety of organs is not unusual. It is frequently used to identify the presence of the virus in patients with progressive generalised lymphadenopathy. It is highly specific and sensitive. Identified positivity indicates diagnosis of HIV infection and has equivalent value to the serological HIV test.

PAX5: The PAX5 gene is essential for B-cell differentiation. Four isoforms of the gene are known and PAX5a has been most studied. It is expressed by immature and mature B-cells but is downregulated during terminal differentiation into plasma cells. Expression is nuclear. PAX5 influences the expression of other B-cell specific genes, including CD19, CD20 and CD79a and precedes the expression of CD20. Reactive lymphoid tissue shows the same distributions of expression of PAX5 as CD20. They are both positive in germinal centres, mantle zones, marginal zones, monocytoid B-cells (weak) and intra-epithelial lymphocytes in extranodal sites. Haematogones in the bone marrow are also PAX5 positive. In diagnosis of precursors B-cell lymphoma/leukaemia (ALL) it is used as a surrogate of CD19, and positivity in this context outlines commitment to B-cell lineage, in the absence of any other B-cell marker. It is also helpful in differentiating H/RS cells of cHL (where it typically shows weak nuclear expression) from T-cells and 'null' cells of ALCL lymphoma. However, expression of PAX5 is rarely seen in ALCL of T/null cell phenotype as well. It may aid distinction of B-cell lymphomas with plasmacytic components (such as CLL/SLL with lymphoplasmacytic differentiation, LPL and MZL of MALT type) from plasma cell myeloma. It is of note that PAX5 is also commonly expressed by Merkel cell carcinoma, small cell carcinoma and alveolar rhabdomyosarcoma, small blue round cell tumours that are often in the differential diagnosis with lymphoma.

PD-1: Similar to ICOS, this is a member of the CD28 family of receptors and is expressed on all CD4+ T-cells and half of CD8+ T-cells. It is expressed by activated T-cells, B-cells and myeloid cells. There are at least 2 ligands for PD-1, PD-L1, and PD-L2, which are expressed on a range of cells. The expression of PD-1 has been widely reported in AITL; it is also expressed by T-cells associated with neoplastic B-cells in NLPHL. It is considered to be a more specific marker for AITL because, unlike CD10 and BCL-6, PD-1 is expressed by few B-cells and appears to stain a higher number of CD3+ neoplastic cells compared with CD10 and BCL-6. It is also found to be useful in diagnosis of cutaneous CD4+ small/medium-size T-cell lymphoma.

Perforin: See Cytotoxic molecules.

PU-1: Transcription factor that regulates the expression of immunoglobulin and other genes important for B-cell development. Absence of PU-1 results in a block in the early stage of B-cell development. It is crucial for the expression of CD20, CD72 and CD79a. It is expressed in B-lymphocytes, immature and mature, including mantle cells and most cells of the germinal centre, but not in plasma cells, histiocytes or PDCs. Lack of expression of

PU-1 has been shown in cHL, whereas NLPHL and TCRLBCL retain expression. This is the likely contributing factor to the lack of immunoglobulin expression and incomplete B-cell phenotype characteristic of the H/RS cells in cHL.

S100: This marker identifies interdigitating dendritic cells and their tumours. It is also positive in a small proportion of T-cell lymphomas.

SOX11: This is a transcription factor upregulated in the majority of MCLs where its nuclear expression co-localises with cyclin D1. In the context of diagnosis of MCL, lack of expression of this marker identifies a small subset of MCLs characterised by CD5 lymphocytosis, minimal lymph node involvement and low proliferation. These cases pursue an indolent clinical course and require little therapeutic intervention in comparison to conventional SOX11-positive cases. SOX11 is also expressed in lymphoblastic lymphomas, DLBCL and Burkitt lymphoma but is not seen in other small B-cell NHLs. Thus, its positivity in cases of cyclin D1–negative, CD5-positive small B-cell lymphomas indicates diagnosis of cyclin D1 negative MCL. This should be corroborated by immunostaining for cyclin D2, which is usually detectable in cyclin D1–negative cases of MCL.

TCL1: The TCL1 gene (14q32.1) is involved in the leukaemogenesis of mature T cells and its overexpression is observed in more than 90% of T-PLLs. Chromosomal rearrangements bring the TCL1 gene in close proximity to the TCR-α or TCR-β regulatory elements. In normal T-cells TCL1 is expressed in CD4–/CD8– cells, but not in cells at later stages of differentiation.

TCR: TCR performs an antigen-recognition function on the surface of T-cells, analogous to that of immunoglobulins on the surface of B-cells. TCR is associated with the CD3 complex and comes in two forms: the α/β and γ/δ heterodimers. The α/β heterodimer is present on ~90% of thymocytes and mature peripheral T-cells, ~60% of precursor T-cell lymphoblastic lymphomas and ~70% of PTCLs. The γ/δ heterodimer is present on a small number of thymocytes, a small number of cells in the skin and MALT and in rare types of T-cell lymphomas: hepatosplenic T-cell lymphoma and primary cutaneous γδ T-cell lymphoma.

TdT: Expressed by precursor T- and B-cells, and precursor leukaemias and lymphomas. A minority of these tumours may be negative. This is a nuclear stain; cytoplasmic staining may occur with suboptimal technique and should be ignored. TdT is also expressed in up to 10% of myeloid leukaemias. Presence of TdT-positive haematogones in bone marrow trephine sections should not be mistaken for neoplasia. A rare but diagnostically important setting of aberrant expression of this marker is in a small proportion of small cell lung carcinomas.

TIA-1: See Cytotoxic molecules.

ZAP-70: Member of the Syk family of tyrosine kinases. It is involved in T and NK cell receptor transduction and also plays a role in the transition of pro-B to pre-B cells in the bone marrow. In reactive lymph nodes nuclear ZAP-70 staining is seen in paracortical T-lymphocytes and rare, scattered, small lymphocytes in the mantle zones and germinal centres. Histiocytes can sometimes express ZAP-70 in a granular cytoplasmic pattern. It is expressed by B-precursor lymphoblastic lymphoma and a subset of B-CLL/SLL. It is so far the best surrogate immunohistochemical marker of the B-CLL/SLL mutational status and is an independent prognostic marker in this context. Nuclear positivity correlates with the clinically more aggressive unmutated phenotype. Assessment of expression requires correlation with the immunostains for background T-cells (CD3, CD5).

Index